PRACTICAL RELIABILITY ENGINEERING and ANALYSIS for SYSTEM DESIGN and LIFE-CYCLE SUSTAINMENT

T0203674

PRACTICAL RELIABILITY ENGINEERING and ANALYSIS for SYSTEM DESIGN and LIFE-CYCLE SUSTAINMENT

William R. Wessels

CRC Press
Taylor & Francis Group
Boca Raton London New York

CRC Press is an imprint of the
Taylor & Francis Group, an **informa** business

CRC Press
Taylor & Francis Group
6000 Broken Sound Parkway NW, Suite 300
Boca Raton, FL 33487-2742

First issued in paperback 2019

© 2010 by Taylor and Francis Group, LLC
CRC Press is an imprint of Taylor & Francis Group, an Informa business

No claim to original U.S. Government works

ISBN-13: 978-1-4200-9439-8 (hbk)
ISBN-13: 978-0-3673-8425-8 (pbk)

Library of Congress Cataloging-in-Publication Data

Wessels, William R.
   Practical reliability engineering and analysis for system design and life-cycle sustainment / William Wessels.
         p. cm.
   Includes bibliographical references and index.
   ISBN 978-1-4200-9439-8 (hard back : alk. paper)
   1. Reliability (Engineering) 2. Product life cycle. I. Title.

TS173.W45 2010
620'.00452--dc22   2009051339

**Visit the Taylor & Francis Web site at
http://www.taylorandfrancis.com**

**and the CRC Press Web site at
http://www.crcpress.com**

To my O. A. O. and the love of my life, Tudor.

Contents

Preface

Product reliability is becoming a global competitive discriminator. It is a concept that people relate to even if they do not know what must be done to achieve it. Reliability engineering is a young discipline that evolved following World War II. It has its roots in military and commercial aviation use of electronics and digital components to control and monitor aviation systems. Defined as the probability that a system will perform its function without failure for the mission duration under stated conditions of use, "reliability" has been treated as a statistical analysis used to audit a design rather than as a design analysis that is part of system design. Time to failure (TTF) is the dominant parameter of reliability. Reliability failure models seek to characterize part failure rate, failures per hour, and mean time between failure (MTBF).

Few universities offer reliability degree programs. Reliability courses taught at universities introduce probability and statistical methods to characterize point estimates of failure rate and MTBF; they present methods to compute the reliability of serial and redundant design configurations. Reliability engineering books describe the same topics taught in universities with the same emphasis on probability and statistics. Reliability tutorials and seminars emphasize selected reliability topics without context to the whole system.

The bathtub curve is the common thread of these instructional programs. This curve assumes that a system goes through three phases in its life cycle: infant failure, useful life, and wear out. A system is assumed to experience constant, random failure during the useful-life phase. This approach to reliability analysis works well when the system is an electronics circuit board constructed of digital components and solder connections.

However, reliability analysis of mechanical design for structures and dynamic components demands a different approach. Mechanical design must mitigate failure mechanisms acting on components as well as achieve functional specifications. Mechanical structures and dynamic components experience wear out that does not manifest constant failure rates over the useful life. Reliability of mechanical design is based on the relationship between stress and strength over time.

This book departs from the mainstream approach to TTF-based reliability engineering and analysis. As a mechanical engineer, I enjoyed a career in mining before becoming a defense contractor. The most enlightened reliability specialists that I have known are maintenance technicians, foremen, and engineers. They deal daily with the demand to keep mobile machinery and process equipment operating. What I learned from them is first to understand why a part fails, then learn how to fix it, and finally learn how to prevent it from failing. This reliability book seeks to blend those lessons with mechanical engineering design, systems integration, and sustainment in order to enable organizations to achieve world-class reliability in products.

I remain a student of the reliability engineering discipline and welcome comments, criticism, and case studies from everyone who works in reliability. I hope this book generates interest in expanding the body of reliability knowledge.

Best regards,

Bill Wessels
wesselsw3@mchsi.com

The Author

Bill Wessels has been a reliability engineer since 1970. He did not know that he was a reliability engineer until 1989, when he was first tagged with that label. Bill discovered that field engineering assignments were tasked to maintain mobile machinery and process equipment far more often than to design or build them. Maintenance employees in mining and process plant organizations provided Bill with several decades of education in system maintainability and insights into how design for reliability can influence system sustainability. Business managers taught him that cash flow is the lifeblood of every organization, that capital assets only earn revenue when performing their function, and that unscheduled downtime causes lost-opportunity costs that can never be recovered.

Bill wrote this book to convey what he has learned to fellow design and field engineers. His formal education has equipped him to learn from those who perform maintenance and engineering work. It began with a BS in engineering from the U.S. Military Academy at West Point (1970), followed shortly after with an MBA (1975), and culminating in a PhD in systems engineering in 1996. Along the way, Bill earned his professional engineer license in mechanical engineering from Pennsylvania (1982) and ASQ certification as a reliability engineer (CRE).

Bill is currently a reliability, maintainability, and sustainability researcher at the University of Alabama in Huntsville, where he also teaches in the College of Engineering. Bill and his wife, Tudor, live on a small farm with seven dogs; ever changing numbers of horses, donkeys, goats, geese, turkeys, and chickens; and who knows how many barn cats. The beasts ignore the findings of statistically rigorous algorithms, defy any statistical level of significance, behave in a random walk manner, and ridicule his attempted analytical methods when they are applied to them. Tudor finds his attempts to predict their behavior to be a lack of common sense and a waste of time. But he continues to try.

List of Tables

List of Figures

1

Requirements for Reliability Engineering: Design for Reliability, Reliability Systems Integration, and Reliability-Based System Sustainment

> If you don't know where you are going, you might wind up somewhere else.
>
> **Yogi Berra**

Introduction

The objective of this chapter is to define what a reliability engineering and analysis program must include in order to achieve

- design for reliability
- reliability systems engineering
- reliability-centered system sustainment

The scope of this chapter is to inform the reader of the "what" of a reliability program. The subsequent chapters describe the "how."

Reliability engineering is the understanding of failure mechanisms. Failure mechanisms form the basis of

- design analysis, design art, and the bill of materials for a part
- part integration into assemblies, assembly integration into subsystems, subsystem integration into systems
- sustainment of fielded systems

Reliability engineering is a multidiscipline body of knowledge that enables engineers to characterize the reliability of a part, the maintainability of a part, and the availability of a part. Part reliability determines part maintainability, and the combination of the two determines part availability (see Figure 1.1). Reliability engineering is applied to mechanical, electrical, electronics, civil, and chemical engineering disciplines.

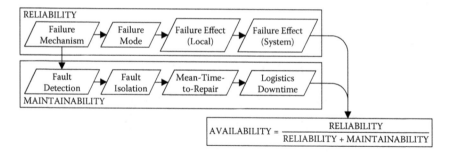

FIGURE 1.1
Reliability, maintainability, availability.

Part reliability describes and measures the failure mechanisms acting on the part, the part failure modes caused by the failure mechanisms, the effects of the failure modes on the part, and successive effects of the part failure mode on the higher design hierarchies up to and including the system. Part maintainability describes and measures fault detection of the failure mechanisms acting on the part, the isolation of the fault to a part, the mean time to repair the part, and the logistical aspects of part repair.

Part availability describes and measures the relationship between part reliability and part maintainability as the percent of time that a part is capable of performing its function.

Part Reliability

> Part reliability is defined as the probability that a part will perform its function without failure for the specified mission duration under stated conditions of use.[1,2]

Reliability is stated as a percent, when the period of time is a mission, as the mean time between failure (MTBF), when the period of time is the life cycle. Conditions of use have two sources:

- operating environment: stresses (thermal, vibration, and mechanical loads; chemical, corrosion, or biological reactivity) caused by the functioning of the part, its next higher assembly, and proximate parts rarely known to the design engineer
- ambient environment: stresses (thermal, vibration, and mechanical loads; chemical, corrosion, or biological reactivity) caused by the weather and proximate systems

The books of Kapur and Lamberson[1] and O'Conner[2] show that this definition has not changed much over the last several decades. The definition

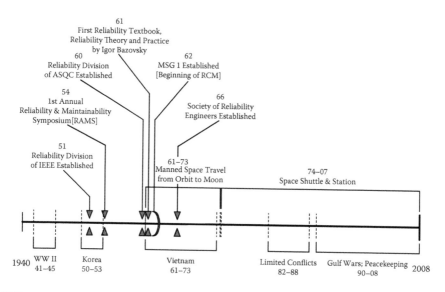

FIGURE 1.2
History of reliability engineering discipline.

passes the test of time. Otherwise, much that we know, or think we know, about reliability has been evolving over the same period of time. This is an important point because the reliability discipline is relatively young (indeed, I am older than the discipline). Civil engineering can trace its practice over several thousands of years. Mechanical engineering was born with the Industrial Revolution a few hundred years ago, reaching the status of a discipline in the late 1800s. Electrical and chemical engineering emerged in the early 1900s. The first reliability technical society was formed by the Institute of Electrical and Electronics Engineers (IEEE) in the early 1950s; the first reliability textbook was published in 1961 by Igor Bazowsky (see Figure 1.2).

The inclusion of the space programs and wars on the time line shows that the creation and evolution of the reliability engineering discipline in the United States has been closely related to National Aeronautics and Space Administration (NASA)[3] and military systems development programs. Indeed, even today the majority of reliability engineers are employed by NASA, the Department of Defense (DoD), and their contractors. Aviation has been and continues to be the driving motive for their huge investment in reliability engineering. But NASA and DoD have expanded their use of reliability engineering to weapons systems, ground vehicle systems, and radar systems, to name a few. One can posit that commercial aviation has shared the benefits of government applications in reliability engineering because the same companies operate in both sectors.[4]

Private sector application of reliability engineering has been less extensive. The U.S. automobile industry was forced to apply reliability engineering in the 1970s as a result of Japanese and German cars that were quickly known to deliver higher reliability. The global economy now demands part reliability as a competitive factor.

Fortunately, reliability engineering is not a mutually exclusive analysis to design analysis; only the orientation differs. The part design analysis is traditionally oriented to meet the functional specification flowed down from the system specifications through the work breakdown structure (WBS). This approach results in a part design that can be expected to function without failure the first time the system is used. How many times it will function without failure after that first time is anybody's guess. Part reliability design analysis applies the same functional specification flowed down from the system specifications through the WBS to determine the failure mechanisms that will cause the part to fail.

Failure Mechanisms

Material failure theories assert that part failure occurs when a stress exceeds one of the uniaxial material strength properties in tension or in compression (Rankine's failure theory) or when stress exceeds the shear material strength properties (Tresca–Guest failure theory). A failure mechanism is a stressor that weakens a part.[5–7] The effects of a failure mechanism are the changes to the geometry and material properties of a material.

The source of a failure mechanism can be functional, environmental, or the interaction of the two. A functional failure mechanism is a stress that occurs from the use of the part. For example, a rotating shaft delivering torque to a pulley drive experiences a torsion stress that causes torsion strain, a bolt fastening a pressure cap to a pressure vessel experiences tensile stress that causes tensile strain, and an electronics circuit board experiences thermal stress on solder connections that can cause thermal strain.

An environmental failure mechanism is a stress that occurs from external conditions of use: for example, temperature of air surrounding the part based on seasonal climate conditions and altitude, vibration from the system acting on the part, and exposure to corrosive materials.

The functional and environmental sources of stresses are often well known during the part design analysis, but less understood is the interaction of the two. Often the stresses acting on a part are not mitigated when the sources are evaluated to be an insignificant failure mechanism. Part failure can occur when two such stresses create a combined failure mechanism. For example, the O-ring failure on the *Challenger* space shuttle resulted from lack of understanding of the combined stresses of low temperature (environmental) and internal pressure loads (functional) on the seal. Another example is the failure mechanism from the combination of moisture (environmental) and thermal stress (functional) yielding corrosion.

Mechanical failure mechanisms are represented by the principal categories of stresses: *force, temperature,* and *reactivity.* Forces act as loads on the geometry of the material and are graphically described in six

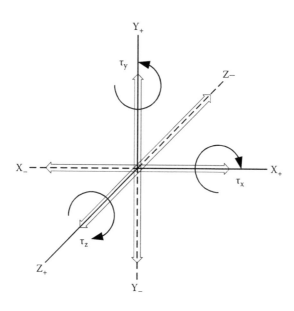

FIGURE 1.3
Six degrees of freedom.

degrees of freedom. The most basic uniaxial forces load a material in tension ("+" axis orientation), compression ("–" axis orientation), or in torsion ("+" or "–") about each axis. A force vector that does not align along one of the three axes can be reduced to its x-, y-, and z-axes resultants (see Figure 1.3).

Force can be described and measured as static and dynamic loads. Static loads are constant in magnitude and fixed in location. Static loads can be fixed at a point on or distributed across a segment of one or more axes (see Figure 1.4).

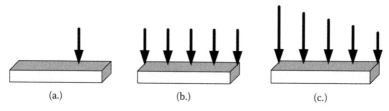

(a.)   (b.)   (c.)

Example Static Loads:
(a.) Point Load
(b.) Uniform Equal Loads
(c.) Uniform Decreasing Loads

FIGURE 1.4
Static loads.

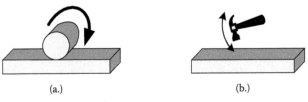

(a.)   (b.)

Example Dynamic Loads:
                    (a.) Uniform Dynamic Loads
                    (b.) Random Cyclical Loads

FIGURE 1.5
Dynamic loads.

Dynamic loads are variable in magnitude, direction, and location. The variation can be patterned or random in force magnitude and frequency of occurrence (see Figure 1.5). Forces acting on the joining or interface between two parts cause shear and bending loads (see Figure 1.6).

Material strain is the most basic response to a force; the magnitude and effect of strain are a function of the elastic property of the material. Material strength is the ability to survive each application of stress. The stress failure theory assumes that failure occurs when stress exceeds strength. This can happen in two ways:

- A single stress is applied that exceeds the material strength.
- A single stress that exceeds the strength is applied repeatedly and fatigues the material.

The former is the more straightforward design analysis method for material specification and is applied more frequently. It is easier to teach, quicker and cheaper to apply, and supported by ample published material properties information. It is the "safety factor" approach to design analysis.

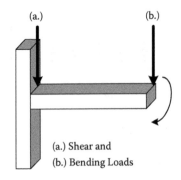

(a.)   (b.)

                    (a.) Shear and
                    (b.) Bending Loads

FIGURE 1.6
Shear and bending loads.

The latter is more challenging to teach, time consuming and expensive to apply, and supported by limited published material properties information. It is the damage tolerance, or fatigue life, approach to design analysis.

Temperature acts on a material to cause thermal strain in more complex ways than forces. Thermal strain results from a material's interatomic spacing localized at the location of the concentration of heat. For this reason, use of the coefficient of thermal expansion provides a less precise understanding of the magnitude and location of material thermal strain.

Thermal strain for a material acts in all three axes with different magnitudes, unlike force-induced strain, which can be evaluated in each axis individually. For example, a tensile load elongates a material in one axis; a thermal strain elongates a material in one axis and swells the material in the other two axes. Thermal strain is most complex to understand at joining and interfaces between disparate materials where the magnitudes differ for each material.

Thermal strain occurs as a steady-state, cyclical, and shock load, often for the same material during one cycle of operation. Steady-state thermal loads are relatively constant magnitudes over time. Cyclical thermal loads vary in consistent pattern or random magnitudes over time. Thermal shock loads are significant changes of magnitude over very short periods of time. For example, the internal combustion engine experiences thermal shock when it is started as the magnitude of thermal load on the structural materials and dynamic components changes from ambient temperature to operating temperature very quickly. Then the thermal load achieves a steady-state magnitude during constant speed operation, and the thermal loads vary cyclically during stop-and-go, accelerate/decelerate operation.

Thermal strain complicates design analysis further by the variability in the conditions of use of a material specified for a system that is widely distributed geographically. Consider the previous example of the internal combustion engine that can be found operating in subfreezing or very hot locations (International Falls, Minnesota, and Death Valley, California, respectively).

Reactivity—the change in chemical composition of a material—is the least understood mechanical failure mechanism, although its effects are well known. Oxidation (rust) is the most common occurrence of reactivity, and design analysis mitigates the failure mechanism through specification of nonoxidizing materials when such materials are both available and applicable. Reactivity includes corrosion of materials that are exposed to chemical and biological contaminants from both operational and environmental sources. Reactivity is often a combined effect of force and thermal failure mechanisms. Combined force and thermal loads change material properties of metals and elastomers from elastic to plastic (embrittlement), changes material chemistry of solder and welds (intermetallic compounds), and changes material properties of lubricants (viscosity), to name a few.

Failure mechanisms act on materials that form the structure of parts. Curiously, there is no standard engineering definition for a part, sometimes called a component. The concept of a part is important to understanding reliability engineering.

The definition of reliability engineering specifies that the probability of failure applies to a part. This is important because only a part can fail. A system does not fail, nor does any subordinate design level in the system design configuration; only the part fails. Reliability engineering investigates the effects of a part failure on the functionality of the next higher design configuration and the functionality of the next higher design configuration above that through to the functionality of the system. This relationship is illustrated in Figure 1.7.

Part selection by the design engineer is the first reliability decision that will influence system reliability, maintainability, and availability and, by extension, customer satisfaction with the system. A part is defined here to be the design unit that is replaced to restore the system to full functionality—a replaceable unit, which is distinguished by the maintenance practice established by system operations and maintenance (O&M). A line replaceable unit

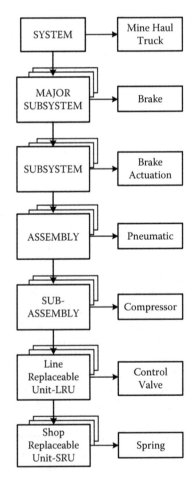

FIGURE 1.7
Design configuration hierarchy.

(LRU) is a part that is replaced on the system. An LRU is repairable if it can be restored to full functionality after it is removed from the system. A shop replaceable unit (SRU) is a part that is replaced in the LRU, usually in a shop or depot. Repairable LRUs are returned to the spare parts inventory to be used as replacements for failed LRUs. An LRU is not repairable if it cannot be restored to full functionality and is discarded following its removal from the system.

The systems engineering WBS allocates technical and functional requirements to the design engineer, who subsequently performs design analysis that defines the part-level design configuration. The design engineer will decide between two part alternatives: make or buy. The "make" decision allows the design engineer to designate the LRU and SRU parts. The "buy" decision allows the design engineer to designate the LRU parts but often not the SRU parts. Systems integration typically addresses LRU parts to the exclusion of SRU parts. The lack of SRU part understanding is foisted on field O&M sustainment engineers, where logistical support analysis determines maintenance practices and spare parts inventory. The SRU parts disconnect has resulted and will continue to result in vastly disparate O&M system reliability, maintainability, and availability results.

Failure Modes

Only a part (LRU, SRU) can have a failure mode, just as only a part can fail. The logical progression is that a failure mechanism causes a failure mode. The failure mode can be expressed as (1) functional or (2) design.

The functional failure mode is described for the LRU during the design concept stage of part design analysis. A functional failure mode can be— indeed, should be—developed prior to part selection. A functional failure mode describes the loss of LRU functionality that results from a failure mechanism. Recording simply that an LRU loses functionality is insufficient. An LRU may have two or more failure modes that experience different lost functionality events; therefore, each must be described. For example, a hydraulic control valve that is subject to the reactivity (oxidation) failure mechanism can fail open, fail closed, or fail intermittently.

Development of the design failure mode follows the functional failure mode and is used to understand LRU failure as part of the design analysis. The design LRU failure mode is described in terms of the SRU material responses to failure mechanisms. For example, the reactivity failure mechanism acting on the valve (LRU) changes the material properties of the diaphragm (SRU) (continuing with the example hydraulic control valve and the reactivity failure mechanism).

Failure Effects—Local

The local failure effect is the result of the SRU failure mode on the LRU. For example, the changes in the SRU material properties of the diaphragm result in the LRU failure effect of shear failure of the diaphragm at the joining with the valve body.

Failure Effects—Next Higher

LRU failure effects are realized in varying degrees by the higher design configuration levels of the system. The preceding design configuration hierarchy figure offers a notional view of the relationships of a spring (SRU) in a control valve (LRU) to the next higher design configuration—the compressor assembly—through to the top design configuration, the haul truck system. The functional and design failure effects' logic converge at the next higher design configuration. The "failure effect—next higher" describes the loss of functionality for each failure mode. For example, the design failure mode, "valve fails open," results in "compressor cannot maintain operating pressure as the failure effect—next higher."

Failure Effects—System (End Effect)

The system failure effect is the response of the system to the LRU failure effect. The description of failure effects as one climbs the design configuration hierarchy morphs from functional descriptions to more general statements of function loss. The system and all design configuration levels above "next higher" can exist in one of three states:

- A *full functioning system* is a state in which no LRU is in a failed state.
- A *degraded functioning system* is a state in which one or more LRU parts have failed causing one or more assemblies to be in a down or degraded state that by extension degrade the system function. For example, dual rear tires on a haul truck are LRU parts that are part of the wheel assembly (next higher design configuration) that is part of the rear suspension subsystem (second next higher) of the truck suspension major subsystem (third next higher), which is part of the haul truck system (end effect). Failure of one tire causes its wheel assembly to be in a degraded mode, causing the rear suspension subsystem to be in a degraded mode that causes the truck suspension major subsystem to be in a degraded mode, which causes the haul truck system to be in a degraded mode.
- A *down system* is the state in which an LRU fails, causing its next higher assembly to be in a down state that, by extension, causes the system to be in a down state.

For example, the starter motor on a haul truck is an LRU that is part of the starter assembly that is part of the electrical major subsystem of the haul truck system. Failure of the starter motor causes the starter assembly to be in a down state, that causes the electrical major subsystem to be in a degraded mode, that causes the haul truck system to be in a down state. Note that the LRU failure effect extended to the electrical major subsystem only puts it into a degraded mode. Other assemblies under the electrical major subsystem are not down and remain functional (e.g., the lighting assemblies).

The progression of an LRU failure effect on the next higher design configuration hierarchy structure must include each level. Doing so recalls a many centuries old ditty:

> For want of a nail the shoe was lost.
> For want of a shoe the horse was lost.
> For want of a horse the rider was lost.
> For want of a rider the battle was lost.
> For want of a battle the kingdom was lost.
> And all for the want of a horseshoe nail.[8]

However, the best-practice guidelines for LRU failure analyses recommend a leap from the next higher design configuration level immediately above the LRU to the system. That guideline shortens the ditty to nonsense:

> For want of a nail the shoe was lost: LRU.
> For want of a shoe the horse was lost: next higher.
> For want of a battle the kingdom was lost: system.

Failure Modes and Effects Analysis

The functional failure modes and effects analysis (F-FMEA), followed by the design failure modes and effects analysis (D-FMEA), is the first and most important analytical tool that provides an understanding of system failure. The FMEA is the structure that combines the failure mechanisms, failure modes, and failure effects into a medium that facilitates the understanding of failure on the system. The FMEA is a bottom-up analysis that uses LRU parts as the line item. Higher level design configuration levels are identified in the title section of the FMEA.[9–11]

The various FMEA formats are defined by standards promulgated by industry and discipline-governing bodies, as well as by corporations. MIL-HDBK-1629 provides a Department of Defense FMEA format used by government contractors as well as private sector corporations that have not created their own formats. Standard-promulgating organizations include the Society of Automotive Engineers (SAE), ASTM International (one of the largest voluntary standards development organizations in the world, originally known as the American Society for Testing and Materials), and American National Standards Institute (ANSI). Corporations that promulgate their internal FMEA formats include global giants like Boeing and Raytheon. Functional and design FMEA formats are consistent in content regardless of the standard followed. This is notionally presented in Table 1.1 using an automobile starter motor for illustration.

Part identification includes part number and nomenclature, at a minimum. Drawing number, reference to WBS code, and vendor identification number can be used in lieu of part number. Consistent use of baseline part identification must be observed. Failure understanding is jeopardized when different part identification is used. A common example is when vendor nomenclature is used to identify a part that has different nomenclature in the systems engineering WBS.

Failure mechanism describes all of the force, thermal, and reactivity stresses as well as their interactions. Failure mechanisms must include operational and environmental sources. A common error is to focus too narrowly on operational and environmental sources of failure mechanisms that are directly attributed to system functionality and those specifically identified by the scope of work by the customer. Limiting the failure mechanism entry to just the expected most prevailing source of stress is another common error (used to simplify the line item entry).

Failure mode describes all of the modes for each failure mechanism. The common errors listed for failure mechanisms are compounded for understanding of failure by preventing identification of failure modes for excluded failure mechanisms.

Failure effects for the local (LRU), next higher (assembly), and end (system) are further affected by the common errors listed for failure mechanisms. Additionally, the lack of intermediate design configurations renders the end effect confusing to those who use the FMEA without benefit of participating in its development.

Most FMEA formats add columns for other information including criticality analysis parameters, actions proposed or taken for mitigation of failure mechanisms, assignment of responsibility for mitigation, dates of actions taken, and updates for criticality analysis resulting from mitigation actions. The impact of adding all the additional information is to make the document unmanageable. Such formats yield a document that must be printed on ANSI-E sized sheets to use a readable font size. Large-sheet FMEAs are unwieldy to use. Consider an FMEA that runs to 50 sheets; there is not enough table-top area or wall space to place the document! An effective FMEA is manageable and meaningful.

FMEA formats have been little changed since their inception. An FMEA is an excellent candidate for a database application. Until a better database FMEA is available, a proposed FMEA format is presented in Table 1.2. The title section includes the design hierarchy in which the LRU line item is located from system down to assembly. Part number and nomenclature, failure mechanism, failure mode, and local and next higher failure effects remain unchanged from the notional FMEA format. The failure effects for all of the design configuration levels above next higher to the end effect are added. The functional state is provided or all failure effects above next higher. This

TABLE 1.1

Failure Modes and Effects Analysis—Notional

System: Automobile

Subsystem: Starter

Part No.	Part Nomenclature	Failure Mechanism	Failure Mode	Failure Effects		
				Local (Part)	Next Higher (Assembly)	End (System)
ABC-01-123	Starter motor		Loosens wire connector	Short circuit	Starter subsystem fails to function	System does not start
		Vibration	Mounting bracket fractures	Starter motor fasteners fail	Wiring stressed by displacement of starter ne'er	None
		Corrosion	Wire connections lose connectivity	Intermittent loss of circuit	Starter subsystem erratic	Intermittent system starts
		Low-temperature moisture	Mechanism locks up	Start fails to function	Starter subsystem fails to function	System does not start

TABLE 1.2

Failure Modes and Effects Analysis—Proposed

System: Automobile
Major Subsystem: Power Plant
Subsystem: Electrical
Assembly: Starter

| | | | | | | Failure Effects | | |
Part No.	Part Nomenclature	Failure Mechanism	Failure Mode	Local (Part)	Next Higher (Assembly)	Subsystem	Major Subsystem	End (System)
ABC-01-123	Starter motor		Loosens wire connector	Short circuit	Starter subsystem fails to function	Down state	Down state	Down state
		Vibration	Mounting bracket fractures	Starter motor fasteners fail	Wiring stressed by displacement of starter motor	Degraded mode	Degraded mode	Degraded mode
		Corrosion	Wire connections lose connectivity	Intermittent loss of circuit	Starter subsystem erratic	Degraded mode	Degraded mode	Degraded mode
		Low-temperature moisture	Mechanism locks up	Start fails to function	Starter subsystem fails to function	Down state	Down state	Down state

distinction from the conventional shows traceability of the LRU failure effect through the entire design configuration hierarchy. The brief statement of the functional state simplifies the document. All other information is excluded.

Part criticality analysis and all other information associated with mitigation are specific to the LRU and justifiably excluded from the FMEA. All engineering tasks are performed under cost and schedule constraints. Criticality analysis (CA) cannot be performed for every LRU! The FMEA serves its purpose by identification of all LRU failures that result in a system down state. That fact alone determines the engineering actions that must have resources allocated to perform the CA.

Criticality Analysis

The CA evaluates the consequences of the system down state and provides a ranking method to determine the most critical to the least critical. There are two approaches to a CA: qualitative and quantitative.

The most common *qualitative approach* is the risk priority number (RPN) method. The RPN evaluates each failure mode for each part by subjectively characterizing scores for three factors: severity of the failure effects of the failure mode (S), likelihood of occurrence of each failure mode (O), and the perception of detection of the occurrence of each failure mode (D). The RPN for each failure mode is the product of three factors rated from 1 to 10. An ideal RPN is a score of $SOD = 1$ ($1 \times 1 \times 1$) and the most critical RPN is a score of $SOD = 1,000$ ($10 \times 10 \times 10$).

- *Severity* is evaluated on a scale of 1 (least severe) to 10 (most severe). Severity is evaluated for consequences to the system (system damage, lost opportunity cost, maintenance costs, availability), personnel (death, lost time, or permanent injury), and regulation compliance (violations that result in punitive actions, mitigation costs). Severity is a lower-is-best criterion where a score of $S = 1$ is the best case and $S = 10$ is the worst case.

- *Occurrence* is evaluated on a scale of 1 (least likelihood) to 10 (will occur). This scale lends itself to subjective probability assessment in increments of 10% (1 = 10% to 10 = 100%). Occurrence is a lower-is-best criterion where a score of $O = 1$ is the best case and $O = 10$ is the worst case.

- *Detection* is evaluated on a scale of 1 (obviously detectable) to 10 (not detectable). Detection is a higher-is-best criterion, but we reverse the score scale to conform to the overall logic of low scores being least critical. A score of $D = 1$ is the best case and $D = 10$ is the worst case.

The most common *quantitative approach* is the modal criticality number (C_m) method. The C_m method evaluates each failure mode for each part by subjectively characterizing scores for three factors: the failure mode ratio, α; the conditional probability of mission loss given that the failure mode occurs, β; and the part failure rate, λ_p. The modal criticality number for each failure mode is the product of three factors times the mission duration, τ. The equation is expressed as

$$C_m = \alpha\beta\lambda_p\tau \tag{1.1}$$

- The failure mode ratio (α) is the probability distribution of the failure modes. It is the percent of the time each failure mode will occur, ranging from 1 to 100%, given that the part fails. Given a part that has two failure modes and one failure mode (A) that occurs 65% of the time failure, then mode B will occur 35% of the time. The failure mode ratios must sum to 100%. A part that has a single failure mode has a failure mode ratio of 100%.

- The conditional probability of mission loss given that the failure mode occurs (β) is the probability that the end effect of the failure mode is mission loss. The conditional probability of mission loss, ranging from 1 to 100%, is related to the severity of the failure mode. It is not a probability distribution and need not sum to 100% for all failure modes.

- The part failure rate, λ_p, is the only quantitative variable and is a calculated value. The product of the failure mode ratio and the part failure rate characterizes the portion of the part failure rate that applies to each failure mode. For a part that has two failure modes (A and B), a part failure rate of 500 failures per 1,000,000 h and failure rate ratios of $\alpha_A = 70\%$ and $\alpha_B = 30\%$, we can say that 350 failures per 1,000,000 h ($\alpha_A\lambda_p = 0.7 \times 500$ failures per 1,000,000 h) are due to failure mode A and 150 failures per 1,000,000 h ($\alpha_B\lambda_p = 0.3 \times 500$ failures per 1,000,000 h) are due to failure mode B.

One will notice that the subjectivity of the failure mode ratio and the conditional probability of mission loss, given that the failure mode occurs, hardly make the method quantitative; it is just as qualitative as the RPN method. A troubling assumption of the modal criticality number method is that part failure rates are constant and random over the useful life of the part. This assumption is correct for electronic or digital parts, but it does not reflect the reality of mechanical, dynamic, and structural parts that wear out over the useful life. A comparison of the two methods is provided in Table 1.3.

The parameters for the RPN and modal criticality number methods are notionally presented. The part failure rates (λ) are expressed as failure per million hours. The rank ordered values for RPN and C_m are tabulated below the part input table. Note that several part failure modes with high severity are ranked low and several part failure rates with lower severity are ranked

TABLE 1.3

Comparable Criticality Analyses Methods

RPN Method					Modal Criticality Analysis					
Criticality Analysis					Criticality Analysis					
Part (LRU)	Failure Mode	S	O	D	RPN	Part (LRU)	Failure Mode	α	β	γ	τ	C_m
Part 1	1A	4	2	6	48	Part 1	1A	0.6	0.8	550	8	2112
Part 1	1B	3	7	1	21	Part 1	1B	0.4	0.5	550	8	880
Part 2	2A	9	1	9	81	Part 2	2A	0.5	0.3	30	8	36
Part 2	2B	7	1	1	7	Part 2	2B	0.3	0.1	30	8	7.2
Part 2	2C	5	6	7	210	Part 2	2C	0.2	0.1	30	8	4.8
Part 3	3A	1	5	8	40	Part 3	3A	0.7	0.6	10	8	33.6
Part 3	3B	1	7	9	63	Part 3	3B	0.3	0.1	10	8	2.4
Part 4	4A	2	3	9	54	Part 4	4A	0.9	0.2	125	8	180
Part 4	4B	1	2	10	20	Part 4	4B	0.1	0.1	125	8	10
Part 5	5A	5	2	3	30	Part 5	5A	0.6	0.8	25	8	96
Part 5	5B	3	8	3	72	Parts	5B	0.3	0.7	25	8	42
Part 5	5C	1	3	2	6	Parts	5C	0.1	0.3	25	8	6
Part 6	6A	1	4	4	16	Part 6	6A	1	0.1	1	8	0–8
Part 7	7A	6	4	1	24	Part 7	7A	1	0.2	1	8	1.6
Rank Ordered Criticality Analysis					Rank Ordered Criticality Analysis					
Part (LRU)	Failure Mode	S	O	D	RPN	Part (LRU)	Failure Mode	α	β	γ	τ	C_m
Part 2	2C	5	6	7	210	Part 1	1A	0.6	0.8	550	8	2112
Part 2	2A	9	1	9	81	Part 1	1B	0.4	0.5	550	8	880
Part 5	5B	3	8	3	72	Part 4	4A	0.9	0.2	125	8	180
Part 3	3B	1	7	9	63	Part 5	5A	0.6	0.8	25	8	96
Part 4	4A	2	3	9	54	Part 5	5B	0.3	0.7	25	8	42
Part 1	1A	4	2	6	48	Part 2	2A	0.5	0.3	30	8	36
Part 3	3A	1	5	8	40	Part 3	3A	0.7	0.6	10	8	33.6
Part 5	5A	5	2	3	30	Part 4	4B	0.1	0.1	125	8	10
Part 7	7A	6	4	1	24	Part 2	2B	0.3	0.1	30	8	7.2
Part 1	1B	3	7	1	21	Part 5	5C	0.1	0.3	25	8	6
Part 4	4B	1	2	10	20	Part 2	2C	0.2	0.1	30	8	4.8
Part 6	6A	1	4	4	16	Part 3	3B	0.3	0.1	10	8	2.4
Part 2	2B	7	1	1	7	Part 7	7A	1	0.2	1	8	16
Part 5	5C	1	3	2	6	Part 6	6A	1	0.1	1	8	08

higher using the RPN method. A similar statement can be made for the conditional probability of mission loss, given that the failure mode occurs using the modal criticality number method. Neither method effectively ranks safety or reliability issues of severity or the conditional probability of mission loss, given that the failure mode occurrences are considered to be important.

A more distressing feature of both criticality analysis methods is the ambiguity of the subjective values. On many occasions, I have asked engineers in reliability training to assign the parameters of the RPN method (S, O, and D) to the water pumps in their cars and calculate the RPN. The RPN scores ranged from a low of 13 to highs in the 400s. One would think that such wide variations would not happen among engineers who know car design generally and their cars specifically. The variation was more pronounced when the same engineers were tasked to calculate the modal criticality number for the same water pump given three internal part SRUs (the hose connection, pump seal, and pump vanes). The failure mode ratio can be quantitative only when empirical data are available; it is subjective in practice when engineers must estimate the ratios. So too is the conditional probability of mission loss; the range from low to high was in two orders of magnitude in this application. The conclusion is that neither criticality analysis method is effective.

The consequences analysis developed by John Moubray is a far more effective criticality analysis method; it achieves the prime purpose to identify the relevant few from the insignificant many, to paraphrase the nineteenth century economist Alfredo Pareto (see Figure 1.8). Moubray defined the parameters of part criticality to be the interactions of three parameters: (1) end effect on the system (catastrophic, operational, degraded mode, and run to failure), (2) the operator's awareness of the degradation of the part (evident, not evident), and (3) the time from perception of the degradation to the part failed state: perception (P)–failure (F) interval.

System End Effects

The end effect is a straightforward description of what happens to the system if the part fails. The end effect determination is not ambiguous and can be recognized and understood by design and system engineers, project and organization managers, system operators and maintainers, and maintenance planners and supervisors. The direct consequences of the part failure are related to the severity parameter (S) of the RPN CA method, and the conditional probability of mission loss (β) of the modal criticality method. The engineers asked to calculate the criticality of water pump failure were unanimous in their assessment that the end effect is operational.

P–F Interval

The time between the perception of part degradation and occurrence of part failure is Moubray's P–F interval. Part degradation ranges from instantaneous to a life cycle of thousands of hours. System operators have the P–F interval to mitigate part failure. Consider a tire on the left front wheel of a car; a "hot" spot on the sidewall can cause a part failure that has a catastrophic end effect: The system operator loses control of the car, the car veers to the left into oncoming traffic, and then a head-on collision kills the occupants of both cars and

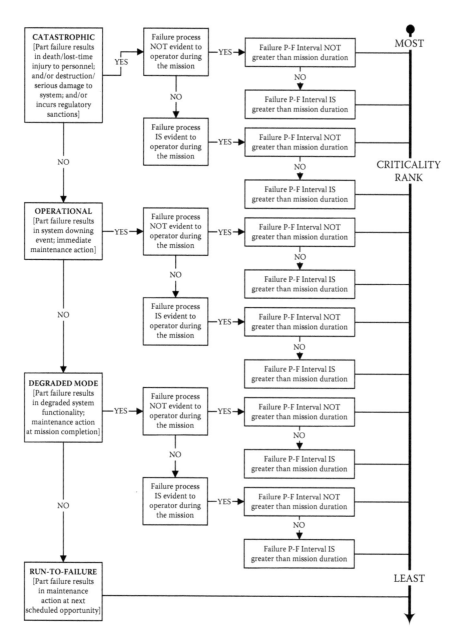

FIGURE 1.8
Consequences analysis logic.

destroys the cars. The part failure was not evident to the operator, and the P–F interval was instantaneous.[12] Compare the tire failure to the water pump; the system end effect consequences are operational, the operator perceives the part degradation leading to failure, and the P–F interval of many minutes provides the operator with sufficient time to implement mitigation. The P–F interval of part failure has no relationship with the RPN CA or the modal criticality method. The engineers were unanimous on the duration of the P–F interval.

Design engineers need to determine whether the P–F interval is greater than or less than the mission duration. A P–F interval that is less than mission duration means that a system will experience a down state with catastrophic or operational end effects for a part failure that begins its degradation after the mission begins. A P–F interval that is greater than mission duration means that a system will complete the mission for a part failure that begins its degradation after the mission begins.

The relationship between P–F interval and mission duration is a threshold metric that can influence

- design
  - active redundant dual tires on tractor-trailer systems located where loss of a tire will have a catastrophic end effect on high-speed highways
  - overdesigned rotor blades on a helicopter where redundancy is not technically feasible and blade failure will have a catastrophic end effect
- maintenance
  - daily inspections of conveyor belt roll-cylinders to schedule part replacement where wear-out is visually evident and the P–F interval is much longer than a day-long shift
  - scheduled maintenance intervals to implement replacement, inspection, and repair of mining haul truck air brake canisters where wear-out is not evident, the P–F interval is instantaneous, and the end effect can be catastrophic
- operations
  - reduced loading on the system to extend the probability of completing the mission or improving the system downing situation, as illustrated by a loaded mine haul truck experiencing a rear dual tire failure that drives away from the haul road traffic, dumps its load, and travels out of the mine to the maintenance facility
- will not interfere with scheduled events

Operator Awareness of Degradation

Operator awareness of degradation requires the design engineer to assume that the operator is trained and is able to use functional performance information available to determine if a deviation from normal part condition indicators has occurred.[13] System downing events are potentially far more critical when the cause is hidden from the operator. Surprises in system functionality can be confusing, cause panic or rash reactions, and are the worst situation for fault detection, isolation, and mitigation.

The contrast is where it is evident to the operator that degradation has occurred. Fault detection and fault isolation can be determined in a calm and deliberate atmosphere. If mitigation exists, it will be potentially more likely to be recognized and implemented. There is a relationship between the operator's perception of the part failure to the detection parameter (D) of the RPN CA method and that there was not a corresponding parameter of the modal criticality method. The engineers asked to calculate the criticality of water pump failure were unanimous in their assessment that the part degradation (fault detection) was evident from the water temperature gauge in the car (fault isolation to the cooling subsystem) and that the fault mitigation was to exit the road safely.

Maintainability and Maintainability Engineering[14,15]

Maintainability is defined to be the probability that a system can be restored to functionality; it forms the basis for the development of the maintenance concept employed by the system operator and maintainer. Maintainability engineering is a dependent element of reliability engineering. It is intuitively obvious that part failure must occur to create the need for a system to be restored to full functionality. Logically, maintainability is the response to failure that is described by reliability engineering. Design for maintainability is the extension of design for reliability where the likelihood of failure is decreased by design mitigation and followed by definition of features that make failures detectable, locatable, and manageable.

Effective system maintainability is achieved when part failure mechanisms acting on every part with associated failure modes are understood. Understanding part failure modes provides guidance to developing maintenance practices that are best suited to remove and replace parts. Understanding maintenance practices provides guidance to defining maintenance skills, specialty tools, and facilities required to perform the work. Understanding how failure mechanisms behave provides guidance to developing spare parts scenarios and policy.

Maintainability also includes servicing tasks that are essential to sustaining system functionality. It is not a response to part failure but rather a periodic replenishment of consumable materials required for a part to function. Consider the personal car: Parts require lubricants to function. A periodic change of engine oil is an essential maintenance task that is not initiated by part failure. Servicing tasks are operational conditions of use of the system, a critical element of the definition of reliability. Servicing tasks can provide a point of confusion. Some maintenance actions can be service tasks (e.g., cleaning an air filter at the same time that engine oil is replaced) or a repair task (e.g., replacing an air filter that fails).

Fault Detection

Fault detection is the accepted term for detection of a degraded mode or down state—not a part failure mode. Fault detection identifies failure effects: the symptoms of part failure. Design for maintainability uses the failure modes and effects analysis to document opportunities for fault detection.

Failure effects are detectable at design configuration levels ranging from the failed part to the next higher assembly and up to the system. An illustrative example is a car that will not start. The operator has detected a fault that is manifested by the end effect of a part failure causing the down state of the system. The same car is driving normally when it begins to overheat. The operator has detected a fault that is manifested by the effect of a part failure causing the down state of the cooling subsystem, which may cause a system down state.

In the same car, the brakes are sluggish and do not stop the car in the expected time and distance based on the speed of the car. The operator has detected a fault that is manifested by the effect of a part failure causing the down state of an assembly in the braking subsystem that may cause a catastrophic system down state. The car experiences a blowout failure of a tire (the part), causing an immediate system down state, including down states for all of the design configuration levels from the assembly up to the system.

Fault Isolation

Fault isolation is the accepted term for isolating and identifying the failed part that caused the failure effect. Maintainability best practices for fault isolation use a fault tree analysis (FTA), which is a systems integration tool. Fault tree analysis is a top-down failure analysis that starts with a system down state and analyzes the design configuration levels to drill down to identify the failed part. The result of this practice posits that the design engineer does not have a role in fault isolation. This view is changing with the implementation of reliability-centered maintenance that recognizes that fault isolation is achieved by recognizing the failure condition of the failed part. Condition indicators are the manifestation of failure mechanisms acting on the part and can be characterized by the design engineer. Condition indicators serve to identify a potential system down state prior to the occurrence of the down state.

Fault detection and isolation that is driven by condition indicators can range from operator perception of degraded system functionality and system downing events, built-in test equipment, off-system test equipment, and visual inspection. Most cars have battery volt meters, engine temperature and oil pressure gauges, and RPM tachometers that measure and report vital system condition indicators. Car maintenance service providers have computer-based diagnostic equipment that measures and reports additional vital system condition indicators. Visual inspection of drive belts, air filters, and tire tread depth also measures and reports vital system condition indicators. All of these tangible means to measure and report condition indicators are provided by design engineers.

Operator "feel" is an intangible measure of system condition. Anecdotal evidence reports that many process equipment and mobile machinery operators can perceive a departure from normal system functionality based on the senses of sound, sight, and smell. This is one aspect of fault detection and isolation that is beyond the scope of design engineers.

Part Mean Time to Repair

Part mean time to repair (MTTR) is the fundamental part maintainability design metric. Repair is defined to be all of the direct maintenance actions required to replace a failed part. Repair actions are those that the design engineer can define and describe that will differ little from the repair actions that are performed on the part by the operator and maintainer. Repair actions include accessing the failed part, removing the failed part, and installing the replacement part.

Part MTTR is characterized in design by analysis and maintainability experiments. Characterization of a part MTTR is an expense justified by the end effect of the part failure: the findings of the FMEA and criticality analysis. Analysis includes the maintenance history of the part, or a similar part, from prior field experience and math modeling and simulation. Maintenance data from field use are a sample of prior repair records. The characterization of the MTTR is the arithmetic average of the time to repair (TTR) and its standard deviation. Math modeling and simulation of the MTTR is a Monte Carlo approach that uses the triangular distribution from an Adelphi survey to characterize the minimum, modal, and maximum TTR for the part. Maintainability experiments are performed on parts under controlled conditions where the TTR is calculated from the empirical results of the tests.

Administrative and Logistical Downtime

Repair actions must not be confused with prerepair logistics actions, postrepair logistics actions, or administrative actions. Each is an element of the total time during which a system is down following a part failure.

Repair actions are performed under a system of management controls that requires documentation of direct labor, direct materials, and direct overhead expenses. The allocation and scheduling of repair actions are essential administrative actions that add time to the task of restoring a system to full functionality. Administrative actions vary between organizations and cannot be anticipated by design engineers.

Logistical actions also add time to the task of restoring a system to full functionality. Prerepair logistical actions include isolating the location of the failed part, determination of the maintenance practice to be performed and skills required, acquiring the replacement part, and accessing specialty tools and facility. Postrepair logistical actions include verifying the repair actions performed to restore the system and disposal of the residue of the repair actions. Logistical actions vary between organizations and cannot be anticipated by design engineers.

Part and System Availability[16,17]

Part availability (A) is the probability that a part is in a full functional state when it is scheduled for use; it is determined by the reliability and maintainability of the part design. Assembly, subsystem, and system availability (A) is the probability that the design is not in a degraded mode functional state when it is scheduled for use and is a function of the availability of all subordinate design configuration levels. Availability is the ratio of reliability (uptime) to reliability plus maintainability (downtime) and can be expressed in general terms as

$$A = \frac{\text{Uptime}}{\text{Uptime} + \text{Downtime}} \tag{1.2}$$

Availability expressed as a constant is typically referred to as the part's or system's steady-state availability. Availability is calculated for four phases of system design and operation; each is based on how uptime and downtime are measured:

- inherent availability: part to system design phase, predictive of operational performance
- instantaneous availability: part to system design phase, predictive of operational performance
- operational availability: system integration design and system sustainment phases, predictive of operational performance
- achieved availability: system sustainment phase, descriptive of operational performance

Reliability in an Organization[18,19]

Reliability, when it is present in an organization, is located in quality assurance. The role of quality assurance is to assure conformance to design and functional specifications. How that role is achieved must change in the highly competitive global economy.

Quality assurance was intentionally isolated from design, production, and sustainment engineering until the advent of total quality management (TQM) and six-sigma (6σ) quality. Internal barriers were created that placed physical and cultural walls between departments. Quality, safety, and reliability engineering departments within quality assurance were in separate locations and viewed their respective missions as distinct from each other. The interface between quality, safety, and reliability engineering departments and design, systems, and sustainment engineering occurred at design reviews. These reviews emphasized whether conformance to functional specifications had been achieved and how to mitigate those occasions where they were not achieved (redesign or waiver) (see Figure 1.9).

The barriers between departments made sense in an adversarial way: Quality assurance would answer to top management to avoid being influenced by the departments that it audited. This discredited approach to quality assurance was referred to as inspecting quality into the system.

Organizational structure is a tactical approach to implementation of an organization's mission and must be based on the organization's strategic plan. Organizational structure defines the allocation of work and resources that produce value for the organization. That value is traded on the open market for cash flow or its equivalent.

The fundamental objective of any organization's strategic plan is economic survival. Private sector for-profit organizations (private business as large and global as General Electric and as small and local as a sole proprietor) achieve economic survival by maximizing the rate of return on owner's investment. Private sector not-for-profit organizations (private charitable and social organizations as large and global as the International Red Cross and as small as the local Boy Scout troop) achieve economic survival by sustaining a positive cash flow—solvency. Public sector organizations achieve economic survival by spending no more than the budget provided by the respective governing authority. Cash flow is the common thread that determines the success (positive cash flow) or failure (negative cash flow) of every organization.

The engineering design process in a barrier-based organization is allowed to proceed to a milestone review where quality assurance review evaluates its acceptability. The majority of the design tasks (design analysis, design art, and bill of materials) were completed long before the design review and the responsible engineers had worked on successive tasks leading up to the design review. The rework feedback loop on planning flow charts looks good

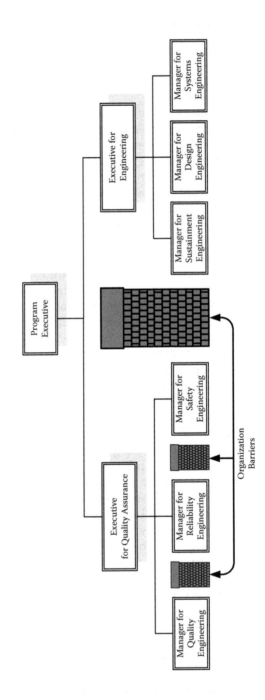

FIGURE 1.9
Conventional organizational structure with barriers.

on paper, but it is doomed to wreak havoc on a design process that must meet technical, cost, and schedule requirements. Such rework should not occur.

TQM and 6σ introduced a management theory where the owner of the work performed was responsible for its quality, as well as the concept of internal customers, where all work performed in an organization is a process of services performed by employees. Each process has one or more inputs and one or more outputs. An employee performing a task is the customer of the work-in-progress item inputs and is the vendor of the work-in-progress item outputs passed on to the next process task. TQM and 6σ empower and delegate system process control and capability to the task owners. The empowerment is accomplished by training employees to understand, use, and analyze the findings of quality engineering methods. The rework feedback loop in TQM and 6σ organizations is instant and is performed by the same employee who performed the work in need of rework.

The Need for Change in Conventional Organizational Structure

The transition emerging in organizations that have achieved world-class quality has not engaged reliability and safety engineering. Industrial, institutional, and system safety engineering continues to be viewed as an enforcement organization that integrates design and sustainment engineering solutions to meet regulatory requirements. Reliability engineering is viewed as a specialty that is not understood in its entirety. Ask a system program manager to identify the role of reliability engineering in design, systems integration, and sustainment; he or she will rarely mention maintainability and availability and will probably stammer about failure modes and effects analysis.

World-class high-quality, reliability, and safety systems are the result of the integration of quality, reliability, and safety engineering analysis in the design analysis performed by the engineering department.[20] The design engineer applies functional discipline, reliability, quality, and safety design analysis from the concept through the validation stages in such organizations. Many analytical tools used to perform functional design analysis are common to the reliability, quality, and safety design analysis. For example, the stress-strength analysis leads to a functional, reliability, quality, and safety materials selection. Many reliability analysis tools are common to quality and safety. For example, the FMEA identifies and ranks parts that demand investment in quality process control and capability and also identifies and ranks that demand investment in safety hazards analysis.

The systems integration engineer applies functional discipline, reliability, quality, and safety design analysis from the concept through the validation stages in such organizations. Systems integration analytical tools used to perform functional integration analysis are common to the reliability,

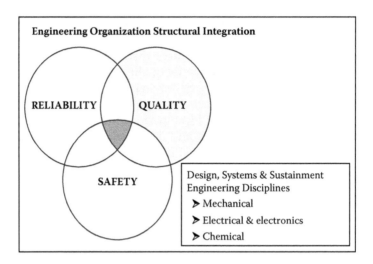

FIGURE 1.10
Reliability, quality, and safety interactions.

quality, and safety integration analysis. For example, system configuration trade studies to evaluate make–buy and design configuration decisions lead to a functional reliability, quality, and safety decision. Many reliability integration tools are common to quality and safety. For example, the reliability block diagram (RBD), math model, and simulation evaluate reliability, quality, and safety model response parameters.

The sustainment engineer applies functional discipline, reliability, quality, and safety sustainment analysis to optimize maintenance policy in such organizations. The impact is more evident than in design and systems engineering. Sustainment engineers see immediate feedback from the application of reliability, quality, and safety to increase system effectiveness and reduce system operating costs (see Figure 1.10).

Proposed Organization Structure

The proposed organization structure removes barriers in two ways. First, the performance of reliability, quality, and safety engineering actions is shifted to the sustainment, design, and systems engineering departments through personnel assignment and training. Second, the quality assurance department changes its mission to training engineering employees to perform reliability, quality, and safety engineering tasks from performing reliability, quality, and safety engineering tasks. The engineering department becomes the customer and quality assurance the vendor of reliability, quality, and safety engineering knowledge infrastructure. Quality assurance responds to changes in engineering demand for reliability, quality, and safety engineering knowledge infrastructure (see Figure 1.11).

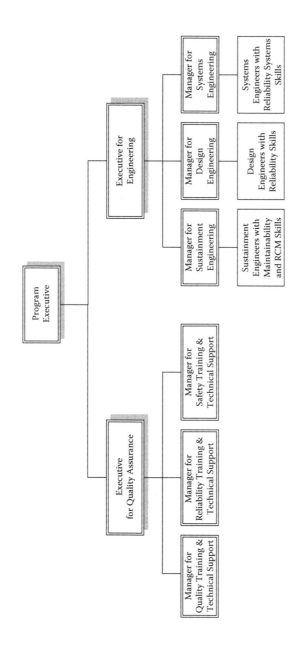

FIGURE 1.11
Proposed organizational structure.

Design for Reliability: Reliability Engineering Requirements for Part Design

A guideline is presented for an organization to restructure the quality assurance and design engineering departments to allocate tasks and resources to achieve design for reliability for part design. The burden is for reliability engineering to fit the part design process, rather than to create a new design process. An accepted part design flow chart is presented in Figure 1.12 that shows the assignment of reliability tasks. The objective of the guideline presented is to describe what must be done and the order in which the reliability tasks are performed in the design process. The methods used to perform the reliability tasks are provided in subsequent chapters (see Figure 1.12).[21]

Design Requirement for a System

The design requirement for a system is defined in a contract agreement between an organization and its customer and is the starting point for all part design. System design requirements provide the functional requirements for the system and the environment that the system will experience during operations. System requirements include functional and environmental metrics that must be met and applicable standards to assure that accepted procedures are used to evaluate the metrics.

Functional requirements can be as vague as the requirements for an unmanned vehicle that can detect enemy motion in poor or low visibility and retrieve a wounded soldier or as specific as the requirement for an unmanned vehicle that has forward-looking infrared sensors and is capable of dragging a 200-lb dead weight.

System requirements identify contract deliverable documents and milestone events. Contract deliverable documents include but are not limited to the design analysis, design art, bill of materials, test plans and test reports, and operator and maintenance manuals.

Milestone events include design reviews (concept, preliminary, critical, and final), qualification and demonstration test events, and technical interchange meetings.

Systems Engineering Work Breakdown Structure

The system design requirement is translated into a WBS by the systems engineering department. The work breakdown structure identifies the *concept* design configuration hierarchy down to the assembly and, to a lesser degree,

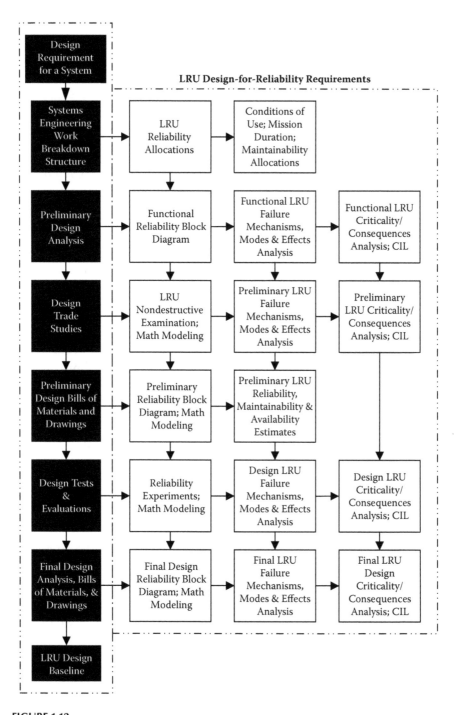

FIGURE 1.12
Design-for-reliability logic. (Adapted from Kerzner, H. 1992. *Project Management,* 4th ed. New York: Van Nostrand Reinhold.)

to the part/LRU design configuration level. Concept design configuration levels define the function that will be performed but do not identify the part. The engineering disciplines required at each level of the design configuration hierarchy are defined in parallel to the concept. For example, an unmanned ground vehicle concept design specifies an electric-drive power supply and electrical, mechanical, and controls engineers.

Functional and environmental requirements metrics are allocated down through the design configuration hierarchy to the part/LRU. Allocations are treated as budgets for limited capability. For example, the electrical power, weight, and volume available for the electric-drive power supply are allocated on the basis of the system concept design limits.

Lowest Replaceable Unit, LRU, Reliability Allocations

The system WBS defines the part/LRU design configuration level. The system design requirement is the source for reliability requirements. A system reliability requirement can be stated in terms of probability percent (e.g., 90% reliability that the system will function without failure for a mission duration of 8 h) or as a statistic (e.g., mean time between downing events of 360 h). Stating a confidence limit on the two statements for a system reliability requirement is a superior statement of a system reliability requirement. The system reliability requirements are allocated down to the LRU.

Systems engineering must provide a system reliability requirement goal in the absence of a reliability requirement from the customer. The lack of a system reliability requirement is as wrong as lacking a system functional requirement. The responsibility to understand the needs of the customer is the responsibility of the vendor regardless of how helpful customers are in articulating their needs.

The LRU design configuration level is an input over to the conditions of use, mission duration, and maintainability allocations and down to the functional reliability block diagram.

Conditions of Use, Mission Duration, and Maintainability Allocations

The conditions of use must be understood as much from what information is not included in the system design requirement as from what information is included in the requirement. Conditions of use include understanding stresses resulting from system operation that part material strength is expected to resist and understanding stresses resulting from external sources (e.g., vibration and shock, temperature and humidity, and exposure to corrosive agents caused by the location of system operation) that part material strength is expected to resist. Mission duration is an estimate for how long the system will be scheduled to operate between opportunities for maintenance actions.

The system maintainability requirements must be allocated down to the part/LRU design configuration level. The conditions of use and mission duration contribute to the design decisions that determine the ability of the system to conform to maintainability requirements. Maintainability system design requirements include some or all of the following: MTTR (e.g., 2 h), fault detection (FD) probability (e.g., 99% of critical LRUs), fault isolation (FI; e.g., 95% to one LRU, 90% down to a serial path of several LRUs), and a false alarm rate (e.g., less than 5%).

The conditions of use and mission duration contribute to the design decisions that determine the ability of the system to conform to availability requirements (e.g., *A* greater than 90%).

Functional Design Analysis

The preliminary design analysis is the transformation of the WBS to provide a system structure that will achieve the system functionality. Functional allocations are assigned to lower levels of the design hierarchy. A formal design review (FDR) provides the opportunity to verify the efficacy of the preliminary design specifications.

Functional Reliability Block Diagram

The preliminary design analysis initiates the development of a functional reliability block diagram. The RBD translates the WBS into a functional process flow chart that describes the design configuration, starting at the system level down the hierarchies to the part level. The blocks are functionally labeled because actual part definition has not yet occurred. An example of the functional RBD is presented in the Figure 1.13.

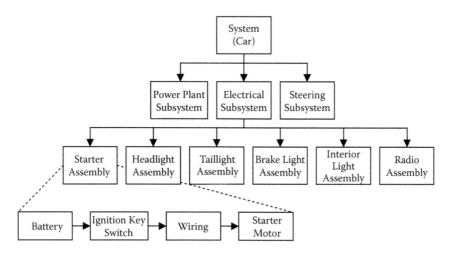

FIGURE 1.13
Functional RBD: car.

The functional RBD illustrates a partial WBS for the design of a car that shows the system, a car, subsystems, power plant, electrical and steering, assemblies below the electrical subsystem, starter, headlight, taillight, brake light, interior light, and radio. The functional RBD departs from the WBS format at the LRU design configuration level under the starter assembly, where the LRUs, battery, ignition switch, wiring, and starter motor are depicted as a process flow chart. This example describes the starter assembly as a serial design of parts. Note that the functional RBD does not specify the selection of LRUs.

The functional RBD inputs over to the functional LRU failure mechanisms modes and effects analysis and down to the LRU nondestructive examination and math modeling.

Functional LRU Failure Modes and Effects Analysis

The functional LRU failure modes and effects analysis is constructed from the functional RBD. The failure modes and effects analysis is the first reliability analysis of the design and enables an understanding of LRU failure effects on each design configuration hierarchy up to the system. Each LRU begins as a single point of failure that will cause the starter assembly to be in a down state. The down state of the starter assembly will cause the electrical subsystem to be in a degraded mode that will put the system in a down state. The F-FMEA inputs over to the preliminary LRU consequences analysis and critical items list (CIL) and down to the LRU reliability, maintainability, and availability estimates.

Functional LRU Criticality/Consequences Analysis and Critical Items List

The functional criticality/consequences analysis (F-CA) for LRU expands on our understanding of failure by describing the impact of LRU failure on the system. Not all failures are equal in system impact; some need to be mitigated and some do not. Scarce resources and limited time demand that we be able to distinguish the differences quickly and accurately. The consequences of the failure of any starter assembly LRU put the car in a down state. All the starter assembly LRUs have an operational critical consequence for loss of system function.

Contrast the understanding of the down state of the starter assembly to the down state for the interior light assembly. The interior light assembly also puts the electrical subsystem in a degraded mode, but the end effect is not a down state for the car. LRUs in the interior light assembly (light bulb, switch) do not have an operational critical consequence. Next, evaluate the down state for the brake light assembly. The brake light assembly also places the electrical system in a degraded mode, but the end effect is a degraded mode of the system that can have a catastrophic consequence by causing an accident or a regulatory consequence from being cited by the police.

The functional critical parts are listed in the first version of the CIL. The CIL identifies all of the parts that require failure mitigation. Ideally, the CIL will be reduced to zero parts, but this is often an unrealistic outcome. Consider critical LRUs in the brake hydraulic assembly on a car. LRU failure always has the catastrophic consequence of loss of control and collision at high speed. Failure mitigation cannot prevent LRU failure, but design decisions can make the likelihood of the event remote and, in combination with maintenance actions, very remote. The F-CA and F-CIL input down to the preliminary (P)-CA and P-CIL.

Design Trade Studies

Design trade studies are performed to identify, evaluate, and select a material or part to fill the functional placeholder of the WBS. Trade studies are analytical evaluations of design, functional, cost, and schedule factors for the alternative materials and parts.

LRU Nondestructive Examination and Math Modeling

Reliability contributions to design trade studies combine physical understanding of material and part characteristics through nondestructive examination (NDE) techniques and understanding of material and part failure through math modeling. NDE techniques are noninvasive methods that measure harmonic frequencies, thermal conduction and expansion, resistance to oxidation and corrosion, surface treatments, mass density, and geometry.

Math modeling includes finite element and reliability math models. Data from the NDE are used to construct mesh and node structure and boundary conditions for the finite element math model. Finite element simulation is performed by introducing stresses and observing the reactions. Finite element math models and simulations sort the statistically significant failure mechanisms from the trivial.

Reliability math models are developed for the statistically significant failure mechanisms identified from the finite element math modeling and simulation. Reliability math models fit failure data to the probability density function (pdf), $f(t)$, with the least error. The pdf leads to the characterization of the survival function, $S(t)$, the hazard function, $h(t)$, and the reliability function, $R(\tau|t)$. The LRU nondestructive examination and math modeling input over to the P-FMEA and down to the preliminary RBD and math modeling.

Preliminary LRU Failure Mechanisms: Modes and Effects Analysis

Failure modes and effects analysis is an iterative process through the design cycle and should be baselined at each design milestone review. The preliminary failure mechanisms modes and effects analysis is updated to replace the functional line items with the specific materials and parts defined in the trade studies. The statistically significant failure mechanisms modes and effects are documented in the P-FMEA.

This is a very important transition for the FMEA process. Complaints about the FMEA process are that it is too time consuming and provides too much information. Reducing the F-FMEA to the P-FMEA focuses on only the materials and parts that have failures that must be understood. Line items that do not have statistically significant failure are not included. The P-FMEA inputs over to the P-CA and P-CIL and down to the preliminary reliability, maintainability, and availability estimates.

Preliminary LRU Criticality/Consequences Analysis and Critical Items List

The P-FMEA narrows the scope of the consequences analysis to the meaningful few. The P-CA of LRU failure is understood in the context of the failure mechanisms and modes. Failure mitigation options are better understood. The P-CIL is more specific. Management and customer support for proposed design mitigation is facilitated from clear understanding of the factors that put a material and part on the P-CIL. The P-CA and P-CIL input down to the design criticality analysis and critical items list.

Preliminary Design Bills of Materials and Drawings

The preliminary (P) design bills of materials (BOM) and drawings are the product of the preceding design and reliability analysis and decisions. A preliminary design review (PDR) is conducted at this point in the system design process, where the vendor communicates the progress and achievements of the system design process to the customer. The reliability design goal at design reviews is to demonstrate a clear understanding of the reliability, maintainability, and availability requirements; conditions of use; and mission duration and to present the allocation of those requirements down to the part design configuration level.

This is not the time to discover the need for major rework. Minor differences typically occur, but such differences should be customer-driven requests resulting in the customer's better understanding of the needs illustrated by the design review. Such changes to the scope of work are rightly funded by the customer and revisions to the program plan are negotiated. The vendor absorbs the costs and schedule penalties to rework the design when the customer assesses that any differences exist because the design did not correctly understand the specification of the system requirements.

Preliminary Reliability Block Diagram and Math Modeling

The preliminary reliability block diagram (P-RBD) is a continuation of the development of the system design. System reliability growth is investigated based on design and maintainability solutions to failure mitigation and evaluated to include changes to the design configuration from serial to parallel and redundant introduction of standby parts and derating materials. Part reliability math models for the design solution are developed and analyzed to assess the reliability growth improvements achieved. The P-RBD and math modeling input over to the preliminary reliability, maintainability, and availability estimates.

Preliminary LRU Reliability, Maintainability, and Availability Estimates

The results of the P-RBD and math modeling generate point estimates and estimated confidence limits for the parameters of LRU reliability, maintainability, and availability. These estimates are compared to the allocations to find differences that must be reconciled. The reconciliation is a technical interchange among multidiscipline, multifunctional design, and management employees. The preliminary reliability, maintainability, and availability estimates input down to the design failure mechanisms modes and effects analysis.

Design Tests and Evaluation

All design activities to this point are by analysis. The analysis is verified by functional test and evaluation. Functional tests range from physical testing to assure that material strength properties are realized to evaluation of input/output functions on a small-scale operational test.

Reliability Experiments and Math Modeling

Reliability and maintainability experiments are designed and performed to verify the findings of the finite element math models and simulations and the reliability math models. Reliability experiments serve three purposes:

- *Use empirical data to estimate parameters of the part reliability functions, including mean time to failure (MTTF) and MTTR.* Accelerated life tests (ALTs) are designed to simulate part life exposure to operating stresses through condensing cycle time or application of acceleration factors. ALTs can range from functional cycling tests to stress cycling tests. Maintainability experiments serve two purposes:
  - *Use empirical data to estimate MTTR.* Maintainability experiments are performed to estimate the mean time to replace an LRU. Experiments are controlled, timed events in which a part repair is performed. The experiment is controlled by fixing repair factors that reduce the complexity of the experiment. The experimental scenario is one source of controlled factors that include but are not limited to ambient temperature, use of a facility protected from the elements rather than a field site, skills and experience of

the maintenance technicians, and lack of wear or induced failure of the part. The objective is to characterize the ideal part TTR.

- *Use empirical data to verify fault detection and fault isolation.* Fault detection, fault isolation experiments are tests of the fault-detection/fault-isolation (FD/FI) method, built-in test equipment (BITE), off-system diagnostic equipment, and inspection methods. Seeded faults are randomly introduced in a selection of parts from a larger sample. The sample parts are integrated into a sample of the next higher assembly. The test fixture is activated and the abilities of the BITE, off-system diagnostic equipment, and inspection methods are evaluated for their effectiveness.

- *Induce failure to identify material and part destruct limits.* Highly accelerated life tests (HALTs) are designed to subject materials and parts to stress factors, including vibration, shock, temperature extremes, thermal shock, humidity, and exposure to reactive agents. Stress factors can be applied individually, as main effects, or combined as interactive effects.

- *Screen parts to verify that they meet vendor reliability parameters.* Highly accelerated stress screening (HASS) exposes parts to ambient and operational conditions of use to verify the part's survivability. Probability ratio sequence tests (PRSTs) are designed to reach an accept/reject decision. The PRST subjects test articles to combined stresses between operational and design limits and constructs an acceptance–rejection region.

The reliability experiments and math modeling input over to the D-FMEA and down to the F-RBD and math model.

Design LRU Failure Mechanisms Modes and Effects Analysis

The findings of the reliability experiments and math modeling provide further understanding of part failure mechanisms, modes, and effects. Successive improvement of that understanding is used to influence design to achieve increasing system reliability, maintainability, and availability. The preliminary FMEA is updated to document the behavior of failure mechanisms on the part, the identification of newly discovered failure mechanisms, and the lack of significance of previously hypothesized failure mechanisms. The D-FMEA inputs over to the design (D)-CA and design (D)-CIL and down to the F-FMEA.

Design LRU Criticality/Consequences Analysis and Critical Items List

The continued understanding of part failure consequences reveals knowledge of the effectiveness of failure mitigation employed. By this time, the

D-CIL should be reduced to parts with catastrophic and operational consequences that cannot be eliminated but rather have been mitigated to an acceptable risk. Part failure analysis and the effects of part failure on higher design levels should eliminate many parts from further analysis. The D-CA and D-CIL input down to the F-CA and F-CIL.

Final Design Analysis, Bills of Materials, and Drawings

The final design bills of materials and drawings are the product of the preceding design and reliability analysis since the preliminary design review. The final design review is conducted at this point in the system design process, where the vendor communicates the completed design of the system to the customer for acceptance. The reliability design goal is to demonstrate a clear understanding of the reliability, maintainability, and availability requirements; conditions of use; and mission duration and to assure that all have been met.

Final Design Reliability Block Diagram and Math Modeling

The final design reliability block diagram (F-RBD) should reflect only those parts that have an impact on the ability of a system to function without failure for a specified mission duration under specific conditions of use and to survive for a specified lifetime. Reliability math models are baselined as contributions to the LRU baseline design document package. The F-RBD and math model input over to the F-FMEA.

Final LRU Failure Mechanisms Modes and Effects Analysis

The final LRU failure mechanisms, modes, and effects analysis should document the understanding of sources of significant failure mechanisms acting on each part in the F-RBD, the significant part failure modes, and the local and higher design level effects. The F-FMEA is baselined and inputs over to the F-CA and F-CIL. Final LRU design criticality/consequences analysis and critical items list (F-CIL) are contributions to the LRU baseline design document package.

Design Reviews

The design review is an iterative step where the vendor communicates the progress and achievements of the system design process to the customer. The reliability design goal at design reviews is to demonstrate a clear understanding of the reliability, maintainability, and availability requirements; conditions of use; and mission duration and to present the allocation of those requirements down to the part design configuration level. This is not the time to discover the need for major rework. Minor differences typically occur, but such differences should be customer-driven requests resulting in the customer's better understanding of its needs as illustrated for them by the design review. Such changes to the scope of work are rightly funded by the customer and revisions to the program plan are negotiated. The vendor absorbs the costs and schedule penalties to rework the design when the customer assesses that any differences exist because the design did not correctly understand the specification of the system requirements.

Reliability Systems Engineering Requirements for System Integration

Everyone agrees that systems engineering is not design engineering, but few agree on the best-practice body of knowledge of systems engineering. Sage suggests that systems engineering is system definition, system design and development, and system operations and maintenance.[22] He describes 22 phases to implement systems engineering. Various authors present other multiphase approaches to implement systems engineering. Organizations have systems engineering practices that work well for them. The approach presented here incorporates system design and development that is implemented in concert with part design engineering to achieve system integration.

System integration is defined as the analysis of combining parts into the next higher assembly, combining assemblies into the next higher subsystem, and combining the subsystems into the system. System integration analysis is defined as the measure of the achieved design and functional requirements of each design configuration level and the comparative evaluation of the achieved requirements to the system specifications.

System integration is implemented by math modeling and simulation of the assemblies, subsystems, and system. The inputs to the next higher assembly math models are the findings of the part/LRU design analyses. The inputs to the design configuration levels above the assemblies are the findings of the lower design configuration levels. The pattern of inputs is illustrated graphically by the WBS from the wide base of all parts/LRUs funneled to each next higher assembly design configuration level, ending at the lone system level (see Figure 1.14).

Comparative evaluation of the achieved requirements to the system specifications is necessitated by and must acknowledge the following:

- *Variances are additive.* Part/LRU design and functional requirements are allocated down from the system design and functional requirements as fixed or point estimate values, but the achieved requirements are point estimates often lacking specification for a measure of dispersion. The variability of a serial combination of two or more parts is

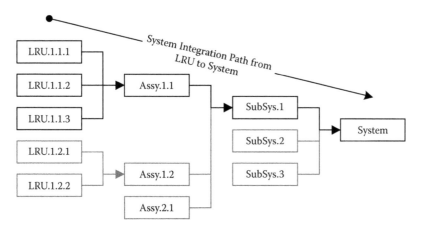

FIGURE 1.14
System integration logic.

the sum of the measures of dispersion of the parts. This fact explains the inability of a system to meet design and functional requirements even though each part/LRU achieved its allocated requirements.

- *Interface structures are often ignored.* The design engineer recognizes that a part/LRU must have a structure that allows for input and output interfaces. However, the interfaces between assemblies and subsystems are beyond the design engineer's control.

The assembly lead engineer can control the input and output interfaces for an assembly, but interfaces between other assemblies and subsystems are beyond control. The upward flow of interface understanding and control by design engineers is compounded by the complexity of the system and by an organization structure that transfers system integration from design to the systems engineers. The disconnection between the two engineering functions is aggravated further by differences in the respective analytical goals. Design engineers use design to achieve functionality. Systems engineering traces functional requirements. The relationship between design and systems engineering appears to make sense abstractly; in practice, it is flawed:

- *The design–systems engineering business model works only when the tasks are strictly serial.* Design–systems engineering reality does not experience a serial business model; it is more likely a waterfall model in which design and systems engineering are performed concurrently when the start of systems engineering for an assembly or higher design configuration level occurs after the start of the design engineering and lags. The following assumptions are made:
  - *All requirements allocations down to part/LRU design engineering are precisely defined and exhaustive for all possible conditions of use,*

operational and ambient. However, the reality is that (1) functional allocations are estimates of functionality parameters that are not precise, (2) operational conditions of use are not understood until after part and material selection occurs following the design concept, and (3) the ambient conditions of use are general estimates for typical locations.

- *All part/LRU design requirements need no feedback loop (i.e., they are done correctly the first time).* It is folly to plan for a perfect part/LRU design that will not have a feedback loop from a design review that will require design modifications to the preliminary design analysis, drawings, and bill of materials. Look where the system design progress is by the time a design modification to a part/ LRU is required—months or more after the original work was performed. The original design team may not be available to perform the modification (e.g., separations, retirement, assigned to another task). Assigning the modification to an engineer who was not part of the original team implies a learning curve that will increase costs, will delay schedules, and may fail to achieve the modification goals. Another response to a design modification feedback loop is to waiver the nonconforming part. This saves the costs and schedule delays, but introduces new variability to the system that is not understood (see Figure 1.15).

  The solution is to perform iterative systems integration at the assembly design configuration level where the primary sources of variability are introduced to the system. The assembly and subsystem modifications will be performed at the earliest, least disruptive opportunity and will be far more likely to be assigned to the original design team prior to their becoming unavailable. The design review feedback loop will not be eliminated; however, the cost, schedule, and design conformance to specifications impact will be profoundly reduced (see Figure 1.16).

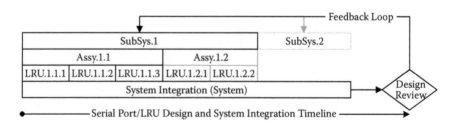

FIGURE 1.15
Systems engineering waterfall.

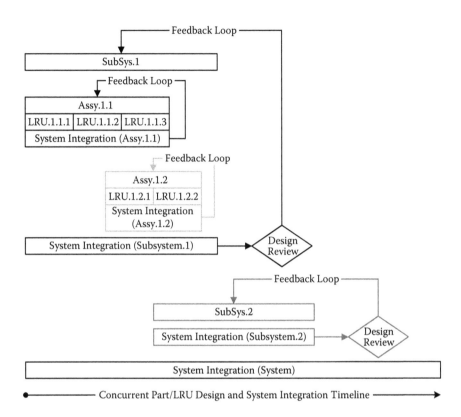

FIGURE 1.16
Iterative systems engineering waterfall.

- *All design engineering disciplines are employed in the part/LRU design (mechanical, electrical, chemical, reliability, quality, safety, cost).* This is an ideal that is not possible. No single engineer is capable of multidisciplined thinking in so many fields. Teaming engineers introduces human interactions and communications issues that will not achieve the singular thinking needed to achieve the ideal applications of engineering design.

  The solution is to define what a design engineer learns from performing the design analysis that can be used by reliability, quality, safety, and cost engineers as the design engineer is doing the work. The failure mechanisms, modes, effects, and consequences analysis for a part/LRU is best performed by the design engineer, who creates that understanding directly from design analysis.

- *All part/LRU design variability sums to an insignificant system variability. This is another ideal that is not possible.* System variability decreases to the lowest common cause source only when the complexity of a system is reduced to a single part/LRU. The sum of part/LRU variability can only increase proportionately to the system complexity. The iterative systems engineering waterfall approach is the best method for understanding the sources of variability at the source and making the determination whether the variability can be mitigated and how the mitigation can be achieved. Not all mitigation need be by design modification; maintainability measures can be identified and documented to reduce the effects and consequences for the system.

- *All interfaces are understood, are independent, and do not provide a source of stress loads not allocated to the part/LRU design engineer.* However, interface failure mechanisms, failure modes, and effects and consequences of those effects are rarely understood. This is due largely to the complexity of the system. Nor are interfaces' failures independent. Consider a mounting bracket and fasteners comprising the interface between a control valve and an assembly housing. Corrosion (failure mechanism) of a mounting bracket fastener causes the fastener to change its material properties (failure mode), resulting in reduced strength of the fastener (local effect), in a degraded mode for the mounting bracket assembly (next higher effect), and in no effect on the system (end effect). Corrosion is not a failure mechanism acting on the control valve; yet, the degraded mode of the mounting bracket allows a vibration failure mechanism to act on the control valve that causes the valve to fail and potentially put the system in a down state.

  The iterative systems engineering waterfall approach can recognize dependent relationships between part/LRU interfaces, as well as the need to understand interface failure mechanisms, modes, effects, and consequences.

The reliability systems engineering approach to implement systems integration is illustrated in Figure 1.17.

Part/LRU-to-Assembly Integration

Part/LRU-to-assembly integration is analogous to building a wall with individual bricks. The only difference is that all bricks are the same and all parts are different. As previously noted, the integration of parts and LRUs is achieved through interfaces, just as the integration of one brick to the other is through cement. The integration of parts and LRUs to the assembly level combines not only hardware to hardware but also design analysis of each part into the assembly.

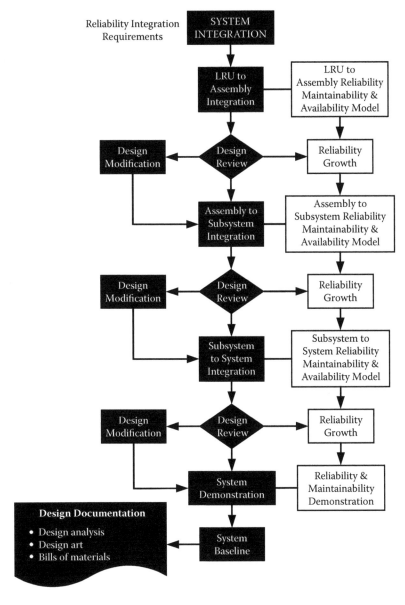

FIGURE 1.17
System integration.

Part/LRU-to-Assembly Reliability, Maintainability, and Availability Model

Part and LRU probability failure math models that were developed in the part design for reliability are the elements that build the assembly reliability, maintainability, and availability math model. The assembly reliability, maintainability, and availability math model is used to calculate point estimates of the reliability, maintainability, and availability parameters.

Simulation trials of the assembly math model provide more than just point estimates; simulation trials provide minimum and maximum estimators and standard deviations that can be used to calculate confidence limits. Two simulation designs are required: mission duration and life cycle. The mission duration simulation solves for the assembly reliability probability. A mission duration simulation trial is run for the time of the mission, τ (e.g., a single mission of t hours, c cycles, x starts). The number of simulation trails, n, is calculated as follows:

1. State the confidence, C%, of the simulation findings (e.g., 90, 95, 99, or 99.9%).
2. Calculate the level of significance, $\alpha | \alpha = 1 - C$, where α is stated as a decimal.
3. Solve for $n|n = \ln[\alpha]/\ln[R]$, where $\ln[R]$ is the natural logarithm, base e, of the allocated reliability (stated as percent) to the assembly.

For example, a simulation with 95% confidence for an assembly reliability allocation of 90% for a mission duration of $\tau = 12$ h must have a minimum of 29 trial runs of 12 h each:

$$n = \ln[0.05]/\ln[0.90] = -2.99573/-0.10536 = 28.4 \text{ rounded up to } 29$$

The life-cycle simulation solves for the assembly maintainability and availability parameters. A life-cycle simulation trial is run for the system useful life, T (e.g., 20 years at 12 h/operating day at 260 operating days = 62,400 h, 1,000,000 cycles, 2,500 starts). The number of simulation trials, n, is the same as for the mission duration simulation. For example, a simulation with 95% confidence for an allocated assembly reliability of 90% for a useful life of $T = 62,400$ h must have a minimum of 29 trial runs of 62,400 h each.

The findings of the assembly mission and life-cycle simulations are compared to the allocated mission reliability, maintainability, and availability. A straightforward test of hypothesis of the equivalence between the sample and the requirement for statistical significance is performed in the following steps:

1. State the assumption that the means of the reliability, maintainability, and availability parameters are approximately normally distributed.
2. Determine whether the reliability, maintainability, or availability parameter criterion is highest is best, nominal is best, or smallest is best (see Table 1.4).
3. Calculate the appropriate confidence limit:

TABLE 1.4

Reliability, Maintainability, and Availability Criteria

Parameter	Criterion
Reliability as a probability	Highest is best
Reliability as a MTBF	Highest is best
Reliability as failure rate	Lowest is best
Reliability as MTBF	Highest is best
Reliability as MTBDE	Highest is best
Maintainability—MTBM	Highest is best
Maintainability—MTTR	Lowest is best
Maintainability—mean downtime	Lowest is best
Maintainability—system failures	Lowest is best
Availability—inherent	Highest is best
Availability—operational	Highest is best
Availability—achieved	Highest is best

a. Use lower confidence limit (LCL) for highest-is-best parameter criteria:

$$LCL = \bar{X} - st_{\alpha,v}$$

b. Use upper confidence limit (UCL) for lowest-is-best parameter criteria:

$$UCL = \bar{X} + st_{\alpha,v}$$

4. Compare the allocated reliability, maintainability, or availability parameter to the appropriate confidence limit:

   a. Compare highest-is-best allocated reliability, maintainability, or availability parameter, X, to the LCL of the simulation value:

      1. IF $LCL < X$, THEN accept the hypothesis that the design meets or exceeds the allocation at $C\%$ level of confidence.

      2. IF $LCL > X$, THEN reject the hypothesis that the design meets the allocation at $C\%$ level of confidence.

   b. Compare lowest-is-best allocated reliability, maintainability, or availability parameter, X, to the UCL of the simulation value:

      1. IF $UCL > X$, THEN accept the hypothesis that the design meets or exceeds the allocation at $C\%$ level of confidence.

      2. IF $UCL < X$, THEN reject the hypothesis that the design meets the allocation at $C\%$ level of confidence.

X is the allocated reliability, maintainability, or availability parameter; $\bar{X}$ is the sample mean of the simulation characterization for the reliability, maintainability, or availability parameter; s is the sample standard deviation of the simulation characterization for the reliability, maintainability, or availability parameter; and $t_{\alpha,v}$ is the Student's t-sampling statistic calculated at a significance and v degrees of freedom, where $v = n - 1$.

Assembly Design Review

The assembly design review is not a formal systems engineering milestone. It is informally performed by the assembly lead engineer and the design engineers to ensure that all aspects of failure are understood at the assembly level. The assembly design review has the most influence on the system reliability. It is the first opportunity to evaluate the achieved assembly reliability, maintainability, and availability with effects of the parts and compare that with the allocated assembly parameters. This is the best time to take corrective action with the least impact on the budget and schedule.

The primary issue is evaluation of the math model simulation findings. The results of the comparison of the allocated reliability, maintainability, or availability parameters to the simulated parameters' confidence limits point directly to the status of the assembly design analysis. The objective is to identify achieved parameters that do not meet the allocations and to propose design mitigation alternatives.

A secondary issue is evaluation of the conditions of use that the parts impose on the assembly. Operational conditions of use will include interface hardware between the parts and the assembly. Ambient conditions of use will include stressors applied to the assembly by the surrounding environment.

Design Modification

Action items from the assembly design review will range from design modification to maintenance solutions for the assembly to preserve its functionality. The authority to approve the design modifications remains with the lead assembly engineer because the results of the part/LRU design are not yet baselined. Proposals to mitigate reliability, maintainability, and availability problems with maintenance can be raised to higher engineering management at the earliest opportunity to be incorporated in the overall systems integration.

Reliability Growth

A design modification that serves to improve the reliability of the assembly is referred to as reliability growth. The reliability math model provides guidance for the need for reliability growth; the weakest-link part is the first source of reliability improvement. The weakest-link part is always the determinant of the assembly reliability. Reliability growth is achieved by improving the reliability of the limiting part or by introducing a redundant design configuration for the limiting part.

Assembly-to-Subsystem Integration

Assembly-to-subsystem integration has similarities to part-to-assembly integration; it can be viewed as evaluating the interfaces between assemblies to form subsystems much as parts form assemblies. The key distinction from part-to-assembly integration is the added complexity of the design. Each assembly is stressed by the interfaces with other assemblies as well as from the operating loads acting on the assembly from its parts.

Another distinction is the interactions and lack of interactions between the assemblies. A subsystem can be a serial design, in which each assembly is dependent on the operation of all assemblies, or a subsystem can comprise assemblies that are independent of each other. Consider the electrical subsystem of a car that is composed of the battery, alternator, starter, headlight, brake light, turn indicator, and dashboard light assemblies. Each of the light assemblies is independent of each other; they are independent of the battery when the engine is operating, and they are independent of the alternator when the engine is not operating. Contrast the car to a mine process plant subsystem that has a feeder, crusher, conveyor, and power supply assemblies. Each assembly is a serial design configuration where each assembly is a single point of failure, dependent on another, for the subsystem.

Assembly-to-Subsystem Reliability, Maintainability, and Availability Model

The subsystem reliability, maintainability, and availability model can take two forms. Ideally, the subsystem math model can comprise the assemblies—simple and elegant. This is an accepted approach when the assemblies have been scrutinized as described in the preceding paragraphs. In this case, each assembly design has been optimized for functionality and reliability and documented by the assembly design review.

The alternate approach is to construct a subsystem math model comprising all of the parts in the subsystem. This model loses the assembly identity and takes on the complexity of the subsystem. Consider a subsystem composed of six assemblies; each assembly is made up of 20 parts/LRUs. The ideal math model will have five math model expressions; the alternate math model will have 120. The alternate approach will not provide information describing the assemblies' reliability, maintainability, and availability parameters.

The simulation of the subsystem reliability, maintainability, and availability model is performed and evaluated just as for the assembly simulation. Mission duration and life-cycle simulations are designed and trials run. The findings of the simulations provide the information that allows calculation of the confidence limits for the achieved reliability, maintainability, and availability parameters, which are then compared with the subsystem allocations at a statistical confidence level.

Design Review

The subsystem design review, like the assembly design review, is an informal event. Like the assembly design review, the subsystem design review is evaluation of the math model simulation findings. The results of the comparison of the allocated reliability, maintainability, or availability parameters to the simulated parameters' confidence limits point directly to the status of the subsystem design analysis. The objective is to identify achieved parameters that do not meet the allocations and to propose design mitigation alternatives. The need for design mitigation should be minimal and not have significant impact on budget and schedule if the assembly design review, design modification, and reliability growth actions were implemented.

The alternative approach, in which the subsystem math model is the first iteration of evaluation of the part design and integration to the assembly level, poses a more time-consuming and complicated task. Many of the part design engineers will be working on other assignments by this time. The tests of hypotheses will be far more complex and prone to human error. Traceability of nonconforming parts to assemblies and the effects of interfaces will be less clear. The weakest-link part will still be identified, but the proposals for design modifications will be less evident.

Design Modification

Design modifications are less significant in the ideal scenario than for the alternate scenario; both will still be driven by improvement of part design or proposals for maintenance solutions. Design modifications in the ideal scenario will be isolated directly to the weakest assembly and, in the alternate scenario, to a part and then the assembly. Reintegration of an assembly to the subsystem will be well structured; reintegration of a part, then the assembly, and then the subsystem has the potential to create hidden problems that will not be evident until the systems engineering design review milestone.

Reliability Growth

Reliability growth in the ideal approach can readily be identified as design improvement of a part in the weakest-link assembly or introducing a redundant assembly design. Reliability growth in the alternate approach will focus on the weakest-link part without direct understanding of the reliability characteristics of the weakest-link assembly.

Subsystem-to-System Integration

Subsystem-to-system integration is the first opportunity to put everything together to evaluate the achieved system parameters to the system requirements. The subsystem design configuration provides a complexity of dependent and independent relationships that compounds that of the assemblies to subsystems integration. Every part in the system is stressed in varying degrees by all of the operating and ambient environments.

Subsystem-to-System Reliability, Maintainability, and Availability Model

The subsystem-to-system reliability, maintainability, and availability math model can take one of three forms. The ideal scenario is a math model comprising the subsystem math models because the assembly and subsystem math models have been simulated and evaluated iteratively, as proposed in this text as the best practice. A variation on the ideal scenario is a math model that is composed of the assembly math models structured by the subsystem design configuration. This is a good approach when the assemblies are independent within the subsystems and it is necessary to isolate weakest links to the assembly level. This is also the preferred approach when assembly integration crosses subsystem lines. The car electrical and engine cooling subsystems share wiring parts; wire failure modes can have failure effects on assemblies in both subsystems, resulting in one or both being in a down state.

The least desirable alternative is a math model in which the system comprises part failure model elements. Neither the assembly nor subsystem identities are apparent. The math model complexity is at maximum. The alternate approach will not provide information describing the assemblies' reliability, maintainability, and availability parameters.

The simulation of the system reliability, maintainability, and availability model is performed and evaluated just as for the previous design configuration simulations. Mission duration and life-cycle simulations are designed and trials run. The findings of the simulations provide the information that allows calculation of the confidence limits for the achieved reliability, maintainability, and availability parameters, which are then compared with the subsystem allocations at a statistical confidence level.

Design Review

System design reviews are formal systems engineering milestones, ranging from preliminary to critical to final. This is the worst opportunity to discover that achieved reliability, maintainability, and availability parameters do not meet the allocated values and, by extension, do not meet the system requirements. The weakest-link parts are still identified, but proposed mitigation is masked by the complexity of the system and is less effective. Assembly and subsystem interface effects on the design are very difficult to identify. Availability of the part and assembly design teams is at its minimum. The design modifications have their greatest impact on budget and schedule. Some design modifications are no longer economically or technically feasible; waivers of requirements may require negotiation with the customer at increased costs and delivery delays.

Design Modification

Design modifications for the ideal and the assembly-driven variation on the ideal scenario are focused on the subsystem or assembly design configurations, respectively. Information is provided by the system math model and simulations for opportunities to consider redundant design modifications of assemblies as well as identification of weakest-link parts. Functional and design impacts from all of the combined operational and environmental environments are understood.

The alternate approach provides only parts-level understanding of the achieved system reliability, maintainability, and availability parameters. Impacts of part design modification on the assemblies, subsystems, and system are masked and cannot be understood.

Reliability Growth

Reliability growth in the ideal approach, and its variation, can readily be identified as design improvement of a part in the weakest-link assembly or introducing a redundant assembly design. Reliability growth in the alternate approach will focus on the weakest-link part without direct understanding of the reliability characteristics of the weakest-link assembly and its subsystem.

System Demonstration

System demonstration is performed following the final design review. All aspects of the achieved system functionality are empirically evaluated and compared to the system requirements.

Reliability and Maintainability Demonstration

Formal reliability and maintainability demonstrations are part of the systems engineering system demonstration; they are designed to prove empirically that the achieved reliability and maintainability parameters have been met. It is a mistake to wait until the system demonstration to prove the achieved reliability and maintainability parameters. Reliability and maintainability experiments should be performed in conjunction with part design and reliability growth, part-to-assembly integration, and assembly-to-subsystem integration. The purpose of reliability and maintainability experiments is to validate the findings of the math model simulations. The logic of reliability and maintainability experiments is illustrated in Figure 1.18.

Part reliability probability is empirically characterized by material stress strength analysis and time-to-failure experiments. Material and part failures are induced by HALT. Reliability growth is validated by probabilistic reliability acceptance test (PRAT). Part maintainability MTTR is empirically

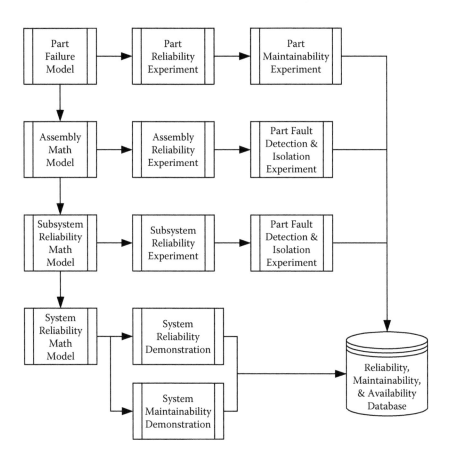

FIGURE 1.18
Reliability and maintainability experiments.

characterized by TTR experiments. Part failure math model parameters that are characterized by design analysis are replaced by empirically characterized reliability and maintainability parameters. The transition from analysis-based to empirically characterized parameters provides greater credibility to the achieved part reliability and maintainability estimations.

Assembly reliability MTBM (mean time between maintenance) is empirically characterized by time-to-failure experiments using ALT. Reliability growth is validated by PRAT. Assembly maintainability is empirically characterized for fault detection and isolation experiments. The MTTR is confirmed in the assembly integration. Assembly reliability and maintainability math models are validated. Reliability design modifications by maintenance solutions are validated. Operating environment conditions of use are empirically determined.

Subsystem reliability mean time between downing events is empirically characterized by time-to-failure experiments using ALT. Reliability growth is validated by PRAT. Subsystem maintainability is empirically character-ized for fault detection and isolation experiments. Reliability design modi-fications by maintenance solutions are validated. Subsystem reliability and maintainability math models are validated. Operating environment condi-tions of use are empirically validated.

System reliability and maintainability are empirically demonstrated with the outcome known from the subsystem reliability and maintainabil-ity experiments.

System Baseline

The system design analysis, drawings, and bills of materials are baselined, reviewed, and approved for fabrication, manufacture, field sustainment, operating and maintenance manuals, training, and warranty terms and conditions.

Configuration Management

The system baseline documentation and the reliability and maintainability database are controlled.

Reliability Engineering Requirements for System Sustainment

System sustainment is the least supported part of a system's life cycle in the engineering disciplines. Standards, best-practice methods, and guidelines exist for design and systems engineering; however, little exists for sustain-ment: system O&M. The principal source of sustainment documentation is the operator and maintenance manuals provided to the customer by the system vendor. These manuals provide the guidance the vendor thinks the customer needs rather than delivering the guidance the customer requests. Such manuals are therefore "one size fits all": Each customer uniquely oper-ates and maintains its system. Consider a specific model car from a specific car company; the use of the car ranges greatly:

- a personal automobile of a family that lives in the suburbs of a major metropolitan area that accumulated 300 miles per week in stop-and-go traffic
- a personal automobile of a retired couple that lives in a small town that accumulated 150 miles per month and 2,000 miles once annually
- a fleet of business and government automobiles that are assigned to employees for random trips of 300–500 miles per week under speci-fied use guidelines

- a fleet of rental cars used by business and vacation travelers for random trips of 50–500 miles per week under few if any specified use guidelines
- a fleet of taxicab or delivery automobiles assigned to one or more employees for continuous daily use ranging from 100 to 300 miles per day

The car company sees each car that is shipped as an identical system and provides each car with the same O&M manuals. The customer sees each car that is received as an asset unique to the customer's use of the car. The O&M manuals more than likely do not reflect the customer's use of the car.[23]

Organizations pursue O&M methods that conform to their needs of a system. Operation is focused on producing the maximum functionality constrained only by the limits of proximate systems that control the use of a system. A haul truck in a surface mine may be capable of hauling 100 tons of ore twice an hour between the excavator and the process plant, unless the excavator can only load the truck once per hour or the process plant can only receive one load per hour.

Similarly, maintainability is focused on sustaining operations at maximum functionality. The most common sustainment approach is repair maintenance: performing maintenance actions when a system downing event occurs to repair the failed part/LRU (often in the place where the failure occurred) and to return the system to service. Repair maintenance is reactive maintenance when randomly occurring part failure causes all system downing events and stimulates all maintenance actions. One may conclude that repair maintenance is system sustainment that is performed as if design-for-reliability and systems-reliability engineering had not occurred. The elements of repair maintenance include:

- *Reacting to part/LRU failure.* Managers and field engineers react to part/LRU failure based on the consequences of the system effect. Degraded functionality may be tolerated to prevent a total loss of productivity. Such decision making is described as "putting out fires" and is viewed as a normal component of the job using systems to generate productivity. Another fire always erupts as or shortly after the current fire is extinguished. It is exhausting, frustrating, and inefficient.
- *Use of fault detection and fault isolation methods and subsystems to point maintenance operations to the right course of action.* Managers and field engineers measure repair maintenance efficiency by how quickly maintenance teams can find and replace the failed part/LRU. System FD/FI methods and subsystems are viewed as a way to react more efficiently to part/LRU failure.
- *Reporting MTBM to trend future O&M costs.* MTBM is not always initiated by a system downing event; maintenance actions are also

performed following a system's scheduled use to restore degradation to full functionality and to perform "preventive maintenance"
(PM)—another term that means different things throughout organizations. Some organizations classify PM as daily or periodic servicing tasks (check and replenish engine oil, inspect and replace worn
v-belts, etc.). Others classify PM as periodic inspections, calibration,
adjustments, and other maintenance tasks that are expected to prevent system downing events.

Well-managed organizations implement good recordkeeping for system
maintenance actions and functionality—not to perform failure analysis, but
rather to measure and control productivity and costs. The reporting period
interval for review of costs ranges from daily to weekly to monthly to quarterly
to annually by various levels of organizations. Planning productivity and cost
forecasts for the next reporting period use trend analysis of past actual productivity and cost metrics. The accuracy of such forecasts is a function of how far
in the future the plan forecasts. Forecasting tomorrow's productivity and costs
is less likely to be wrong than forecasting those for next month, the quarter, or
the year. The farther out in time one forecasts, the greater the errors will be.

Forecasting can be described as driving a car with the windshield painted
black using the rear-view mirror to steer. The slower one travels, forecasting
one day to the next, the less likely one is to miss a turn and drive off the road;
the faster one drives, forecasting longer into the future, the more likely one
is to drive off the road. Therefore, well-managed organizations do not com-
pensate for the inefficiency of repair maintenance as a system sustainment
approach; they just do a better job of measuring the lost production opportunity and maintenance costs—the symptoms of inefficient sustainment.

Reporting system mean time between downing events (MTBDE). Mean time
between downing events is the sum of the three preceding elements. Reaction
to system downing events is analogous to a swarm of people coming to the
aid of a stricken person. All are well meaning, all have different skills, and
every random event involves a different swarm of people reacting to a different ailment.

Maintainability is performed in infrastructure logistical support, ranging
from a backyard shade tree to a state-of-the-art depot. The elements of logistical support include:

- *Spare parts strategies.* Spare parts inventory and consumption repre-
  sent the single largest direct materials and indirect overhead system
  sustainment costs. The ideal spare parts strategy is to receive the
  spare part at the same time that it is required to repair the system;
  it is impossible to achieve. Spare parts can either be ordered when
  needed or stored on site in inventory. Availability of an ordered part
  at the down system location experiences delays due to time to place
  the order, time needed by the vendor to fill the order, time required

to ship the order, and time needed by the customer to accept and distribute the part to the system.

Parts stored in on-site inventory have administrative delays of much shorter duration but have inventory carrying costs. The cost of the part is a direct material cost for either method of making the part available for use. Buying a part on an as-needed basis can be more expensive if quantity discounts are offered for larger quantities. Shipping an as-needed part is a direct cost that can also be lower for quantity discounts. Both spare parts purchasing strategies carry indirect overhead expenses: the ordering and receiving employees and facility expenses, the inventory facility, the inventory control system, etc. Any delay in delivering the spare part to the down system causes a lost opportunity cost for the system due to lost productivity.

- *Facility requirements.* Maintenance events require a variety of facility capabilities, including control of ambient conditions (protection from rain, wind-blown sand, excessive temperatures, etc.), overhead lifting devices, storage and access to bulk materials and fasteners, maintenance and parts list documents, meeting rooms, safety stations, database terminals, tool storage, and property security.

- *Mobile maintenance requirements.* Dispersed systems require that maintenance is performed at the system location. Mobile maintenance requirements include delivering the right skills, tools, and replacement parts to the system. For example, process equipment and large excavation machinery cannot be towed to a fixed facility; maintenance must be taken to the system location.

- *Special tools requirements.* Part/LRU repair maintenance actions can require special tools that are not standard tools used by skilled trades' employees. Some special tools are designed and fabricated to perform a single maintenance task that is unique to the system and they are not available for purchase through tool vendors. Other special tools are available through tool vendors and are expensive beyond the means of skilled trades' employees. Many hoist and leverage tools used in heavy industry are fabricated on site and are controlled by the maintenance organization. Computer-based diagnostic tools for automobile inspection, fault detection, and fault isolation are expensive and also controlled by the maintenance organization.

- *Maintenance trade specialties.* All maintenance organizations are composed of people who have trade skills to perform maintenance tasks. Trade skills include mechanics, electricians, and welders, to name a few. Mechanics specialize in diesel engines, transmission and power train, industrial wheels and brakes, hydraulics, and pneumatics, for example. Maintenance managers are resource constrained on selecting which, and how many, trade skills employees can be hired and which trade skills will be contracted on an as-needed basis. The decision criterion is to minimize lost opportunity costs that result from waiting for a needed skills trade that is in use or finding and contracting a skills trade versus having a skills trade on the payroll that is not required at the quantity employed.

- *Database requirements.* Using repair maintenance records to trend and forecast maintenance costs requires a database system that can range from written records to organizationwide computer-based systems. Database requirements require an effective data acquisition methodology that is accurate, measures the right metrics, and provides meaningful information.

The root causes of maintenance actions—part/LRU failure—are the common thread for repair maintenance and logistical support. Trend analysis of system maintenance actions and spare parts consumption draws samples, in the statistical sense, from complex populations of data. This means that no useful inferences can be drawn from the trend analysis. Lack of understanding of part/LRU failure mechanisms, failure modes and effects, and consequences means that all decisions made to develop a system maintenance concept are "educated" guesses that are cost inefficient and force an organization always to be reacting to circumstance.

The corollary is that understanding the root causes of maintenance actions enables accurate decisions for the operating and capital investments in a maintenance concept that give an organization control over system O&M. Reliability-centered maintenance (RCM) is a time-proven method to transform an organization's maintenance concept from repair maintenance to proactive maintenance—from reacting to part/LRU failures to preserving system functionality. The transformation to RCM is illustrated in Figure 1.19.

System Sustainment

System sustainment in RCM is influenced by the reliability, maintainability, and availability models developed in design and systems integration. The system O&M organization can develop meaningful and manageable reliability, maintainability, and availability models in the absence of vendor models or when the organization's conditions of use differ from the assumptions used by the vendor. RCM reliability analysis is performed by the system O&M to identify the CIL with the benefit of maintenance records to identify actual failures.

The system O&M has firsthand knowledge of part/LRU condition indicators, combined with CIL elements of failure consequences, perception of failure, and the P–F interval, that enables development of a condition-based maintenance (CBM) approach to RCM. The system O&M also has firsthand

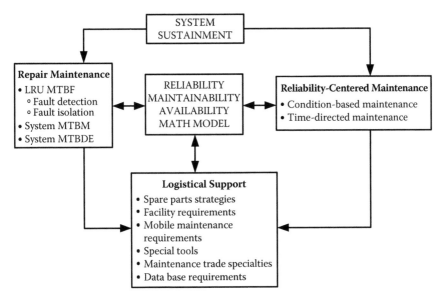

FIGURE 1.19
Reliability-based system sustainment.

knowledge of the part failure and LRU hazard function that enables development of a time-directed maintenance (TDM) approach to RCM.

Repair Maintenance

Repair maintenance is too frequently a reaction to part failure during scheduled system operations. This leads to maintenance actions performed under less than ideal working conditions, excessive downtime, and the costs associated with lost production. The key benefit of an RCM sustainment program is transformation to proactive maintenance for the part/LRU failures that are deemed cost effective by the CIL. Proactive maintenance preserves system functionality by scheduling part replacement prior to its failure.

LRU Mean Time between Failure

The use of MTBF to schedule maintenance actions will be eliminated. CIL part/LRU failure analysis is based on understood degradation of individual parts, rather than from historical trends of MTBF-based part consumption rates or statistical and probability analysis of a population of parts. The forecast of part failure over a reporting period is replaced by CBM and TDM prognostics of scheduled maintenance actions that support the operations demands on the system availability.

Fault Detection and Isolation

CIL part/LRU fault detection and isolation is integrated with condition indicators and probabilistic risk assessments using the hazard function to schedule imminent maintenance actions that support the operations demands on the system availability.

System Mean Time between Maintenance

System MTBM will become controlled by the organization to reflect scheduled maintenance actions that are determined by condition indicators and hazard rates. System MTBM will probably decrease as CIL part/LRU reliability-centered maintenance is expanded to exhaust the list. However, more maintenance actions scheduled to accommodate operational demands on the system will have significantly shorter logistics downtime and eliminate induced failures from field corrective actions.

System Mean Time between Downing Events

System MTBDE will increase significantly as CIL part/LRU reliability-centered maintenance is expanded to exhaust the list. Unscheduled system downing events will decrease, approaching but realistically never reaching zero.

Logistical Support

System logistical support will be optimized as CIL part/LRU reliability-centered maintenance is expanded to exhaust the list to match maintenance investment and resources to actual needs.

Spare Parts Strategy

System spare parts are a major cost of doing business that has the potential to tie up operating capital for parts that sit in inventory for long periods of time, emergency acquisition of spare parts required by an unscheduled maintenance action, and spare part stockpile degradation. Condition indicators and hazard analysis enable optimum spare parts acquisition and storage costs and reduce spare part degradation in stockpiles.

Facility Requirements

System facility requirements are a limited resource that provides value only when in use. Unscheduled off-site maintenance actions waste facility resources through lack of use. High demand for facility resources resulting in a queue also carries negative value. Condition indicators and hazard analysis enable optimum scheduling of facility resources.

Mobile Maintenance Requirements

System mobile maintenance requirements are a special case of facility requirements. Unscheduled off-site maintenance actions are inefficient uses of a mobile maintenance unit. Condition indicators and hazard analysis enable optimum scheduling of mobile maintenance resources.

Special Tools

System special tools are maintenance task specific and often demand is random and unpredictable. Special tools range from fixed lifting machinery to portable diagnostic equipment. Unscheduled off-site maintenance actions are removed from the former and are often inefficient uses of the latter. Condition indicators and hazard analysis enable optimum scheduling of special tools.

Maintenance Trade Requirements

System maintenance trades skills are the lifeblood of a maintenance organization. The literature and common field experience teach that reactive and inefficient demand for services and allocation of resources diminishes the effectiveness and morale of skilled employees. Unscheduled off-site maintenance actions are poor working environments that frustrate employees when ideal facilities are available but not accessible. The opportunity to induce another part failure from exposure to off-site environmental conditions frustrates employees who take pride in their work. Condition indicators and hazard analysis enable optimum scheduling of employees and improve their effectiveness and job satisfaction.

Database Requirements

Data are absolutely required to understand, control, and manage part failure and maintenance actions. Condition indicators and hazard analysis demand continuous information to prompt condition-based and time-directed maintenance actions.

Notes

1. Kapur, K. C., and L. R. Lamberson. 1977. *Reliability in engineering design.* New York: John Wiley & Sons.
2. O'Conner, P. D. T. 2002. *Practical reliability engineering,* 4th ed. New York: John Wiley & Sons.

3. National Aeronautics and Space Administration.
4. For example, Boeing, Lockheed Martin, and McDonnell-Douglas concurrently have designed and developed commercial aircraft, military aircraft and missiles, NASA missiles, and the space lab, space shuttle, and space station.
5. Shanley, F. R. 1967. *Mechanics of materials.* New York: McGraw–Hill.
6. Shigley, J. E. 1977. *Mechanical engineering design,* 3rd ed. New York: McGraw–Hill.
7. Collins, J. A. 1993. *Failure of materials in mechanical design,* 2nd ed. New York: John Wiley & Sons.
8. Anon. For want of a nail.
9. MIL-STD-1629. 1992. Procedures for performing a failure mode, effects, and criticality analysis.
10. Carlson, C. S. 2007. Lessons learned for effective FMEAs. *Proceedings of the Reliability and Maintainability Symposium 2007,* tutorial notes.
11. Ireson, W. G., C. F. Coombs, Jr., and R. Y. Moss. 1996. *Handbook of reliability engineering and management,* 2nd ed. New York: McGraw–Hill.
12. The tire located on the right rear wheel on an untraveled road would have a less critical end effect (operational); the operator perception and P-F interval remain the same. Criticality analysis must define the worst case when such choices are known.
13. This assumption is a necessary statement of the conditions-of-use element to the reliability definition. Assignment of an unqualified operator is a special cause of variability that is beyond the scope of responsibility of the design engineer.
14. MIL-STD-721. 1983. Maintainability.
15. Ireson, W. G., C. F. Coombs, Jr., and R. Y. Moss. 1996. *Handbook of reliability engineering and management,* 2nd ed. New York: McGraw–Hill.
16. Ibid.
17. O'Conner, P. D. T. 2002. *Practical reliability engineering,* 4th ed. New York: John Wiley & Sons.
18. Crosby, P. B. 1979. *Quality is free.* New York: McGraw–Hill.
19. Brocka, B., and M. S. Brocka. 1992. *Quality management.* Homewood, IL: Business One Irwin.
20. Wessels, W. R. 2008. Unpublished reliability benchmarking study of aviation and mobile machinery systems to be completed in 2010, Huntsville, AL.
21. Kerzner, H. 1992. *Project management,* 4th ed. New York: Van Nostrand Reinhold.
22. Sage, A. P. 1992. *Systems engineering.* New York: John Wiley & Sons.
23. There are some very effective exceptions to this situation, such as vendors who take great pains and effort to meet their customers' specific O&M needs; Caterpillar Tractor is an excellent example.

2

Part/LRU Reliability Modeling for Time-to-Failure Data

Life is like a sewer. What you get out of it depends on what you put into it.

Tom Lehrer

Introduction

The objective of this chapter is to characterize the parameters of the failure model for a part and to specify the failure model. The application is design for reliability.

Time-to-failure (TTF) data are the workhorse of reliability engineering and analysis. The common thread for reliability, maintainability, and availability parameters is part/line replaceable unit (LRU) failure. Part through system reliability is defined as either the probability that a failure will not occur or the mean time between failure (MTBF). Part through system maintainability is defined as the probability that a failure can be repaired or the mean time to repair (MTTR). Part through system availability is defined as the probability of failure-limited functionality.

Time-to-failure data are not so straightforward as presented in many reliability texts and seminars. Too many reliability references present methods to acquire and analyze TTF data for assemblies and higher level design configurations, up to and including the system. Yet, assemblies and higher design configurations do not fail; rather, they experience a down state from the failure effects of part/LRU and materials.

A single part/LRU failure occurs in a cauldron of multiple factors that must be understood to characterize the math model parameters for reliability, maintainability, and availability. To do less is a cursory approach that does not provide understanding of part/LRU failure sufficiently to influence design analysis, systems integration, and field sustainment. This chapter presents a proposed approach that will provide understanding of part/LRU failure sufficient to influence system design, systems integration, and field sustainment. The reliability, maintainability, and availability procedures in this chapter meet the design-for-reliability requirements presented in Chapter 1 (see Figure 2.1). There are two distinctions between the logic figures:

- The design for reliability requirements includes procedures that are inputs from the systems integration (i.e., part/LRU reliability allocation) that are addressed in Chapter 8 on reliability systems integration.
- Many of the design-for-reliability requirements include procedures that are iteratively performed; each successive procedure that iteratively builds on new or modified information is addressed only once in this chapter.

The cursory and proposed approaches share some common methods, but the implementation differs significantly; the logic flow is distinctly different and emphasizes that new procedures requiring different organizational

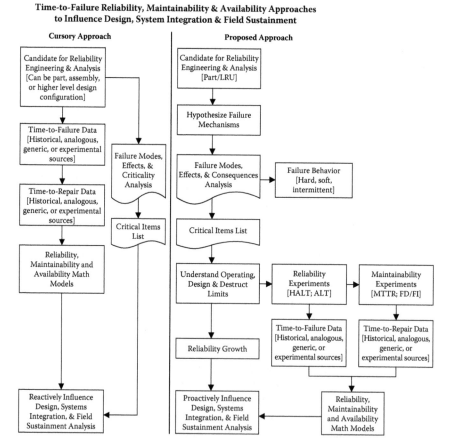

FIGURE 2.1

TTF reliability, maintainability, and availability math modeling logic.

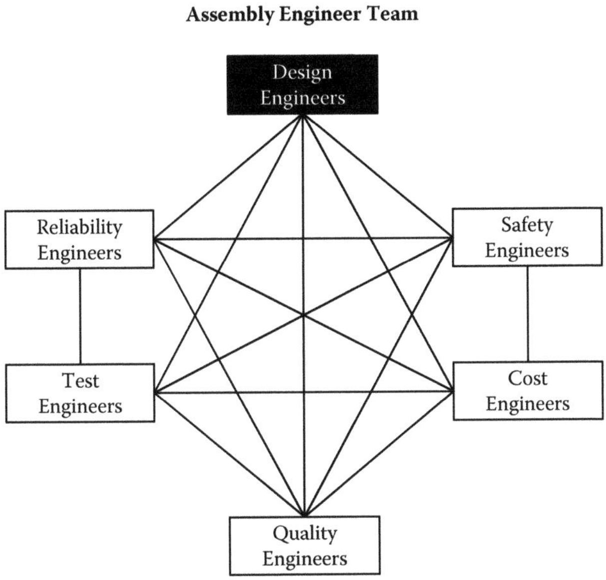

Assembly Engineer Team

FIGURE 2.2
Proposed multidiscipline team/matrix.

teaming approaches are required. Organizations that separate engineering disciplines will assign all of the cursory approach to a reliability engineer located in the quality assurance or specialty engineering department, where he or she functions without direct contact with design and systems engineers. The proposed approach requires a multidiscipline team (or matrix project team) structure where design, systems, and reliability engineers work together (see Figure 2.2).

The design engineers are purposely the apex of a design team; all work performed by the reliability, quality, safety, test, and cost engineers influences the design objective, which is the responsibility of the design department. The work performed by these engineers is initiated by the design engineers.

Part Candidate for Reliability Engineering and Analysis

The candidates for reliability engineering and analysis are all parts/LRUs identified by the design engineers and the respective interfaces between the parts and LRUs. A concept design is developed from the work breakdown structure (WBS) and translates the system requirements into functional blocks that should reach down to the assembly design configuration level and may identify a specific part. We can illustrate the evolution from system requirement to WBS to concept design with a notional statement of work (SOW), provided by the customer, to design a process plant.

SOW: Design, operate, and maintain a process plant system on the customer's property that blends three liquids in equal proportion in a reaction vessel that can be stored for up to 48 h prior to bulk shipment by truck and rail or loaded into barrels for longer storage periods. Reaction time is 5 h/batch – 20 min + 10 min. Customer production is scheduled for 24 h/day for 6 days/week, Monday through Saturday, for 50 week/year. Bulk shipping is scheduled for 24 h/day, 6 days/week for 50 weeks/year. Barrel loading is scheduled for time between bulk vehicle availability, off-shipping days, Sunday, and as an emergency measure for storing the finished product for unscheduled downing events that prevent bulk shipping. Recurring plant sustainment maintenance is scheduled for 1 day/week, Sunday.

Plant overhaul maintenance is scheduled for 8 days in August, Sunday through Sunday. Fluid A is a cryogenic liquefied gas at 3 atm pressure and 40°F. Fluid B is a stable, nonhazardous fluid mixture at ambient atmospheric pressure, 0.9–1.2; humidity, 10–90%; and temperature, –10 to 110°F. Fluid C is a hazardous, caustic fluid at ambient atmospheric pressure, 0.9–1.2; and humidity, 10–90%, and it is stable in storage at extreme temperatures, –40 to 160°F. The bulk finished product must be delivered to trucks at 25 gal/min and to rail cars at 35 gal/min. The reaction vessel must be flushed with recycled-water-process water from a joint three-stage settling pond and pump system between batches.

The customer will provide pipeline supply of the raw materials from bulk delivery tanks and the recycled water. Waste-process water from the process plant will be pumped to the first settling pond. The system design boundary at input is the connection of the supply pipeline to the process plant work-in-process (WIP) tanks. The system design boundary at output is the bulk product loading at the truck and rail yard, the customer product barrel storage facility, and the first stage of the settling ponds for process waste water. All utilities and containment structures will be included as part of the system.

Systems engineers develop a WBS from the SOW that expands the system definition to create a top-down design logic that describes the system design configuration hierarchy by engineering discipline, as shown in the notional WBS shown in Figure 2.3.

Mechanical engineering takes responsibility for "1.01.00.00.00.000 Mechanical Major Subsystem" and further devolves the design hierarchy to as low as is technically feasible. A mechanical major subsystem notional design concept is developed in Figure 2.4.

Reliability and maintainability engineering and analysis should commence at this point. The reliability system integration engineer translates the mechanical WBS into the concept reliability block diagram (RBD), as shown in Figure 2.5.

Concurrently, the design engineers expand the functional related design analyses to include reliability failure analyses. The best time to develop an understanding of part/LRU failure is concurrent to the performance of design

Work Breakdown Structure - Plant Design					
1.00.00.00.000	Process Plant System				
1.01.00.00.000	Mechanical Major Subsystem	1.02.00.00.000	Electrical Major Subsystem	1.03.00.00.000	Civil Major Subsystem
1.01.01.00.000	Raw Materials WIP Tank Subsystem	1.02.01.00.000	Production Electrical Utilities	1.03.01.00.000	Structures
1.01.01.01.000	Fluid A WIP Tank Major Assembly	1.02.02.00.000	Maintenance Electrical Utilities	1.03.01.01.000	Fluid A Structure
1.01.01.01.000	Pressure Tank Assembly	1.02.03.00.000	Administration Electrical Utilities	1.03.01.02.000	Fluid B Structure
1.01.01.01.001	Pressure Tank			1.03.01.03.000	Fluid C Structure
1.01.01.01.003	Input Connection			1.03.01.04.000	Reaction Vessel Structure
1.01.01.01.004	Tank Check Valve			1.03.01.05.000	Rail Bulk Structure
1.01.01.01.005	Tank Control Valve			1.03.01.06.000	Truck Bulk Structure
1.01.01.02.000	Pump & Pipe Assembly			1.03.01.07.000	Control Structure
1.01.01.02.001	Pump			1.03.01.08.000	Maintenance Structure
1.01.01.02.002	Pipe and Fittings			1.03.01.09.000	Administrative Structure
1.01.01.02.003	Pressure Gauge				
1.01.01.02.004	Flow Indicator				
***					
1.01.02.00.000	Fluid B WIP Tank Major Assembly				
1.01.03.00.000	Fluid C WIP Tank Major Assembly				
1.01.04.00.000	Reaction Vessel Major Assembly				
1.01.05.00.000	Truck Bulk Delivery Major Assembly				
1.01.06.00.000	Rail Bulk Delivery Major Assembly				
1.01.07.00.000	Barrel Delivery Major Assembly				
1.01.08.00.000	Recycled Water Major Assembly				

FIGURE 2.3
Notional system WBS.

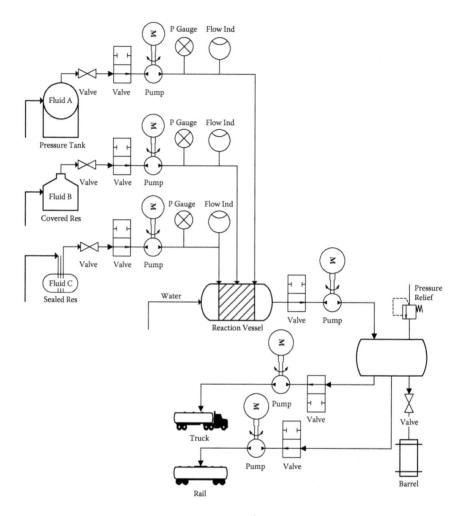

FIGURE 2.4
Notional process system.

analysis. The design engineer can adapt functional design thinking to failure thinking. He or she can ponder the causes of part/LRU failure at the same time as conducting analysis to determine how to achieve functionality.

Hypothesize Part Failure Mechanisms

The design concept is not too early to begin the hypotheses of failure mechanisms; rather, it is the best time. The logic of failure mechanism hypotheses starts with a dichotomy between identification of those the design engineer

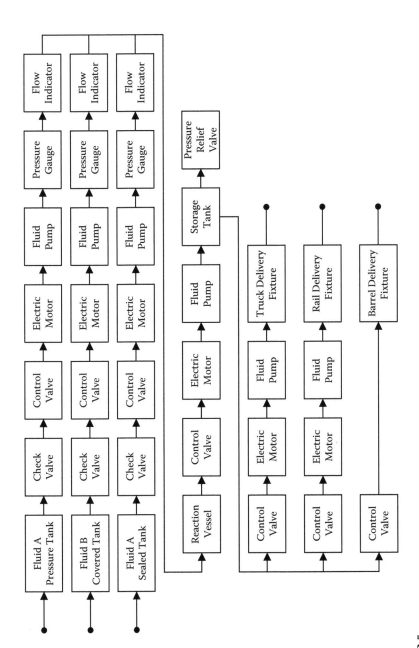

FIGURE 2.5
Notional RBD.

can control and those that are uncontrolled—the operating conditions of use and the ambient conditions of use, respectively. Consider the fluid pump for the caustic fluid: The operating conditions of use define an expected failure mechanism from temperature loads acting on the pump from exposure to the fluid, but the ambient conditions of use provide little information for failure mechanisms from variable and cyclical ranges of temperature loads acting on the pump from exposure to the surrounding environment. Compounding the complexity are failure mechanisms induced by the interaction of the controlled and uncontrolled temperature loads.

A concept or functional failure mechanism analysis translates the customer's statement of the conditions of use with investigation of the possible operating and ambient sources of loads that will act on the part/LRU. The failure mechanisms hypotheses evolve as the system concept evolves into design decisions, part by part.

The logic of failure mechanisms acknowledges that loads acting on parts/ LRUs are not deterministic values, but rather are characterized by a measure of central tendency—mean or median—and a measure of dispersion—standard deviation or range. Another element of the logic of failure mechanisms acknowledges that loads acting on parts may occur in phases throughout the mission duration in constant or cyclical magnitudes or at specific times during the mission. The load on a car battery to energize the starter motor is a single discrete event that does not continue after the car begins its mission. The load on the car electrical system to operate the various light assemblies varies with operational events—use of brake lights when the brakes are applied or use of headlights when the car operates in darkness or during rain. The loads on the water pump occur constantly while the engine is operating.

Developing hypotheses of failure mechanisms is a prime opportunity to show the value of multidiscipline engineering teaming at its best. A working group is assembled for a technical interchange meeting (TIM) in a conference room with a whiteboard and computer. Representatives of the customer and remotely located team members can call in on a conference call or participate in a Web meeting. The responsible part/LRU design engineer prepares for the TIM with an agenda that defines the part/LRU and outlines the known conditions of use. The responsible part/LRU design engineer serves as the TIM facilitator while an administrative assistant or one of the engineers records the findings of the TIM on the computer.

The TIM should last no more than an hour and less time is often the case. The objective is to brainstorm the failure mechanisms that act on the part/ LRU. There is only one rule for a brainstorming session: No team member is allowed to critique an idea proposed by any member.[1] The structure of the TIM failure mechanisms brainstorming session is a worksheet that is included in the agenda. A workable notional format is shown in Figure 2.6. The structure of the TIM part failure modes brainstorming session is a worksheet that is included in the agenda.

			Failure	Mechanics Brainstorming Worksheet		
Lead Assembly Engineer					TTM Date:	----------
Design Engineer					TTM Time:	------------
Reliability Engineer						
Quality Engineer						
Safety Engineer						
Test Engineer						
Cost Engineer						
Customer Representative						
Part:		1.00.00.00.00.000	Process Plant System			
		1.01.00.00.00.000	Mechanical Major Subsystem			
		1.01.01.00.00.000	Raw Materials WTP Tank Subsystem			
		1.01.01.01.00.000	Fluid A WTP Tank Major Assembly			
		1.01.01.01.02.000	Pump & Pipe Assembly			
LRU		1.01.01.01.02.001	Pump			
SRU			Seal			
			Housing			
			Connections			
			Shaft			
			Blades Impellar			
			Bearing			
Conditions		of Use				

Design, operate and maintain a process plant system that blends three liquids in equal proportion in a reaction vessel that can be stored for up to 48-hrs prior to bulk shipment by truck and rail or loaded into barrels for longer storage periods. Reaction time is 5-hr/batch -20-min + 10-min. Customer production is scheduled for 24 hrs/day for 6 days/week, Monday - Saturday. for 50 weeks per year. Recurring plant sustainment maintenance is scheduled for 1 day per week. Sunday. Plant overhaul maintenance is scheduled for 8 days in August. Sunday through Sunday. Fluid A is a cryogenic liquefied gas at 3 atmospheres pressure and 40°F, The bulk finished product must be delivered to trucks at 25 gal/min; to rail cars at 35 gal/min. All Utilities and Containment structures will be included as part of the System.

Failure Mechanism	Operational	Ambient	Interaction	Comment
Mechanical				
Vibration				
Shock				
Temperature				
Steady State				
Thermal Shock				
Cyclical				
Corrosion				
Oxidation				
Contaminants				
Reactivity				
Abrasion				

FIGURE 2.6
Notional TIM part failure mechanisms brainstorming worksheet.

The completed format includes the following information:

- identity of the part including its place in the design hierarchy
- identity of the components/shop replaceable units (SRUs) and materials on which the failure mechanism acts
- detailed description of the conditions of use for the part/LRU
- identity of the failure mechanisms by the environmental sources (operations, ambient, and interactions)
- hypothesis of the interactions of the main factors
- specific comments that address the logic of the part/LRU failure mechanisms (measures of central tendency and dispersion)
- specific comments that address phases of the exposure of the part/LRU to failure mechanisms

Part Failure Modes Analysis

The concept or functional failure mode is a statement of loss of part functionality caused by the hypotheses of failure mechanisms. It forms the skeleton of the design failure modes analysis and evolves as design decisions are made, part by part. A design failure mode is a statement of the symptom of an imminent or immediate part failure resulting from exposure to a failure mechanism by describing changes in the material from with which the part is constructed and changes of interface material. Brainstorming is the best approach to identify the changes in the part and interface materials[2]; it should include:

- change in material geometry (elastic and plastic deformation, physical separation)
  - strain: elongation, compression, bending, torsion, physical and thermal deformation
  - fatigue: high, low, and variable cycles, thermal, steady state
  - fracture: physical and thermal shock
  - surface: pitting, wear, spalling, galling, buckling
- change in material properties (chemical reactivity and physical wear out)
  - corrosion: oxidation, chemical, biological
  - embrittlement: oxidation, chemical, biological
  - surface: erosion, leaching
  - bonding: intermaterial compounds

Each failure mechanism from the brainstorming TIM is analyzed to evaluate expected failure modes. The failure modes TIM will screen out insignificant failure mechanisms when the engineering judgment of the team members is that the total feasible range of magnitudes of operational and ambient environmental loads will not cause a failure mode. The TIM will identify significant failure mechanisms and document all of the failure modes.

Pairing failure mechanisms to failure modes is a critical element of part failure analysis and should not be omitted.[3] Knowing a symptom is an incomplete understanding of failure. Consider a common human "failure mode": pain in the chest. The symptom is meaningless and cannot justify "end effects" and "consequences analyses" or appropriate "mitigation." The "failure mechanisms" include causes such as (1) indigestion, (2) hiatal hernia, (3) cracked rib, (4) muscle strain, and (5) arteriosclerosis. The respective end effects, consequences, and mitigation are (1) discomfort, no consequence, self-medicate with over-the-counter antacid; (2) discomfort, no consequence, prescription medication; (3) discomfort, no consequence, self-treatment with restricted movement; (4) discomfort, no consequence, self-treatment with heat; and (5) heart disease, death, major surgery.

The structure of the TIM failure modes brainstorming session is a worksheet that is included in the agenda. A notional format is shown in Figure 2.7.

The completed format includes the following information:

- identity of the part including its place in the design hierarchy
- identity of the component/SRU and materials on which the failure mechanism acts
- detailed description of the failure mechanism that acts on the component/SRU and materials
- identity of the failure modes caused by the failure mechanism
- specific comments that address the logic of the failure modes (measures of central tendency and dispersion)
- specific comments that address phases of the exposure of the part to failure mechanisms

Part Failure Effects Analysis

The concept or functional failure effect is a statement of downing events of assemblies, subsystems, and the system due to loss of part functionality caused by the hypotheses of failure mechanisms. It forms the skeleton of the part design failure effects analysis and evolves as design decisions are made, part by part. A design failure effects analysis is a two-step procedure.

Failure Mechanics Brainstorming Worksheet				
Lead Assembly Engineer			TTM Date:	-----------
Design Engineer			TTM Time:	-----------
Reliability Engineer				
Quality Engineer				
Safety Engineer				
Test Engineer				
Cost Engineer				
Customer Representative				
Part:	1.00.00.00.00.000	Process Plant System		
	1.01.00.00.00.000	Mechanical Major Subsystem		
	1.01.01.00.00.000	Raw Materials WTP Tank Subsystem		
	1.01.01.01.00.000	Fluid A WTP Tank Major Assembly		
	1.01.01.01.02.000	Pump & Pipe Assembly		
LRU	1.01.01.01.02.001	Pump		
SRU		Seal		

Failure Mechanism				

Failure Modes	Operational	Ambient	Interaction	Comment
1				
2				
n				

FIGURE 2.7
Notional TIM part failure modes brainstorming worksheet.

Step 1

The part team identifies the ways in which the part will lose functionality for each failure mode; step 2 identifies the ways in which each part failure effect will cause the next higher assembly, subsystem, and system to lose functionality; experience degraded modes of functionality; or have no significant effect. Team members who perform the two steps must understand the customer SOW; the engineering part functionality and design analysis; assembly functionality, design analysis and integration; subsystem functionality, design analysis, and integration; and the system functionality, design analysis, and integration. Consultants and segregated "reliability" engineers cannot provide this understanding. The composition of the two teams will differ.

The part team performs step 1 and is made up of the same engineers who performed the hypotheses of failure mechanisms and the failure modes analysis. The systems integration team performs step 2 and is composed of systems integration engineers representing the higher design configuration levels, reliability, quality, safety, maintainability, sustainability, production, and logistics.

The structure of the step 1 TIM part failure modes brainstorming session is a worksheet that is included in the agenda. A notional format is shown in Figure 2.8. The completed format includes the following information:

- identity of the part
- detailed description of the failure mode that acts on the component/ SRU and materials
- identification of the failure effect caused by the failure mode

Step 2

The part team initiates the step 2 TIM and presents the findings of the part failure effects analysis to the system integration team including the step 1 TIM part failure effects brainstorming worksheets. The system integration team brainstorms the effects on each successive design configuration design level including the consequences analysis for each successive design configuration design level (see Figure 2.9).

The consequences analysis screens out all part failure mechanisms and modes that do not require further analysis. The criteria for eliminating further analysis are based on the understanding of the customer's SOW. The completed format includes the following information:

- identification of the part and each effect
- identification of the effect for each successive next higher design configuration level up to the system
- identification of the part failure effect consequences for each successive next higher design configuration level up to the system

Failure Mechanics Brainstorming Worksheet				
Lead Assembly Engineer		TTM Date:		-----------
Design Engineer		TTM Time:		------------
Reliability Engineer				
Quality Engineer				
Safety Engineer				
Test Engineer				
Cost Engineer				
Customer Representative				
LRU	1.01.01.01.02.001	Pump		

Failure Mode

Failure Effect

FIGURE 2.8
Notional step 1 TIM part failure effects brainstorming worksheet.

Critical Items List

The critical items list (CIL) is the end product of the part failure mechanisms, modes, effects, and consequences analysis. The CIL identifies the need for mitigation. Mitigation is achieved through a systems engineering approach of the whole system within the project constraints of technical, functional, budget, and schedule factors for design, development, production, delivery, and sustainment and the customer's cost, operation, and maintenance factors.

Mitigation approaches that require[4] quantitative understanding of part failure that characterizes the reliability, maintainability, and availability parameters include:

- part design analysis and specification revision
- assembly design analysis and configuration modification
- logistical support analysis and revision

Failure Mechanics Brainstorming Worksheet			
Lead Assembly Engineer		TTM Date:	-----------
System Reliability Engineer		TTM Time:	------------
System Quality Engineer			
System Safety Engineer			
Production Engineer			
Logistical Engineer			
System Cost Engineer			
Customer Representative			

		Effects	Consequences			
			Catastrophic	Operational	Degraded Mode	Run-to-Failure
1.01.01.01.02.001	Pump					
1.01.01.01.02.000	Pump & Pipe Assembly					
1.01.01.01.00.000	Fluid A WTP Tank Major Assembly					
1.01.01.00.00.000	Raw Materials WTP Tank Subsystem					
1.01.00.00.00.000	Mechanical Major Subsystem					
1.00.00.00.00.000	Process Plant System					

FIGURE 2.9
Notional TIM system failure effects brainstorming worksheet.

A notional format for a CIL is provided in Figure 2.10. The CIL should include part identification, each significant failure mechanism and mode, and the respective consequences rating for each successive design configuration level.

Critical Item List						
Part nomenclature		Failure Mechanism	Failure Mode	Assembly	Subsystem	System
1.01.01.01.02.001	Pump					

FIGURE 2.10
Notional critical items list.

Part/LRU Reliability Analysis: Understanding Failure of a Part/LRU

That part design analysis is the understanding of part success and part reliability analysis is the understanding of part failure should be ingrained in the mind of the reader. Understanding that part failure mechanisms will act on the part, of the failure modes that will occur, of the corresponding local through system effects, and of the expected consequences of failure has led to the creation of the critical items list. Qualitative and quantitative investigations performed for each CIL part serve to expand the understanding of failure mechanisms and measure their magnitudes in terms that relate to the likelihood of the occurrence of the consequences. Time to failure is the best metric that provides insight into the behavior of failure mechanisms acting on parts/LRUs over time. The independent variable, time (t), from which the sample variable (TTF) is drawn is stated in units of chronological time in hours, operating time in hours, continuous cycles of operation, and discrete cycles of operation (e.g., starts).

Qualitative Part/LRU Investigation

Qualitative investigations of a part/LRU serve to validate that each failure mechanism will cause the expected failure mode. Qualitative findings are achieved from math modeling and simulation, nondestructive physical tests, and destructive physical tests. Consider a design analysis that specifies a bolt, washer, and nut to connect two plates. The statement of work requires that two plates be joined to prevent displacement in the x–y axes (see Figure 2.11).

Part/LRU Design Parameters Fall in One of Three Criteria

- *Operating limits.* The part/LRU operating limits are defined to be loads acting on the bolt and are estimated to be 10 kpsi (kpsi = 1,000 lb/in.^{2}) in shear and 15 kpsi in tension.
- *Design limits.* The part/LRU design limits are based on an engineering decision to use a safety factor of 2 that results in the selection of a bolt, washer, and nut with a materials property that provides 20 kpsi shear strength and 30 kpsi tensile strength.
- *Destruct limits.* The part/LRU destruct limits are estimated to be the yield shear and tensile stress for the material that cause transition from elastic to plastic limits.

The relationship of the operating, design, and destruct limits follows one of three criteria: (1) lower is best, (2) higher is best, and (3) nominal is best (graphically illustrated in Figure 2.12):

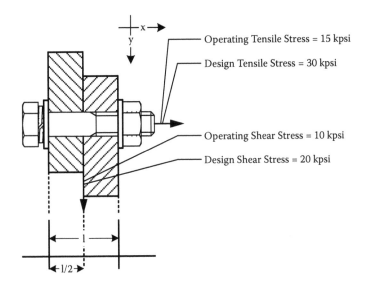

Operating Tensile Stress = 15 kpsi

Design Tensile Stress = 30 kpsi

Operating Shear Stress = 10 kpsi

Design Shear Stress = 20 kpsi

FIGURE 2.11
Notional bolt in tension and shear.

- Loads that cause increased material strain as the magnitude increases are smaller-is-best stressors. Point and distributed forces, vibration, high temperatures, and exposure to reactive agents are examples of smaller-is-best loads.
- Loads that cause increased material strain as the magnitude decreases are larger-is-best stressors. Low temperature is an example of a larger-is-best load.
- Loads that cause increased material strain as the magnitude varies between high and low extreme levels are larger-is-best stressors. Fluid pressure, geometry, and thermal shock are examples of nominal is best.

Part/LRU design for success is the end product of this analysis approach. A notional reliability failure analysis follows the failure mechanisms, modes, effects, and consequences analysis presented in Figure 2.13. The fastener failure mechanisms analysis identifies mechanically and temperature-induced failures under operational, ambient, and interaction environments.

The fastener failure modes analysis identifies failure modes caused by the thermal shock failure mechanism resulting from pump seal failure under operational, ambient, and interaction environments (see Figure 2.14).

The step 1 fastener failure effects analysis identified pump failure effects for the tensile thermal strain failure mode under operational, ambient, and interaction environments (see Figure 2.15). The step 2 fastener failure effects analysis identified pump and pipe assembly, fluid A WIP tank major

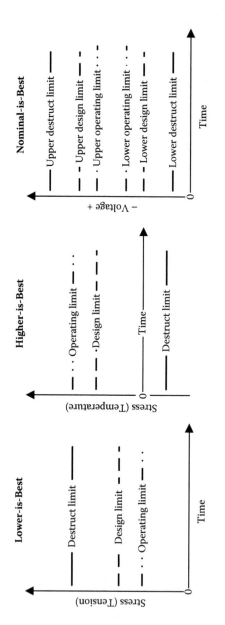

FIGURE 2.12
Operating design and destruct limits.

	Failure	Mechanisms Brainstorming Worksheet
Part:	1.00.00.00.00.000	Process Plant System
	1.01.00.00.00.000	Mechanical Major Subsystem
	1.01.01.00.00.000	Raw Materials WTP Tank Subsystem
	1.01.01.01.00.000	Fluid A WIP Tank Major Assembly
	1.01.01.01.02.000	Pump & Pipe Assembly
LEU	1.01.01.01.02.001	Pump
		Bolt, Washer & Nut

Conditions of Use

Fastener connects pump flange to pump mount on cryogenic WIP tank and is exposed to ambient weather conditions that ränge from -10°F to 103°F seasonally, and up to 30-inches annual rain and 6-inches annual snow.

Failure Mechanism	Operational	Ambient	Interaction	Comment
Mechanical				
Vibration	Normal operating Vibration ränges from 1-5 gRMS			Continuous operating condition
Shock	Vibration shock can occur from nonlaminor flow at start-up, pump bearing failure, and pump cavitation			At least daily; on bearing failure
Temperature				
Steady State	Pump housing conducts -40°F steady-state temperature			Daily
Thermal Shock	Pump seal failure and leak induces - 60°F temperature shock in less than 1-miiute			On bearing failure
	Pump housing conducts -40°F steady-state temperature	Start-up ambient temperature can be greater than 80°F	Hot weather start-up induces 140°F temperature shock	Once per day for up to 40 days per year in summer
Cyclical	Pump housing conducts -40°F steady-state temperature	Temperature ranges from 60°F to 103°F	Bolt temperature varies up to 35°F during a 24-hour operating period	Daily for up to 60 days per year in summer

FIGURE 2.13
Failure mechanisms analysis: fastener.

Failure Modes Brainstorming Worksheet		
Part:		Process Plant System
	1.01.00.00.00.000	Mechanical Major Subsystem
	1.01.01.00.00.000	Raw Materials WTP Tank Subsystem
	1.01.01.01.00.000	Fluid A WIP Tank Major Assembly
	1.01.01.01.02.000	Pump & Pipe Assembly
LEU	1.01.01.01.02.001	Pump Bolt. Washer and Nut

Failure Mechanism _____

Thermal shock caused by pump seal failure induces -60°F temperature change in less than 1 minute.

Failure Modes		Operational	Ambient	Interaction	Comment
1	Thermal strain of bolt in tension	Bolt material experiences thermal strain			On failure of pump seal
2		Connected plates experience thermal strain acting on bolt			On failure of pump seal

FIGURE 2.14
Failure modes analysis: fastener.

assembly, raw materials WIP tank subsystem, and process plant system failure effects for the fastener tensile thermal strain failure mode, as well as the consequences of the fastener failure for each design configuration level (see Figure 2.16).

The CIL lists the pump as a critical item that qualifies for reliability, maintainability, and availability analysis (see Figure 2.17). Qualitative analysis of the CIL failure mechanism and mode, thermal shock, and strain begins with a math model of the bolt material properties for thermal strain. The reliability math model should be the design analysis math model applied to the worst-case thermal load identified in the failure analysis and extend to higher loads to identify the ultimate strain. Nondestructive physical tests apply the worst-case thermal load, followed by measures of actual strain and visual inspection for signs of material changes.

A decision point is reached to determine whether sufficient information is available to reject the failure hypothesis for the failure mechanism. This hypothesis states that the part does not meet the design requirements for the system. Rejection of the failure hypothesis leads to the acceptance of the fastener, justified by documentation of the failure analysis, and to the

Part Failure Effects Brainstorming Worksheet				
LRU	1.01.01.01.02.001		Pump Bolt, Washer, & Nut	

Failure Mechanism _____

Thermal shock caused by pump seal failure induces -60°F temperature change in less than 1 minute.   _____   _____

Failure Mode _____

Bolt material experiences thermal strain

	Failure Effects	Operational	Ambient	Interaction	Comment
1	Thermal plastic limit exceeded	Bolt fractures weakening connection of pump mounting causing increased vibration damage to surviving bolts			Daily
2		Bolt cracks exposing material to water intrusion	Freeze thaw during operations fractures bolt		Daily

FIGURE 2.15
Part failure effects: fastener.

quantitative part/LRU investigation. Two options are available when the failure hypothesis cannot be rejected:

1. Specify a material or geometry for a fastener that will have the properties to eliminate the failure mechanism and repeat the failure analysis procedure.
2. Conduct destructive physical tests.

Destructive physical tests validate failure hypotheses when math modeling and simulation and nondestructive physical tests cannot do so. Destructive physical tests apply failure mechanisms to induce failure. Static design tests induce failure to validate material properties of parts, but the parts are not

	System Failure Effects Brainstorming	Worksheet	Catastrophic	Operational Mode	Degraded Mode	Run-to-Failure
		Effects				
	Bolt, Washer, & Nut	Bolt fractures weakening connection of pump mountain causing increased vibration damage to surviving bolts				
1.01.01.01.02.001	Pump	Loss of pump functionality; immediate maintenance required		X		
1.01.01.01.02.000	Pump & Pipe Assembly	Loss of pump & pipe functionality		X		
1.01.01.01.00.000	Fluid A WTP Tank Major Assembly	Loss of fluid AWIP tank functionality		X		
1.01.01.00.00.000	Raw Materials WTP Tank Subsystem	Loss of raw materials WIP tank functionality		X		
1.01.00.00.00.000	Mechanical Major Subsystem	Loss of mechanical major subsystem functionality		X		
1.00.00.00.00.000	Process Plant System	Loss of process plant functionality		X		

FIGURE 2.16
System failure effects analysis: fastener.

loaded by a failure mechanism. Reliability physical tests load the part to successive magnitudes of the failure mechanism until failure occurs or draw test articles at successive magnitudes of the failure mechanism and conduct static tests to measure material properties.

Highly accelerated life testing (HALT) induces one or more failure mechanisms (vibration, physical shock, temperature extremes, thermal shock, and humidity) at controlled rates. Test articles can be energized, input functions applied, and response variables (functionality parameters, strain gauge data, thermocouple data) acquired as the failure mechanisms are applied. HALT introduces time to physical tests where the failure effects from cyclical loading can be measured and it allows recognition of hard and soft failures. Hard failure is the irreversible part-failed state. Soft failure is a reversible failed state where the part recovers from the applied loads. Intermittent failures during system operations are the result of soft failures and are difficult to

Critical Items List						
Part Nomenclature		Failure Mechanism	Failure Mode	Assembly	Subsystem	System
1.01.01.01.02.001	Pump	Thermal shock caused by pump seal failure induces -60°F temperature change in less than 10 minutes	Bolt material experiences thermal strain	O	O	O
			Bolt cracks exposing material to water intrusion	D	D	D

C—Catastrophic
O—Operational
D—Degraded Mode
R—Run-to-Failure

FIGURE 2.17
Process plant system CIL.

detect and isolate because they cease to manifest failure when the system is shut down for maintenance.

A decision point is reached to determine whether sufficient information is available to reject the failure hypothesis for the failure mechanism. Rejection of the failure hypothesis leads to the acceptance of the fastener justified by documentation of the failure analysis and to the quantitative part/LRU investigation. Only one option is available when the failure hypothesis cannot be rejected: Specify a material or geometry for a fastener that will have the properties to eliminate the failure mechanism and repeat the failure analysis procedure.

Quantitative Part/LRU Investigation

Quantitative part/LRU investigations require data—the information needed to develop reliability, maintainability, and availability math models. Data are the measurement of a metric that correlates to the phenomenon under investigation. Time is the accepted metric used to characterize reliability, maintainability, and availability parameters (i.e., failure rate in failures per hours, MTBF, MTBM, and MTTR).

Reliability math models are constructed from the frequency distribution of part TTF to form the probability density function (pdf) of part TTF, $f(t)$; the cumulative density function (cdf) of part TTF, $F(t)$; the life function, referred to as the survival function, $S(t)$; the instantaneous failure rate, referred to as the hazard function, $h(t)$; and the probability that a part will function without failure for mission time (τ), given that the part has survived to time t, which is referred to as the reliability function, $R(\tau|t)$. Yet, time does not cause failure; failure mechanisms cause failure. Sources of reliability time-to-failure data include historical records of parts' field performance, prototype bench tests, acceptance tests, vendor tests, and reliability experiments. Historical data for like or closely similar parts can be used as an initial estimator of reliability, and engineering judgment of the criticality of the part will determine whether reliability experiments are needed to improve the reliability analysis.

Maintainability math models are constructed from the frequency distribution of part time to repair (TTR) to form the pdf of part TTR, $f(t)$, and the cdf of part time to repair, $F(t)$. Yet, time does not define the requirements for or the performance of repair actions; failure mechanisms do both. Sources of maintainability time-to-repair data include historical records of part field maintenance actions and maintainability experiments.

TTF and TTR Frequency Distribution and Probability Density Function of Part/LRU Failure

TTF and TTR data are plotted as frequency distributions that describe the shape, measure of central tendency, and measure of dispersion of the data. The shape

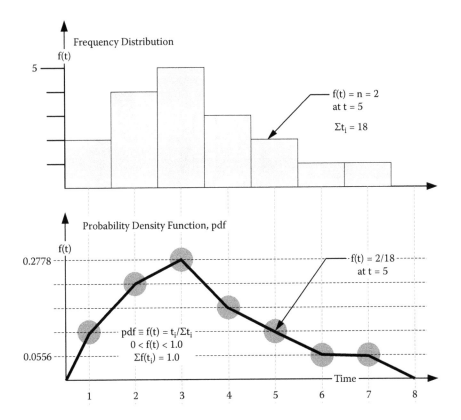

FIGURE 2.18
Frequency distribution to pdf.

of the data can be positively skewed (tail extends to the right toward positive values of time), negatively skewed (tail extends to the left toward negative values of time), symmetrical about the measure of central tendency, exponential, geometrically decreasing over time, and uniform, constant magnitude over time. The data that form frequency distribution are used to fit the probability density functions for reliability and maintainability functions. The pdf takes on the same shape of the frequency distribution but has different scales (see Figure 2.18).

The x-axis for the independent variable, t, does not change. The y-axis differs for the frequency distribution and pdf; the frequency distribution ranges from 0 to n-units of the event measured (failures, maintenance actions) at time t. The pdf describes the probability that failure occurs at time t and ranges from 0 to 100%. The sum of the frequency distribution is the total number of events; the sum, or integral, of the pdf must equal 1.00. (Procedures to fit TTF and TTR data to a pdf are presented in Chapter 3.)

Frequency distribution modes are important information that needs to be evaluated before further math model development is performed. A math model for TTF and TTR that has two or more modes is meaningless; it is a

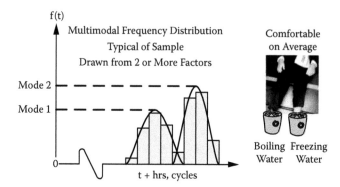

FIGURE 2.19
Multimodal frequency distribution.

measure of more than one failure mechanism. The measure of central tendency, the mean or median, for time to failure does not reflect the behavior of either failure mechanism; the measure of central tendency for time to repair does not reflect the effect of either failure mechanism (see Figure 2.19).

Reliability data plot as frequency distributions for TTF that take two shapes: exponentially decreasing and positively skewed. The exponential frequency distribution describes a failure mechanism and mode that have a constant effect on the useful life of the part, and it is modeled by the exponential probability distribution. The accuracy of the exponential probability distribution to estimate the behavior of failure effects is acceptable for electronic, electrical, and digital parts (see Figure 2.20).

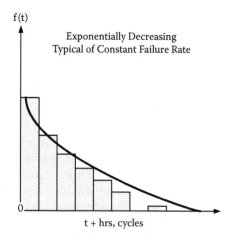

FIGURE 2.20
Exponential frequency distribution.

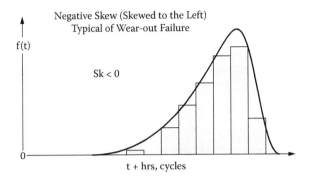

FIGURE 2.21
Weibull frequency distribution for TTF.

The negative, or left-hand, skew frequency distribution describes a failure mechanism and mode that have increasing effect over the useful life of the part, and it is modeled by the Weibull probability distribution. The accuracy of the Weibull probability distribution to estimate the behavior of failure effects is acceptable for structural and dynamic parts (see Figure 2.21).

Maintainability data plot frequency distributions for TTR that take two shapes: symmetrical and positively skewed. The positive, or right-hand, skew frequency distribution describes repair times that are smaller-is-best criteria, and it is modeled by the Weibull probability distribution. The accuracy of the Weibull probability distribution to estimate the behavior of repair time is acceptable for maintainability experiments (see Figure 2.22).[5]

The symmetrical frequency distribution describes repair times that are nominal-is-best criteria, and it is modeled by the Weibull probability distribution. Use of the normal probability distribution for symmetrical frequency

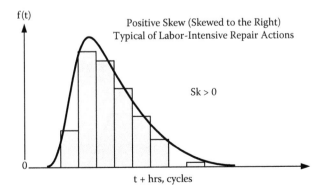

FIGURE 2.22
Weibull frequency distribution for TTR.

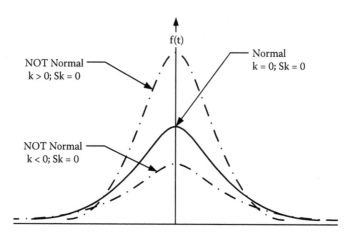

FIGURE 2.23
Symmetrical probability distribution functions.

distributions is a common mistake. The normal probability distribution can be used only when the skew is zero and the standardized kurtosis is zero. Kurtosis measures the degree of peak in the symmetrical plot of the distribution. Consider the three probability density functions in Figure 2.23: Only one can be evaluated as a normal probability distribution; the other two can be evaluated by the Weibull as can the normal.

The accuracy of the Weibull probability distribution to estimate the behavior of repair time is always acceptable for maintainability experiments (see Figure 2.24). Repair maintenance actions that are labor intensive typically describe a positively skewed frequency distribution. Machine-intensive repair maintenance actions typically describe a symmetrical frequency distribution.

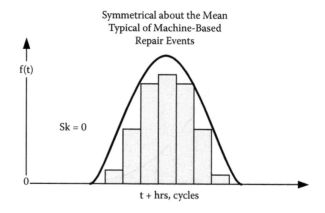

FIGURE 2.24
Symmetrical frequency distribution.

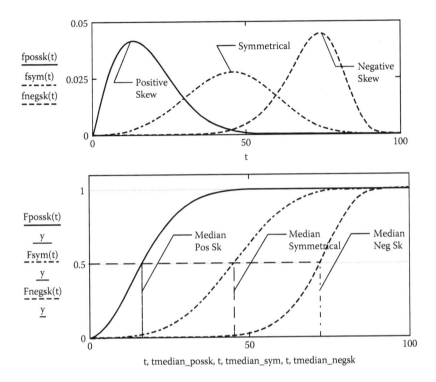

FIGURE 2.25
Cumulative frequency distribution.

Cumulative Frequency Distribution

The cumulative density function distribution is the summation, or integral, of the pdf, $f(t)$ from zero to a specified time, t. The relationship between the three forms of the pdf and the respective cdf, $F(t)$, is illustrated in Figure 2.25.

The median value of a pdf is the center value of the number of the sample data points where one half of the data are less than or equal to the median, and one half of the data are greater than or equal to the median. The cdf of the median time is 0.5. The mean equals the median only when the pdf is symmetrical. The mean time for a positively skewed and exponentially distributed pdf is greater than the median and less than the median for a negatively skewed pdf. Indeed, the sign of mean minus median defines the sign of skew (see Figure 2.26).

The cdf of TTF is the cumulative percent of the population of parts that will fail by time t. The cdf provides percentiles of failure (e.g., the part B10 life is the time to 10% of population failures). The cdf of TTR is the cumulative expectation in percent of the time that a repair is expected to take. The maximum expected repair time for a part is estimated as the nth percentile (e.g., the repair time calculated for the 90th percentile).

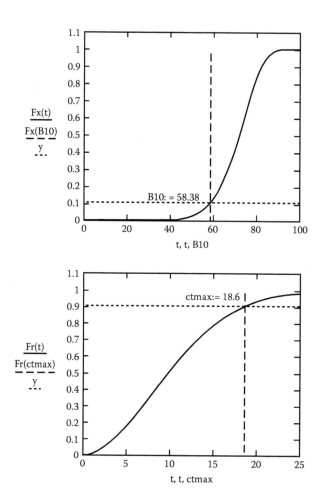

FIGURE 2.26
Percentiles from the cdf for TTF and TTR.

TTF Survival Function of a Part/LRU

The survival function, $S(t)$, is the life function of a part and is often expressed as $R(t)$—the reliability function over the range of the independent variable time, t. Using $R(t)$ creates confusion and should be discontinued. The survival function is the complement of the TTF cdf, or $1\ F(t)$. The survival function describes the cumulative percent of the population that will survive by time t. The characterization of reliability is illustrated by the succession of the part failure frequency distribution to the TTF pdf, $f(t)$; to the TTF cdf, $F(t)$; and to the survival function, $S(t)$, as shown in Figure 2.27 for both the exponential and Weibull probability functions.

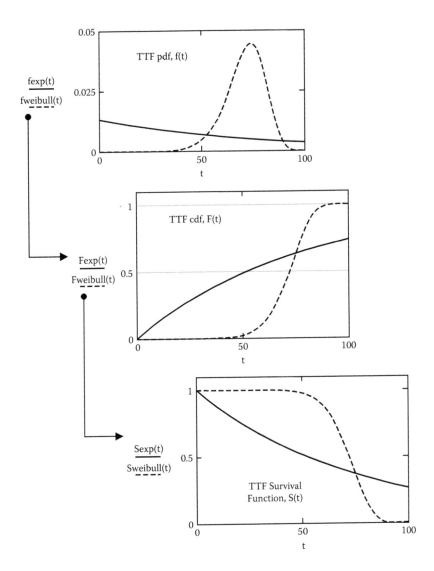

FIGURE 2.27
Progression from pdf to cdf to $S(t)$.

TTF Instantaneous Part/LRU Failure Rate: The Hazard Function

The instantaneous failure rate, referred to as the hazard function, $h(t)$, is the number of failures, r, divided by the number of surviving parts, $S(t)$, at time t. The hazard function is expressed as

$$h(t) = \frac{f(t)}{S(t)} \qquad (2.1)$$

Constant instantaneous failure rate,[6] $h(t) = \lambda,$[7] means that the failure mechanism acts on the part with equal effect at any time in the useful life from new to well past the mean TTF. The exponential probability density function, expressed as

$$f_{exp}(t) = \lambda e^{-\lambda t} \tag{2.2}$$

describes the constant instantaneous failure behavior of a failure mechanism.

The exponential cumulative probability density function, $F_{exp}(t)$, is expressed as

$$F_{exp}(t) = \int f_{exp}(t)dt = 1 - e^{-\lambda t} \tag{2.3}$$

The survival function, $S_{exp}(t)$, is expressed as

$$S_{exp}(t) = 1 - F_{exp}(t) = e^{-\lambda t} \tag{2.4}$$

The exponential pdf, $f_{exp}(t)$, and constant hazard function, $h_{exp}(t)$, are illustrated in Figure 2.28.

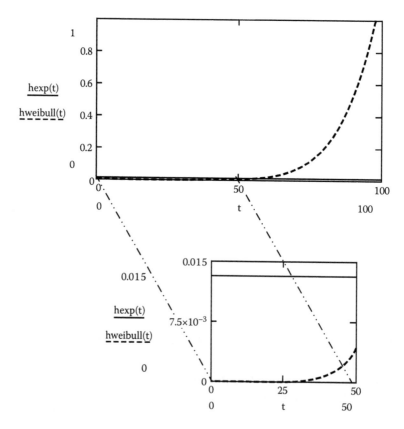

FIGURE 2.28
Exponential $f(t)$ and $h(t)$.

The exponential probability distribution is memory-less and does not measure part degradation. It assumes that there is no degradation and that failure occurs without wear out.

As time in service increases, the occurrence of part failure increases until the number of survivors dwindles to zero. The Weibull probability function, expressed as

$$f(t) = \frac{\beta}{\eta}\left(\frac{t}{\eta}\right)^{\beta-1} e^{-\left(\frac{t}{\eta}\right)^{\beta}} \tag{2.5}$$

describes the increasing instantaneous failure rate behavior of a failure mechanism. The Weibull cumulative probability density function, $F_{\text{weibull}}(t)$, is expressed as

$$F_{\text{weibull}}(t) = \int f_{\text{weibull}}(t)\,dt = 1 - e^{-\left(\frac{t}{\eta}\right)^{\beta}} \tag{2.6}$$

The survival function, $S_{\text{weibull}}(t)$, is expressed as

$$S_{\text{weibull}}(t) = 1 - F_{\text{weiball}}(t) = e^{-\left(\frac{t}{\eta}\right)^{\beta}} \tag{2.7}$$

The Weibull probability density function is a family of distributions that can take the shape of any TTF or TTR frequency distribution. The parameters of the Weibull probability density function are the location parameter, η, that characterizes the measure of central tendency, and the shape parameter, β, that characterizes the measure of dispersion. The Weibull distribution's power to describe the shapes of frequency distributions is illustrated in Figure 2.29 for various values of β with η held constant.

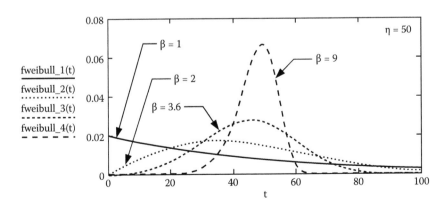

FIGURE 2.29
Weibull probability density function.

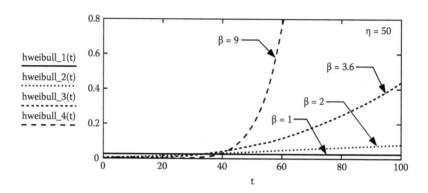

FIGURE 2.30
Weibull instantaneous failure rate.

The Weibull shape for $\beta = 1$ is the exponential distribution, for $\beta = 2$ is an example of a positive skew distribution, and for $\beta = 3.6$ closely approximates the normal probability distribution. An example of a negative skew distribution is $\beta = 9$.

The Weibull hazard function,[8] $h(t)$, is expressed as

$$h(t) = \frac{\beta}{\eta}\left(\frac{t}{\eta}\right)^{\beta-1} e^{-\left(\frac{t}{\eta}\right)^{\beta}} \tag{2.8}$$

It describes an instantaneous failure rate that increases as time increases. The corresponding hazard functions, $h_{weibull}(t)$, for the preceding four examples are illustrated in Figure 2.30.

The accuracy of the Weibull probability distribution to estimate the behavior of failure is always acceptable for structural and dynamic parts that wear out over time. It is preferred over the exponential distribution for TTF that appears to have a constant failure rate. In Figure 2.30 of various values of the shape parameter, β, we observe the following:

When $\beta = 1$, the hazard function is constant over time and the Weibull distribution reduces to the exponential distribution because the factor $(t/\eta)^{\beta-1}$ reduces to 1, $[(t/\eta)^0]$, and $e - (t/\eta)^{\beta}$ reduces to $e - (t/\eta)$.

When $\beta = 2$, the hazard function increases with a constant slope, β/η, because the factor $(t/\eta)^{\beta-1}$ reduces to 1, $[(t/\eta)^1]$.

When $\beta > 2$, the hazard function increases exponentially.

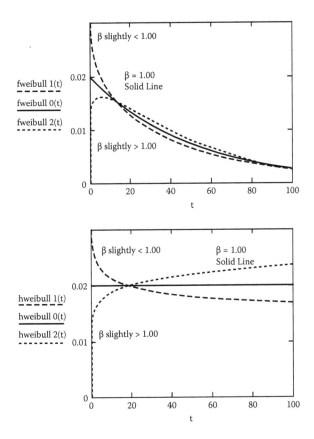

FIGURE 2.31
Nearly and exactly exponential $f(t)$ and $h(t)$.

Figure 2.31 shows the plots for a failure rate that is slightly less than exponentially distributed, $f_{\text{weibull1}}(t)$, exactly exponentially distributed, $f_{\text{weibull0}}(t)$, and slightly more than exponentially distributed, $f_{\text{weibull2}}(t)$. The shapes of the probability density functions and hazard functions show that the difference exists and should not be ignored.

TTF Reliability Function of a Part/LRU

The reliability function, $R(\tau|t)$, is the conditional probability that a part will function without failure for a mission duration, τ, given that the part has survived to time t. The reliability can be expressed as

$$R(\tau|t) = \frac{S(t+\tau)}{S(t)} \tag{2.9}$$

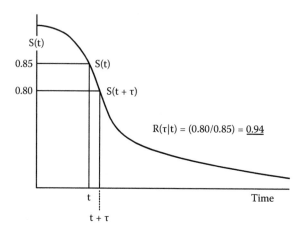

FIGURE 2.32
Reliability calculation: graphical solution.

where $S(t + \tau)$ is the survival function for the time to survival to time t, plus the duration of the next mission, τ, and $S(t)$ is the survival function for the time to survival to time t, as illustrated in Figure 2.32.

The reliability expression using the TTF exponential probability distribution is shown to be

$$R_{exp}(t \mid \tau) = \frac{S_{exp}(t+\tau)}{S_{exp}(t)} = \frac{e^{-\lambda(t+\tau)}}{e^{-\lambda t}} = \frac{e^{-\lambda t}e^{-\lambda \tau}}{e^{-\lambda t}} = e^{-\lambda \tau} \qquad (2.10)$$

Note the similarity between the expressions for the exponential survival function, $S_{exp}(t) = e^{-\lambda t}$, and the exponential reliability function, $R_{exp}(\tau \mid t) = e^{-\lambda \tau}$. The two expressions are very different, as illustrated in Figure 2.33.

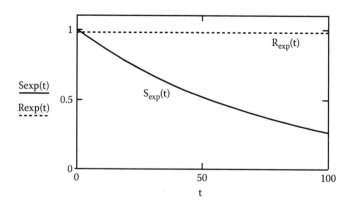

FIGURE 2.33
Exponential survival and reliability functions.

The part reliability for an exponentially distributed failure mechanism is constant over its useful life for a specific mission duration, τ. The assumption of constant mission duration is useful for comparative evaluations of part design alternatives but does not reflect reality when mission duration varies between customers. Consider two mines that use the same haul truck: One operates one 12-h shift/day, 5 days/week, and the other operates two 8-h shifts/day, 6 days/week.

Part/LRU Time-to-Failure Characterization of Reliability Parameters

Data that describe time to failure for a specific failure mechanism are not equally descriptive. All data are small, unbiased samples of randomly selected observations of a failure mechanism metric taken from a vast population—with very few, rare exceptions. The exception is a 100% census of all parts measured for a critical failure mechanism. The exception must meet one of two requirements:

- The sample is drawn from records of historical records and is sufficiently large to be treated as "complete" sample data. The historical records include accurate information for time to failure, root cause failure analysis that accurately describes the failure mechanisms and modes, and conditions of use.
- The part demonstrates that it does not fail or exceed its design limits for an accelerated life screening test duration greater than or equal to its useful life. The part does not experience degradation that prevents its distribution to the end user. The 100% census is cost effective to the end user.

Characterization of reliability parameters using small, random, unbiased samples drawn from the population of all parts is a more reasonable approach in the majority of cases. The very nature of characterizing failure requires that a part must experience a failure mechanism over a suitable time to induce failure. Survivors of reliability tests are degraded sufficiently to prohibit distribution to the end user and must be scrapped. Part TTF reliability tests have an impact on system design and development costs and schedule plans:

- cost:
  - Direct labor expenses are incurred to
    - design, perform, and analyze the findings of TTF tests
    - design and fabricate test fixtures and prototype test articles
  - Indirect labor expenses are incurred to
    - perform purchasing, shipping and receiving, and personnel activities
    - conduct safety and test training

- Direct materials expenses are incurred to
  - acquire test articles
  - contract test services
  - acquire, operate, and maintain test equipment and chambers
- Indirect materials expenses are incurred to
  - provide engineering, test, and administrative materials and supplies
- Direct overhead expenses are incurred to
  - manage and administer the infrastructure for engineering and test activities
  - operate and maintain test equipment and chambers
- Indirect overhead expenses are incurred to
  - manage and administer the organization infrastructure
  - operate and maintain engineering and test facilities

- schedule:
  - Time must be budgeted to design, perform, and analyze the findings of TTF tests.
  - Time must be budgeted to acquire, receive, and fabricate test fixtures and prototype test articles.
  - Time must be budgeted to perform failure mitigation.

Part TTF reliability test data cost resources; data are not free or inexpensive. But the cost and schedule impact on system design and development plans associated with failure mitigation are optimally incurred during the earliest phases of part design analysis and only become prohibitively high in later phases and sustainment.

Unmitigated part failures that doomed systems in use by end users and possibly caused ruin, or near ruin, of the organizations that designed and distributed them are documented in the literature. One needs only to look at the U.S. automobile and consumer products industries since the 1960s, when more reliable foreign imports reduced market share, impugned organizations' reputations in the market place, and necessitated changes in organizations' engineering practices for them to survive. Government-funded systems have experienced higher than forecast design, development, and sustainment costs; extended delivery schedules; and dismal field performance.

Consumer Reports magazine frequently grades automobiles, electronics, and consumer products for cost and reliability. Foreign products are ranked higher than U.S. products, and high-cost, low-reliability lists are populated by U.S. products more frequently than by foreign products. The costs to society associated with low reliability are a lower standard of living (measured by unemployment rate, median income, and gross national product [GNP]) due to the reduction in competitive advantage in the global economy. Successful organizations design system reliability at part design, where the return on investment is highest, the impact on the cost of goods sold is lowest, and the satisfaction of the end user is assured.

Part/LRU Historical Part Failure Data

Characterization of reliability parameters is a statistical estimate that is performed at specified confidence levels. Absolute 100% confidence does not exist. Sources of error in reliability parameters include sampling error, measurement error, and intrinsic material variability. Confidence levels are typically stated at 90, 95, and 99%, with 80 and 99.5% and other levels occasionally observed. Statisticians and common sense claim that the accuracy of the estimate of the mean for a population—the sample mean—is improved by increasing the size of the sample. The accuracy of the estimate is composed of (1) proximity of the sample estimate of the mean to the true population mean, and (2) minimization of the measure of dispersion of the sample estimate. Tests for statistical significance of sample means are used to calculate the size of a sample to approximate the true population mean better. Reliability test samples have two size parameters: (1) number of test articles, and (2) time on test for the test articles.

Complete data approach 100% census sample sizes in terms of number of test articles and time in service. Part MTTF is equal to the total time that all test articles operate (Σt_i) to the failed state divided by the total number of failed parts (r): MTTF = $\Sigma t_i / r$. The total test time required to claim that a part has an MTTF = 10,000 h for only one failed part at 95% confidence is 156,000 h. This test would take a test duration of 2,501 h for a sample size of 50 test articles, assuming that the failure occurs at the last moment of test time (see Table 2.1). More total test time, more sample test articles, or a combination of the two is required for a statistically significant claim that MTTR = 10,000 if the number of failures is greater than one.

The relationship between total test time and test duration for MTTR ranging from 10 to 1,000,000 h for selected sample sizes (10–100 test articles) is seen in Table 2.1.

The time and expense for statistically significant test results for complete data are unrealistic for a part design project. Historical data come close to providing the magnitudes for time in service and number of field failures. The issue is whether the data are available, sufficient to isolate the failures to specific failure mechanisms, and credible. Consumer and commercial tires are an example of a part that has the potential to provide these magnitudes. Tires are capable of being tracked through service organizations that sell, maintain, and replace them. Consumer and commercial car batteries are an example where the magnitudes are present but the credible data acquisition is lacking. Organizations that support their systems after delivery to the end user can

TABLE 2.1

Statistically Significant Test Time

r = 1			Total Test Hours for Selected MTTF Targets						
C	α	χ^2_{av}	10	100	1,000	10,000	100,000	1,000,000	
90%	0.1	6.25	1.25E + 02	1.25E + 03	1.25E + 04	1.25E + 05	1.25E + 06	1.25E + 07	
95%	0.05	7.81	1.56E + 02	1.56E + 03	1.56E + 04	1.56E + 05	1.56E + 06	1.56E + 07	
99%	0.01	11.34	2.27E + 02	2.27E + 03	2.27E + 04	2.27E + 05	2.27E + 06	2.27E + 07	
					Test Direction				
Number of test articles		10	13	125	1,250	12,503	125,028	1,250,278	
		25	5	50	500	5,001	50,011	500,111	
		50	3	25	250	2,501	25,006	250,056	
		100	1	13	125	1,250	12,503	125,028	

develop data acquisition for critical parts that meet the standard for statistically significant estimation of population mean values for time to failure.

Part/LRU Reliability Experiments

Time-to-failure data are acquired through reliability experiments in the absence of historical data. Real-world constraints on resources and time make large sample sizes impossible. Statistically significant estimators for reliability parameters are still possible when experimental design is properly performed.

The components of a reliability experiment are the failure mechanisms, the operating and ambient environments, and the response variables. The basic reliability experimental design specifies a single factor experiment in one failure mechanism, with one or more operating and ambient factors, and response variables that measure magnitude and time to the failure modes. Controlled factors are independent variables set at specified levels and measured to document the magnitudes in real time. Uncontrolled factors are independent variables that occur due to the controlled factors and are measured to document the realized magnitudes in real time. Response variables are the reaction of the part to the levels of the independent variables and are measured to document the realized magnitudes in real time.

Consider the previous bolt example: The controlled failure mechanism is tension at levels from design equilibrium to the destruct limit in increments of 5 psi. The independent controlled factors are the applied installation torque, bolt size (diameter), and bolt material properties. The independent uncontrolled factors are friction-generated heat, environmental temperature, temperature shock, and vibration. The response variables are strain, crack initiation, and fracture.

Time-Censored Experimental Part/LRU Failure Data

Reliability test data are described by the censoring of the data. A censored test is one that is terminated prior to achieving the magnitudes of complete data. All reliability experiments are either time censored or failure censored. Time-censored tests are defined by a limited test duration—a plan that allows use of the test facility and equipment for 1 week, 10 h/day, for 50 test hours. A total time on test of 500 hours is possible, assuming a test capacity of 10 test articles on test. The total time on test is equal to the sum of the time on test for each test article: $T = \Sigma t_i$. A time-censored test without replacement occurs if only 10 test articles are available and failures during the test result in less total time on test, as illustrated in Figure 2.34; the total test time will fall below the potential 500 h.

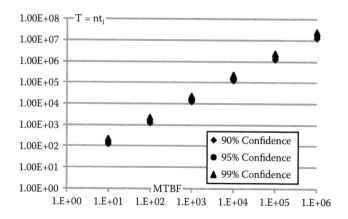

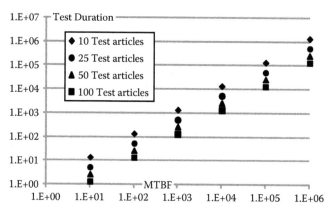

FIGURE 2.34
Time-censored experiment without replacement.

A time-censored test with replacement occurs if sufficient test articles are available to use the entire total test time, net time to replace test articles following failure. The total test time approaches the potential 500 h (see Figure 2.35).

Interval-Censored Experiment

A variation on the time-censored test occurs when test monitoring is less than real time. Consider a field test for pumps that is designed to run 24 h/day at a remote site. The response variable is pump operation. The pumps are checked on a daily interval at the same time. A failed pump is recorded at the end of each time interval. The exact time to failure is not known because the pump could have failed 1 min following the preceding daily check, 1 min before the current daily check, or any time in-between.

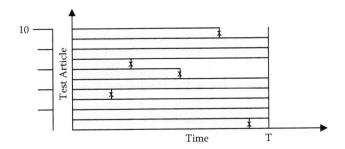

FIGURE 2.35
Time-censored experiment with replacement.

The time to fail will be recorded at the preceding daily time on test, for a conservative estimate, or at the midtime in the failure interval in which it failed.

Failure-Censored Experimental Part/LRU Failure Data

Failure-censored reliability tests are defined by a limited test duration that extends until all or some specified number of test articles fails. Small sample sizes are total failure-censored tests that are not limited by time constraints. Time to the nth failure is used when time is limited and the unit costs are high. For example, consider an expensive test article and limited time. Ten test articles are made available but only one can be destroyed by the test. Placing all 10 test articles on test and running the test to the first failure builds up total test time for characterization of the mean TTF. The test is terminated and censored following the first failure and the surviving test articles are removed from test (see Figure 2.36).

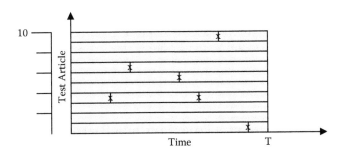

FIGURE 2.36
Failure-censored experiment.

Failure-Free Experimental Part Data

Both time-censored and failure-censored tests can end without part failure. The time-censored example can end with 10 test articles remaining in a functional state when the test duration ends. The failure-censored example can run out of time with no failures. Information is insufficient to calculate a mean TTF, but sufficient to calculate a statistically significant lower confidence limit for the mean time to failure for continuous response variables or lower confidence limit for the discrete probability of failure for attribute response variables.

Attribute Failure Metrics

Attribute data are based on a count (e.g., number of bolts that fracture in a level of tension load). Consider 10 bolts in tension at 15 kpsi with no failures. The β distribution provides the lower confidence limit for the discrete probability of failure at various confidence levels. The β distribution table is available in statistics references and on the Internet. The β table for 10 test articles with no failures and 95% confidence gives a discrete probability of failure of 0.762.

Continuous Failure Metrics

Continuous data are based on a measurement (e.g., geometry, stress, temperature) that can take on any level of precision limited only by the measuring device. (There are infinite increments of length within a 1-inch dimension; a ruler can measure 1/64 increments of an inch, and a micrometer can measure 1/1,000 of an inch.) Time to failure is a continuous metric. The statistically significant lower confidence limit for TTF for the preceding example of 10 test articles on test for 50 h accumulating 500 total hours on test, T, is calculated as

$$\theta_{LCL} \geq \frac{2T}{\chi^2_{\alpha,\nu}} = \frac{1000}{5.99} = 166.9 \qquad (2.11)$$

where α, the level of significance, is equal to $1 - C$ $(1 - 95\% = 0.05)$ and ν, the degrees of freedom, is equal to $2r + 2$ $(2(0) + 2 = 2)$.

Maintainability Analysis Functions of a Part/LRU

Maintainability analysis serves two purposes in system maintainability: (1) characterize the part MTTR for two or more design alternative decisions, and (2) characterize the part MTTR for the system logistical support analysis (LSA).

Part maintainability analysis for two or more design alternative decisions is limited to what the design engineer can control—specifically, the MTTR. The conventional approach for MTTR is to calculate and compare the arithmetic mean from a small sample of maintainability experiments. The decision criterion is to select the part with the lowest MTTR. This approach fails to acknowledge the measure of dispersion for MTTR.

The preferred approach for only two alternative parts is to calculate the standard deviation and conduct a test of hypothesis for the difference between the means. Consider part 1 that has an MTTR_1 = 1.33 h with a standard deviation of s_1 = 0.25 h, and part 2, which has an MTTR_2 = 1.5 with a standard deviation of s_2 = 0.5 h. Both experiments ran five runs of the experiment, $n1 = n2 = 5$. Part_1 would be selected based on its lower MTTR. A test of hypothesis provides the following information:

1. Confidence of the test of hypothesis at 95%; a level of significance of $\alpha = 0.05$
2. Null hypothesis, H0: MTTR_1 = MTTR_2[9]
3. Critical statistic, $t_{crit} = t_{\alpha,v} = t_{0.05,(5+5-2)} = 2.31$, using MS Excel™
4. Calculate test statistic:

$$t_{test} = \frac{\text{MTTR}\_1 - \text{MTTR}\_2}{\sqrt{s_p^2\left(\dfrac{1}{n1} + \dfrac{1}{n2}\right)}} = -0.68 \qquad (2.12)$$

where

$$s_p^2 = \frac{(n1-1)s_1^2 + (n2-1)s_2^2}{n1+n2-2} \qquad (2.13)$$

5. Evaluate expression, $|t_{test}| < |t_{crit}|$:
   a. If $|t_{test}| < |t_{crit}|$, then accept the null hypothesis that there is no statistically significant difference between MTTR-! and MTTR_2 at 95% confidence.
   b. Else reject the null hypothesis and the two estimates for MTTR are statistically different and can be rank ordered.

The conclusion is that the experiment did not provide sufficient information to decide one part over the other and the decision criterion must be something else (e.g., cost, functionality parameters, weight, etc.).

The preferred approach for three or more alternative parts is to perform an analysis of variance (ANOVA). The raw data are used for ANOVA and are calculated in a spreadsheet (e.g., MS Excel) or statistical software program (e.g., Minitab™). Consider four parts that are under consideration and three

TABLE 2.2

Time to Repair: Hours

	Part_1	Part_2	Part_3	Part_4
	1.38	1.60	1.23	1.62
	1.30	1.43	1.07	1.68
	1.48	1.37	0.98	1.78
MTTR	1.39	1.47	1.09	1.69
s	0.09	0.12	0.13	0.08

experimental maintainability experiments run for each. The results for the TTR experiment for each part are in Table 2.2.

A cursory view might prompt the decision to claim that Part_3 should be selected because it has the lowest MTTR (MTTR_3 = 1.09 h). The answer lacks statistical significance. A test of hypothesis states the null hypothesis that the MTTR for all four parts is statistically the same at 95% confidence. The alternate hypothesis is that at least one or more MTTR is statistically different at 95% confidence. The ANOVA was run using MS Excel and the results are tabulated in Figure 2.37.

The summary displays the descriptive statistics for each part MTTR. The ANOVA solves the relationship between F_{test} and F_{crit}. The decision criterion is to accept the null hypothesis when $F_{test} < F_{crit}$.[10] The null hypothesis is not accepted and the MTTR values can be rank ordered and a decision made to select the part with the lowest MTTR.

Resource Requirements for a Part/LRU

Maintainability analysis is used to define part repair requirements for the system logistical support analysis. The maintainability experiments are used to write maintenance procedures, identify tool and facility requirements, and specify labor skills. Design of the maintainability experiment can

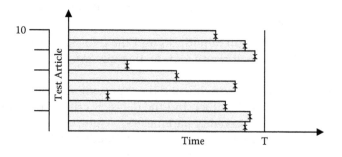

FIGURE 2.37
Maintainability experiment ANOVA.

TABLE 2.3

Inherent Availability

	Hours	A
θ_1	100	
μ_1	3	0.97
θ_2	120	
μ_2	5	0.96

optimize this opportunity to support the system LSA by including logistics engineers on the team.

Inherent Availability of a Part/LRU

Availability is the probability that a part will be in a functional state at the beginning of a mission. The inherent availability (A_i) is the ideal part availability and is a function of mean time between failure (θ) and mean time to repair (μ):

$$A_i = \frac{\theta}{\theta + \mu} \qquad (2.14)$$

Inherent availability is used to evaluate alternative parts in design analysis. Consider two parts that have the reliability and maintainability parameters in Table 2.3. The reliability parameter, MTBF, supports selection of part 2; the maintainability parameter, MTTR, supports selection of part 1. The inherent availability resolves the confusion and supports selection of part 1.

Notes

1. I have made it known that any criticism of an idea will be met with expulsion, and I have followed up with expulsion when the rule was violated. Criticism will mute introverts. All opinions are needed and many of the best ideas come from introverts, but they are sensitive to criticism in a group. Criticism of one introvert's idea will prevent other introverts from participating.
2. Rothbart, H. A. 1964. *Mechanical design and systems handbook.* New York: McGraw–Hill.
3. Many approaches to failure analysis begin with hypothesis of the failure modes. This approach is typically evident when the failure modes and effects analysis is prepared by a consultant or a "reliability engineer" who is not part of the design team.

4. Bill Hewlett, the late cofounder of Hewlett-Packard, would say "demands" rather than "requires." He often admonished his employees with the statement that "what one can measure, one can control, what one can control, one can manage." There is no doubt that critical part failures must be managed.

5. The lognormal probability distribution is the best-practice math model for TTR data analysis; however, the Weibull probability distribution models skew distributions more precisely.

6. $h(t) = \lambda e^{-\lambda t} / e^{-\lambda t} = \lambda$.

7. The failure rate (λ) of the exponential probability density function is the inverse of the mean (θ) of the exponential probability density function (e.g., $\lambda = 1/\theta$).

8.
$$h(t) = \frac{f(t)}{S(t)} = \frac{\frac{\beta}{\eta}\left(\frac{t}{\eta}\right)e^{-\left(\frac{t}{\eta}\right)^{\beta}}}{e^{-\left(\frac{t}{\eta}\right)^{\beta}}} = \frac{\beta}{\eta}\left(\frac{t}{\eta}\right)^{\beta-1}$$

9. The equal sign means statistically the same as—not numerically equal to.

10. The p-value is a more rigorous evaluation statistic than the F statistic. The value for F_{crit} requires a confidence level. We must recompute F_{crit} if we change the confidence level. But the p-value defines the exact threshold for the accept/reject decision. We accept the null hypothesis if p-value $> \alpha$, else reject. In this example, we can reject the null hypothesis for confidence levels at 90, 95, 99, and 99.9%.

3

Reliability Failure Modeling Based on Time-to-Failure Data

Seek first to understand; then to be understood.

Stephen Covey, *The 7 Habits of Highly Effective People*

Introduction

The objective of this chapter is to describe the current best-practice methods to characterize the parameters of material and part reliability failure models using time-to-failure (TTF) data, which has been the dominant approach to characterize reliability parameters in academe and engineering practice. An engineer must understand the use of TTF data to determine when it is an acceptable method for his or her purposes and when to challenge its use as providing meaningless results. TTF data must be empirical. Empirical data are drawn from records of actual events, previous maintenance documents, and controlled experimentation. Two factors dictate how the data can be analyzed: sample size and censoring.

Large sample size, typically drawn from historical data over several years, can be treated as complete data. Large is a vague term, but engineering judgment can make the determination easily. Sample size that is greater than 30 (or 25, depending on the reference) is deemed large. Sample size is a large proportion of the total population, greater than 2–5%.

Consider an industrial machine that has been in service for 15 years with annual production that has grown from 100 units in the first year to 1,000 units recently. Your organization has 15 units with 10 years of historical data. You calculate a round order of magnitude that your maintenance actions have declined from 15% of the population to 1.5%. The historical data can be assumed to be large enough to be classified as complete data.

Contrast the machine to a generator that has been in production for decades with tens of thousands of units produced annually. You have three generators with 5 years of maintenance data. Engineering judgment leans toward not treating the historical data as complete data.

Data that are not complete are censored. Censored data have small sample size and are typically drawn from controlled experiments. The experiment is stopped by a predetermined condition, almost always due to two resource constraints: time and number of test articles. Censored experiments can take one of the following forms:

- *Time-censored data.* Management demands that the test results be available by a specified date, or a facility and test resources are available for 1 week due to competing demand from others. Time-censored empirical data have two constraints:
  - Time-censored data without replacement: a fixed number of test articles is available and placed on test at the commencement of the experiment; no replacement test articles are available to replace failure test articles
  - Time-censored data with replacement: a fixed number of test articles is available but exceeds the capacity of the test resources to run simultaneously; test articles are replaced as test articles fail
- *Failure-censored data.* The test budget limits the number of test articles, all are placed on test, and there is no test time limit. Failure-censored empirical data have two constraints:
  - Failure-censored exhaustive data: a fixed number of test articles is allowed to run until all have failed.
  - Failure-censored to the nth failure data: a fixed number of test articles is allowed to run until a specified number of failures have occurred, at which time the test ends

Use of TTF data can lead to two major inaccuracies in part reliability failure models for the following reasons:

- Time is not a failure mechanism, but TTF data-based reliability failure models treat it as though it were. Therefore, too many engineers view reliability as a statistics exercise rather than as an engineering design analysis. Only when the design engineer isolates a failure mechanism and mode can time be the independent variable to characterize part reliability failure models.
- The mean time to failure (MTTF) lends itself too easily, almost seductively, to the application and misuse of the exponential probability failure model. Constant random failure over part useful life is not viewed by engineers as a rational description of true failure behavior—further supporting their lack of commitment to reliability models.

Part Reliability Failure Modeling

A reliability failure model is a math expression (a probability distribution) that fits the shape of empirical data. Engineers are intimately familiar with linear math models that define the relationship between an independent variable and a dependent variable. Probability distributions describe the frequency that a number of parts fail over time and the frequency that a number of maintenance actions will be performed over time. Distributions have a measure of central tendency and dispersion. Four distributions make up the workhorses for reliability and maintainability math modeling.

Exponential is used or overused to model part failure, life-cycle survival, mission reliability, and the hazard function. It describes a part that fails at a constant rate from the time a part is put in service, and has a mean (θ) and a standard deviation ($\sigma = \theta$). The mean is calculated as the arithmetic average of the time on test (ToT) for all parts. The parameter of the exponential distribution is the failure rate (λ), which is the inverse of the mean. The exponential probability distribution is expressed as

$$f_{\exp}(t) = \lambda e^{-\lambda t} \tag{3.1}$$

Weibull is used to model part failure, life-cycle survival, mission reliability, hazard function, and time to repair. It describes a part that fails at an increasing rate over time from wear out and fatigue that is skewed to the left (negative skew) and a time to repair that is skewed to the right (positive skew). The parameters of the Weibull distribution are the characteristic life (h) and the shape (β). The Weibull probability distribution is expressed as

$$f_W(t) = \left(\frac{\beta}{\eta}\right)\left(\frac{t}{\eta}\right)^{\beta-1} e^{-\left(\frac{t}{\eta}\right)^{\beta}} \tag{3.2}$$

Normal is used to model part time in use, time to repair, and stress and strength parameters. It describes an event that occurs at a stable value and varies below or above that value symmetrically. The parameters of the normal distribution are the mean (μ) and standard deviation (σ). The normal probability distribution is expressed as

$$f_n(x) = \left(\frac{1}{\sigma\sqrt{2\pi}}\right) e^{-\frac{1}{2}\left(\frac{x-\mu}{\sigma}\right)^2} \tag{3.3}$$

Lognormal is used to model part time to repair. It describes part repair time that is skewed to the right (positive skew). The parameters of the lognormal distribution are the mean of the logarithms of the data and the standard deviation of the logarithms of the data. The natural log (ln) is the maintainability best-practice approach. The lognormal probability distribution is expressed as the normal distribution with the understanding that μ and σ are calculated from the logarithms of the data.

This book proposes that the Weibull distribution be used in all characterizations of reliability and maintainability math models for two reasons:

- The Weibull family of distributions can characterize all frequencies of data, including data shaped like the exponential, normal, and positively and negatively skewed.

- Reliability and maintainability body of knowledge and best practices were developed before engineers had access to computers that now allow use of the Weibull distribution. Reliability engineering models failure with the exponential distribution. Maintainability engineering is used to lognormal distribution to transform positive skewed data to the normal distribution so that the normal distribution and its standard normal distribution can be used to characterize the behavior of repair time. The two distributions have been used because they are convenient, rather than because they are good estimators.

However, use of the exponential, normal, and lognormal distributions is presented in this book along with the Weibull approach to inform the reader how to deal with analyses that have used the offending distributions and to show the discrepancies between them and the Weibull method.

Failure modeling for part TTF and part time to repair (TTR) are referred to as characterization of reliability, maintainability, availability, and sustainability parameters. "Parameter" defines a measurable factor—that is, failure rate (MTTF), mean time to repair (MTTR), etc. Yet the true value of a parameter can never be known because there is always error in the estimate of the value of a parameter. Error has three sources:

- *Common cause/intrinsic variability* means that no two samples of the same material are identical in composition or properties, no two measures of the same part are identical in functionality, and no two measurements of the same factor of the same sample by the same person are identical.

- *Special cause/extrinsic variable* means uncontrolled, incapable processes that produce materials; differing levels of expertise of people performing the measurement; varying wear and calibration of measuring devices; and random ambient environments that have an impact on the measurement.

- *Sampling error* is any deviation from 100% census of the measurement.

Common cause variability is natural for material properties, measurement devices and methods, process control and capability, and human performance.

Common cause variability cannot be eliminated. Understanding failure mechanisms, modes, and effects is achieved by understanding the magnitudes of common cause variability.

Special cause variability is manageable and must be minimized as much as is technically and economically feasible. Bill Hewlett, cofounder of Hewlett-Packard, wrote, "That which can be measured can be controlled; that which can be controlled can be managed." Eliminating all special cause variability is an ideal that can never be achieved. Sampling error is a necessary evil. Performing 100% census of all material properties and all part functionality carries prohibitive costs. Every engineer knows that cash flow is the lifeblood of an organization. All data cost resources: direct labor, materials, and overhead that are visible to the engineering organization, as well as indirect labor, materials, and overhead that are not visible. Cost and schedule constraints demand small representative samples be used to estimate the value of parameters.

The lowest cost to an organization to understand failure and characterize reliability, maintainability, availability, and sustainability parameters is performed at the part design configuration level as the part is designed for functionality. Parts' common cause variability is best understood at part design. Parts' special cause variability is best managed and understood at part design. Only sampling error remains, but the methods presented provide knowledge of the impacts of sampling error on understanding the behavior of failure mechanisms.

Measurements of TTF and TTR are the current best practice to characterize part reliability, maintainability, availability, and sustainability parameters. Measurements of TTF must recognize the failure mechanism that caused the failure. Part failure is understood only when the TTF for a specific failure mechanism is modeled. Part failure can be the result of two or more failure mechanisms. Knowing only that a part failed at 100 h of operation does not provide understanding of failure needed to characterize the part reliability parameters. Sources of TTF and TTR include:

- historical maintenance and failure reports
  - advantage:
    - low cost of data acquisition
    - large sample size that can be treated as complete, uncensored data
  - disadvantage:
    - failure mechanism that may not be known
    - conditions of use that may not be known
- analogous information of like parts, vendor data for material/part
  - advantage:
    - low cost of data acquisition
    - large sample size that can be treated as complete, uncensored data

- disadvantage:
  - failure mechanism that may not be known
  - conditions of use that may not be known
  - degree of differences between analogous and design part that may be too great
  - information that may be limited to mean values with no measure of dispersion
- experimental information
  - advantage:
    - design engineer's total control over experimental design
    - specified failure mechanisms
    - controlled or measured conditions of use
  - disadvantage:
    - option with highest cost
    - censored small sample sizes

Censoring occurs when less than 100% census of all parts is tested. Censoring occurs in the following scenarios in which TTF experiments have two distinct time metrics:

- TTF_i: time to failure for the ith part
- ToT (or total test time): equal to the sum of all TTF_i plus the time accumulated by all parts that did not fail

Failure modeling is performed using descriptive and inferential statistical analyses. Descriptive statistical analyses explain how the sample data behave:

- measure of central tendency: where the data are located; the location parameter
  - mean: the center of the magnitudes of the data
  - median: the middle value of the number of the data
  - measure of dispersion: how wide the data are; the shape parameter
  - standard deviation: the standard width of the magnitudes of the data about the mean
  - range: the total width of the number of the data
  - interquartile range: the width of 50% of the number of data about the median
- shape of the data: the quantity, or frequency, of observations at each value of the data; the sample frequency distribution and probability density function (pdf)

Sample frequency distributions have either one mode or two or more modes. Multimodal sample frequency distributions describe samples that measure more than one failure mechanism and are useless for failure modeling. Historical data should always be reviewed for multiple modes.

Descriptive statistics are used to fit the sample data to a pdf. Inferential statistics[1] describe how well the sample describes the behavior of the failure mechanism:

- Goodness of fit of the sample to the pdf is measured by the coefficients of correlation and determination. Software programs calculate the two coefficients.
  - Pearson's rho (ρ) and the Anderson–Darling coefficient of correlation describe how much the sample data vary from the failure model predicted value. Pearson's rho ranges from –1 to +1, where –1 is a perfect negative correlation, +1 is a perfect positive correlation, and 0 is an absolute lack of correlation. The Anderson–Darling coefficient of correlation ranges from zero to positive values; the closest to zero is the best fit.
  - The coefficient of determination describes how much change in the number of failures is the result of a unit change in TTF where zero is no cause and one is 100%.

The interpretation of the values of the coefficients is subjective, relying on engineering judgment, and demands an understanding of the consequences of failure, material properties, cost and schedule, system functionality, and customer requirements. ($\rho = 0.85$ may be acceptable for a jelly bean process machine, but it is unacceptable for a biomedical implantable device!)

Knowing goodness of fit is important to understanding the behavior of failure on a part. Ignoring goodness of fit can result in using the wrong failure model. Too often, the exponential pdf is used to characterize reliability parameters when it is the wrong pdf. Goodness-of-fit analysis may suggest use of the Weibull pdf.

- Confidence limits measure the sum of the effects of common and special cause variability on the precision of the measures of estimates for central tendency and dispersion. Variability is the key to understanding the risk of past failure. Software programs calculate confidence limits.
  - Lower confidence limits are calculated for larger-is-best parameters (e.g., mean time between failure [MTBF]).
  - Upper confidence limits are calculated for smaller-is-best parameters (e.g., mean time to repair [MTTR]).
  - Confidence intervals are calculated for nominal-is-best parameters (e.g., design tolerances).
  - Confidence limits are calculated using sampling statistics, z, t, F, and χ^2.

The interpretation of confidence limits is also subjective. Engineering judgment is used to determine whether the limits are too wide for the criticality of the part on the system.

Confidence limits provide more valuable information than the point estimate of the mean. A notional customer requirement demands that the system MTBF be 100 h. A deterministically calculated design MTBF of 100 h appears to meet the requirement. But the expectation that the customer will experience the 100 h MTBF is only 67%, and the expectation that the customer will experience less than 100 h is 37% (using the exponential pdf to model failure). A part design lower confidence limit of 100 h (calculated for 90% confidence) expresses the expectation that the customer will experience at least the 100 h MTBF requirement demanded with only 10% risk of nonconformance. The design MTBF will be higher than 100 h in order to meet the requirement:

- Test of hypotheses measures the relationship between a sample's statistical estimators and the customer requirement, as well as the relationship between two or more samples' statistical estimators—both at a specified confidence level (typically 90, 95, and 99%, but sometimes at 75, 80, and 99.5%).
  - The null hypothesis, H_0:, states that the difference between the sample statistical estimator and the requirement is zero (null) and that the difference between two or more samples' statistical estimators is zero null.
  - The alternate hypothesis, H_1:, states that the difference is statistically significant.
  - The sampling critical statistics, z_{crit}, t_{crit}, F_{crit}, χ^2_{crit}, define how much difference in the sample's estimators are allowableto accept the null hypothesis, or reject the null hypothesis and accept the alternate hypothesis, at the specified confidence level.
  - The sampling test statistic, z_{test}, t_{test}, and F_{test}, is calculated for the sample estimator.
- The decision rule[2] is to accept the null hypothesis when

  |test statistic| < |critical statistic|

  ELSE reject the null hypothesis and accept the alternate hypothesis

The p-value provides a preferred method to evaluate the relationships between a sample and a requirement and of samples to each other. The p-value is the cumulative probability distribution for the test statistic that is computed by the software packages described in this section.

Consider two alternatives where the requirement is $\eta = 100$. Two alternatives of equal sample size ($n = 30$) are drawn and the sample means are calculated: $\bar{X}_1 = 98$ and $\bar{X}_2 = 104$. The specified standard deviation is $\sigma = 10$. The confidence level is specified at 90%. The level of significance $\alpha = 1 - C = 0.05$. The null hypothesis is stated as $H0: \bar{X}_1 = \mu$ and has an alternate hypothesis, $H1: \bar{X}_1 \neq \mu$. The normal distribution is selected for the test of hypothesis based on the central limit theorem, which states that a sample distribution of size $n \geq 30$ is normally distributed. The test of hypothesis is two sided because the sample mean can be either greater or less than the requirement (μ). The lower and upper critical statistics for the test of hypothesis ($\pm z_{crit}$) are equal to the standard normal z-score for two tails of area $\alpha/2$:

$$z_{crit_{lower}} = -z_{\alpha/2} = -1.645$$
$$z_{crit_{upper}} = z_{\alpha/2} = 1.645 \tag{3.4}$$

The test statistic for each sample mean is calculated as

$$z_{test_1} = \frac{\bar{X}_1 - \mu}{\sigma/\sqrt{n}} = \frac{98 - 100}{10/\sqrt{30}} = -1.095$$
$$z_{test_2} = \frac{\bar{X}_2 - \mu}{\sigma/\sqrt{n}} = \frac{104 - 100}{10/\sqrt{30}} = 2.191 \tag{3.5}$$

The decision for sample 1 is to accept the null hypothesis that there is no statistically significant difference between the sample mean and the requirement at 90% confidence because the absolute value of the test statistic (1.095) is less than the absolute value of the critical statistic (1.645). The decision for sample 2 is to reject the null hypothesis and accept the alternate hypothesis that there is a statistically significant difference between the sample mean and the requirement at 90% confidence because the absolute value of the test statistic (2.191) is greater than the absolute value of the critical statistic (1.645).

But what would the decisions be if the question were restated to be at 95% confidence? The upper and lower critical statistics must be calculated at the new confidence level. The p-value for the two-tailed test of hypothesis is calculated as twice the cumulative distribution of the tail bounded by the test statistics, p-value 1 = 2(0.219) = 0.438 for sample 1 and p-value 2 = 2(0.036) = 0.072 for sample 2. The acceptance criterion is to accept the null hypothesis for p-value $\geq \alpha$. The p-value provides accept/reject of the null hypothesis for any confidence level where use of the critical statistic provides only an accept/reject option for the null hypothesis at a single confidence level. The example data show that the null hypothesis will be accepted at 56.2% or higher confidence for sample 1 and 92.8% or higher confidence for sample 2.

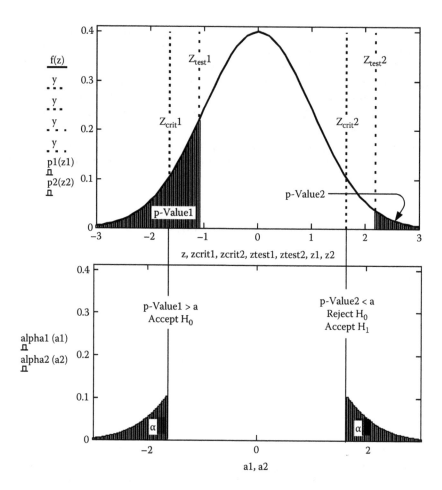

FIGURE 3.1
p-Value relationship to significance, α.

We can subjectively claim that acceptance of the null hypothesis for sample 1 is strong at 90%, and the acceptance of the null hypothesis for sample 2 is weak at 95%. The use of critical statistics and the p-value is illustrated in Figure 3.1.

The p-value simplifies the decision rule: Accept the null hypothesis for p-value $> \alpha$. Failure modeling for TTF is illustrated by example. The process plant example from the preceding chapter is used. Three software tools are suitable for developing failure models:

- spreadsheet (Microsoft Excel™)
  - least expensive; highly available to the design engineer
  - requires step-by-step approaches performed by the engineer

- statistical software (Minitab™)
  - expensive; limited availability to the design engineer
  - performs complete model building with just data entry
- engineering software (MathCAD™)
  - expensive; should be available to the design engineer
  - requires step-by-step approaches performed by the engineer

Failure model parameter estimators calculated from spreadsheet and engineering software approaches will yield the same numerical values because the same characterization methods are performed. Failure model parameter estimators calculated from statistical software use either least-squares fit or maximum-likelihood estimation and will yield numerical values slightly different from those reached from spreadsheet and engineering software.

Candidate for Reliability Engineering and Analysis

Selection of a candidate part is the result of the failure modes, effects, and consequences analysis. The vane pump (see Figure 3.2) for the fluid "A" process is selected from the critical items list.

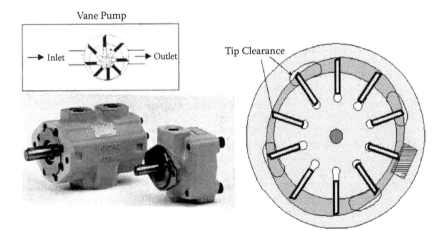

FIGURE 3.2
Vane pump.

Experimental Design for TTF

Experimental design begins with a statement of the general expression for the vane pump failure model, the pdf f_{vp}(TTF), where the independent variable is TTF. The failure model will quantify the failure mechanisms and modes hypotheses from the failure modes and effects analysis (FMEA). A fault tree analysis is used to brainstorm four vane pump failure modes: vane fails, control actuator fails, rotor fails, and bearing fails.

The general expression of the probabilistic math model for the vane pump is expanded by the probability logic of the fault tree analysis (FTA) as follows (see Figure 3.3):

$$P \text{ (vane pump fails)} = P \text{ (vane fails) OR } P \text{ (control actuator fails)}$$
$$\text{OR } P \text{ (rotor fails) OR } P \text{ (bearing fails)} \tag{3.6}$$

The vane pump can be in one of two states: functional and failed. The sum of the probabilities that the pump is functional and failed is unity; the events are exhaustive. The "OR" condition states that the vane pump will fail if the vane fails, OR if the control actuator fails, OR if the rotor fails, OR if the bearing fails. The probability theory states that the probability that an event

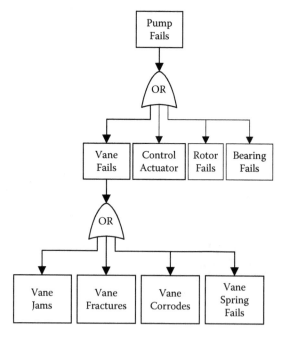

FIGURE 3.3
Vane pump FTA.

"A" occurs OR an event "B" occurs is equal to the sum of the probabilities of events "A" and "B" minus the product of the probabilities of events "A" and "B":

$$P(A) \text{ OR } P(B) = P(A) + P(B) - P(A)P(B) \tag{3.7}$$

The product factor, $P(A)P(B)$, is the probability that events "A" and "B" occur at the same time, but each failure event is independent and is therefore assumed to be zero. The probability that the vane pump fails is the sum of the probabilities that the vane, control actuator, rotor, and bearing fail. The continuous probability of an event "A" is expressed as a pdf, $f_A(t)$, with respect to time. The math model for the vane pump, $f_{vp}(t)$, becomes

$$f_{vp}(TTF) = f_{vane}(TTF) + f_{con}(TTF) + f_{rotor}(TTF) + f_{bearing}(TTF) \tag{3.8}$$

where $t \equiv TTF$.

The FTA shows that the pump vane has four failure modes that define the probability that the vane fails, P (vane fails), expressed as a probability density function, $f_{vane}(TTF)$. The probability expression for pump vane is stated as

$$P \text{ (vane fails)} = P \text{ (vane jams) OR } P \text{ (vane fractures) OR } P \text{ (vane corrodes)}$$
$$\text{OR } P \text{ (vane spring fails)} \tag{3.9}$$

The probability density function is expressed as

$$f_{vane}(TTF) = f_{jam}(TTF) + f_{fract}(TTF) + f_{corr}(TTF) + f_{spr}(TTF) \tag{3.10}$$

What follows is the characterization of the failure model parameters for two pump parts: the pump control actuator and the pump vane. The end result is a reliability math model for the vane pump. The three software programs are used for four sources of TTF data: complete, time-censored without replacement, time-censored with replacement, and failure-censored data. The dominant school of thought claims an order of preference for the sources of TTF data:

1. *Vendor-supplied reliability data.* The advantage is that best expertise is applied to the failure analysis. Disadvantages include lack of applicability to failure modes and conditions of use to the intended conditions of use and the added cost of the part.

2. *Complete historical data.* Advantages include large sample sizes at lowest cost of data and highest time on test. Disadvantages include lack of adequate failure analysis to isolate the failure mode and credibility of the data.

3. *Time-censored data with replacement.* Advantages include control over the failure experiment and failure analysis. Disadvantages include limited time on test at high cost of data.

4. *Tie.* A tie can occur depending on which provides the highest time on test: (1) time-censored data without replacement, or (2) failure-censored data. Advantages include control over the failure experiment and failure analysis. Disadvantages include less time on test that time censored with replacement at high cost of data.

5. *Generic failure data from GIDEP, PRISM, and other TTF databases.* An advantage is the low cost of data. Disadvantages include lack of applicability to the failure modes and the intended conditions of use. Generic data lag development of new parts.

Exponential Probability Distribution Approach

Demonstration of the exponential probability distribution will be applied to the pump control actuator. Historical records from other sites are used to model the control actuator proposed in the design. The vendor confirms the hypothesis that the exponential probability distribution provides the best fit for the failure math model.

Spreadsheet Approach

The spreadsheet approach does not provide a way to differentiate between complete or censored data. Data for historical control actuators TTF for 1999–2007 are tabulated in Excel and presented in Table 3.1. The MS Excel Data Analysis Tool Pak computes the descriptive statistics table, shown in Table 3.2.

The descriptive statistics table explains much about the behavior of failure:

- The measures of central tendency are not equal, suggesting that the distribution is not symmetrical: Mean time to failure[3] is 307 h and median TTF is 311 h. The mathematical mode for a continuous distribution is irrelevant, but the graphical mode defines the location of the distribution's peak.

- The measures of dispersion for the TTF data are the standard deviation of 28.73 h and a range of 170.9 h from a minimum TTF of 192.3 h to a maximum of 363.2 h. The empirical rule states that 99% of sample observations fall within ±3 standard deviations—a width of 172 h in this example. The total range of the sample is 171 h, confirming the empirical rule.

- The shape of the distribution is defined by its peak and its skewness. Kurtosis measures the height of the peak compared to the height of the normal distribution. MS Excel calculates the standard kurtosis

TABLE 3.1

Historical Data: Control Actuator TTF (1999–2007)

				Complete Data				
TTF 1999	TTF 2000	TTF 2001	TTF 2002	TTF 2003	TTF 2004	TTF 2005	TTF 2006	TTF 2007
322.5	303.5	264.1	293.5	296.6	302.7	283.8	285.7	318.8
317.9	296.2	337.4	318.0	315.4	319.5	276.8	284.0	293.2
290.9	324.6	240.7	359.7	319.4	340.0	309.9	320.0	313.0
347.2	355.7	312.9	331.8	314.0	330.0	327.4	336.0	282.3
321.2	293.3	275.9	300.4	357.9	316.6	331.3	302.9	320.7
297.2	322.1	283.0	312.2	333.6	279.2	322.6	277.1	263.5
314.5	352.2	291.9	333.7	316.5	308.3	350.4	240.3	297.1
334.9	320.1	313.4	332.8	273.3	292.6	257.8	312.0	287.0
334.6	318.1	350.8	292.9	331.9	302.5	318.0	308.4	192.3
352.7	347.7	363.2	302.5	290.6	297.8	288.7	297.2	261.6
323.4	315.5	351.8	312.5	283.8	293.1	327.8	322.0	321.5
321.1	337.2	337.4	291.0	287.8	267.8	304.6	288.7	323.4
	327.1	271.4	294.6	202.5	311.5	325.8	284.0	324.8
	337.4	350.5	283.3	303.1	306.9	312.3	293.9	345.5
	292.4	288.8	311.8	320.6	268.2	344.6	264.2	314.4
	309.6	284.4	263.3	247.5	305.3	327.1	287.3	306.4
	314.6	297.1	307.5	347.5	262.8	298.7	308.6	309.4
	315.0	353.6	310.9	330.4	357.8	302.5	318.6	275.8
	275.2	303.4	259.1	328.0	316.9	319.2	297.6	314.9
	280.8	336.4		312.2	342.4	336.1	318.2	281.1
	237.3	253.9		270.6	308.5	294.1	282.4	
	325.3	310.7		295.5		349.4	299.0	
	328.0	345.4		280.9		345.7		
		320.7				294.9		
						243.0		

that equals zero when the sample height fits a normal distribution. The sample kurtosis is greater than 1 at 1.35 and is peaked higher than the normal probability distribution (negative kurtosis describes a flatter distribution than the normal distribution). Skewness equals zero when a distribution is symmetrical; negative skewness describes a distribution skewed to the left from the measure of central tendency and positive skewness describes a distribution skewed to the right. This distribution is negatively skewed.

The summary statistics describes a distribution that is shaped neither like the normal distribution nor like the exponential distribution without looking at the histogram for the frequency distribution.

TABLE 3.2

Summary Statistics: Control Actuator
Complete TTF Data

TTF Historical Data	
Mean	306.96
Standard error	2.09
Median	310.90
Mode	283.78
Standard deviation	28.73
Sample variance	825.29
Kurtosis	1.35
Skewness	(0.73)
Range	170.91
Minimum	192.31
Maximum	363.22
Sum	58,015.26
Count	189

The MS Excel Data Analysis Tool Pak computes the frequency distribution table, and the relative frequency column is calculated from the frequency column, as shown in Table 3.3. Creating a frequency distribution table is part subjective. The number and width of the class interval must be determined. Too few classes will clump all the data in an undistinguishable stack; too many will spread the data too wide to observe a shape. A rule of thumb is seven to ten classes. The class interval or width should be intuitive. Increments of 5, 10, 25, 50, and 100 are more easily perceived and visualized

TABLE 3.3

Frequency Distribution Table: Control Actuator
Complete TTF Data

TTF	Frequency	rf
175	0	0.00000
200	1	0.00529
225	1	0.00529
250	5	0.02646
275	14	0.07407
300	51	0.26984
325	70	0.37037
350	35	0.18519
375	12	0.06349
Sum *f*	1S9	1.00

than increments of 3, 4.5, and 12.635; 25-h increments are selected in this example. The first class interval should include the minimum observation and its starting interval should be the previous interval increment; the first class begins with 175 to include the minimum TTF of 192.3 h.

The relative frequency (rf) is the proportion of the number of observations in a class interval to the total number of observations:

$$rf = \frac{n_i}{\sum n_i} \tag{3.11}$$

The rf is a discrete probability density function for the TTF distribution and sums to 100%, or 1. Exponentially distributed data have a relative frequency that begins at the maximum value at the lowest class interval and decreases at constant rate to the higher class intervals. The rf for this sample does not do that—yet more evidence that the exponential distribution would not ally. The frequency distribution is plotted in MS Excel, as shown in Figure 3.4.

The frequency distribution is exactly the shape suggested by the summary statistics. The peak is located at 325 and the distribution skews to the left, a negative skew. The shape of the distribution does not look like an exponential distribution in any possible way.

The exponential continuous failure math model—pdf, $f_{exp}(t)$—of the control actuator is an estimate of the likelihood of failure occurrences at specific points in time and is expressed as

$$\text{pdf} \equiv f_{exp}(t) = \frac{1}{\theta}e^{-\frac{t}{\theta}} = \lambda e^{-\lambda t} \tag{3.12}$$

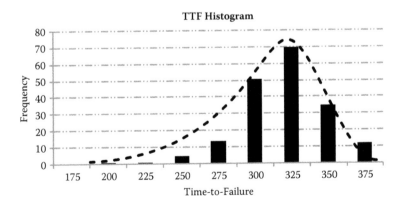

TTF Histogram

FIGURE 3.4
Histogram control actuator complete TTF data.

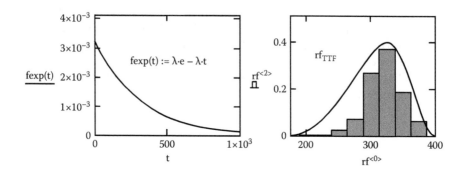

FIGURE 3.5
Exponential pdf and relative frequency distribution control actuator complete TTF data.

The mean TTF (θ) is 306.7 h and the failure rate (λ) is 3.258E-03. The exponential pdf is stated as

$$f_{exp}(t) = 3.258 \cdot 10^{-3} e^{-3.258 \cdot 10^{-3} t} \qquad (3.13)$$

The continuous exponential probability distribution of the TTF data is plotted in Figure 3.5. One can readily see the distinction between the shape of the exponential probability distribution and the shape of the relative frequency distribution. It is not enough that the two shapes are so different; the exponential distribution suggests that a significant portion of the part population fails at TTFs greater than 500 h and the relative frequency distribution suggests that no parts survive beyond 400 h.

Because TTF is a larger-is-best population criterion, the lower confidence limit (LCL) for the MTBF, LCL_θ, of the control actuator is calculated using the χ^2 sampling statistic, as follows:

$$\theta_{LCL} \geq \frac{2T}{\chi^2_{(\alpha,2r+2)}} = \frac{2(58015.26)}{426.45} = 272.082 \qquad (3.14)$$

where
$C = 95\%$
$\alpha = 1 - C = 0.05$
$r = 189$
$2r + 2 = 380$
$\chi^2_{(0.05,380)} = 426.45$

The value for χ^2 was acquired in MS Excel using the paint function command, = CHIINV(α,v), where α is the significance and v is the degrees of freedom, $2r$ + 2. The entry for the example is =CHIINV(0.05,380). The lower confidence limit failure rate (λ_{LCL}) is found as the inverse of the lower confidence limit

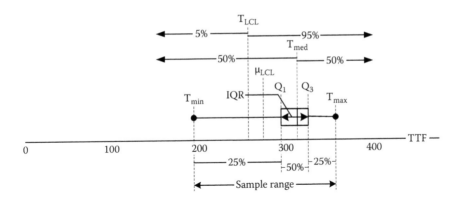

FIGURE 3.6
Control actuator box plot complete data.

for the mean time to failure, θ_{LCL},

$$\lambda_{LCL} = 1/272.1 = 3.675E\text{-}03$$

Another excellent tool for data analysis is the box plot, which allows the visual evaluation of the behavior of the data median and mean, and the total and interquartile ranges, as shown in Figure 3.6. The box defined by the first and third quartiles (Q1 and Q3) shows the interquartile range (IQR), in which 50% of all sample observations fall. The range between the minimum sample value (T_{min}) and Q1 contains 25% of all sample observations. The range between Q3 and the maximum sample value (T_{max}) contains 25% of all sample observations. The sample range contains 100% of all sample values and is defined by the range between T_{min} and T_{max}. The median (T_{med}) is the center of all sample observations; therefore, 50% of all sample observations are less than the median and 50% are greater. The 95% lower confidence limit of the sample is calculated by the mean minus the product of the standard deviation and the standard normal z-statistic:

$$T_{LCL} = \mu + z_{0.05} \times StDev$$

It is plotted on the box plot scale. The value of T_{LCL} (259.75 h) infers that 5% of all population data are less and 95% are greater.

The box plot graphically illustrates the lack of symmetry of the distribution. A very important fact is the location of the lower confidence limit of the mean—between TTF_{min} and the first quartile of the data; this means that over 75% of the observations are greater. The box plot graphically describes the shape of the population data and suggests the appropriate probability distribution that fits the data, as shown in Figure 3.7.

The cumulative exponential probability distribution of the control actuator, $F_{exp}(t)$, is a measure of the likely total proportion of failures of the population

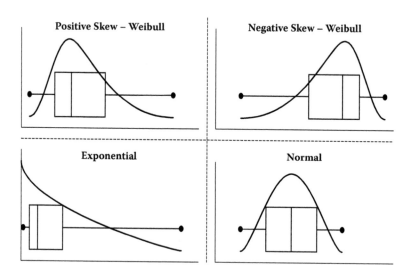

FIGURE 3.7
Comparative box plot shapes and corresponding distribution fit.

of parts over time and is characterized by the indefinite integral of the pdf, $f_{exp}(t)$:

$$F_{exp}(t) = \int_0^t \lambda e^{-\lambda t}\, dt = 1 - e^{-\lambda t} = 1 - e^{3.258 \cdot 10^{-3} t} \tag{3.15}$$

The range of the indefinite integral is from zero to t because there is no minus time in reliability analysis. The cumulative probability distribution is plotted in Figure 3.8.

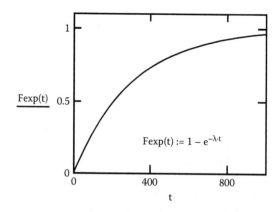

FIGURE 3.8
Exponential cumulative probability distribution control actuator complete TTF data.

The survival function, $S(t)$, is the complement of the cumulative density function (cdf), $F(t)$, where $S(t) = 1 - F(t)$. Recall that a part can be in only two states: functional (survived) and failed. The cdf of failure suggests that 0.593 of the population of control actuators will fail by the end of the second time interval; the survival function suggests that $1 - 0.593 = 0.407$ of the population will survive through the second time interval. The continuous exponential survival function, $S_{exp}(t)$, of the control actuator is expressed as

$$S_{exp}(t) = 1 - F_{exp}(t) = 1 - (1 - e^{-\lambda t}) = e^{-\lambda t} = e^{-3.258 \cdot 10^{-3} t} \qquad (3.16)$$

The lower confidence limit of the continuous survival function, $S_{LCL}(t)$, of the control actuator is characterized using the lower confidence limit of the failure rate (λ_{LCL}) expressed as

$$S_{LCL}(t) = e^{-\lambda_{LCL} t} = e^{-3.675 \cdot 10^{-3} t} \qquad (3.17)$$

The plots of the continuous exponential survival function and its lower confidence limit are illustrated in Figure 3.9. The lower confidence limit states that 95% of population of parts represented by the sample will survive at least at the levels shown in the figure.

The hazard function, $h(t)$, is the instantaneous failure rate evaluated at a point in time during the useful life of the part, and it is characterized as the ratio of the pdf and the survival function:

$$h(t) = \frac{f(t)}{1 - F(t)} = \frac{f(t)}{S(t)} \qquad (3.18)$$

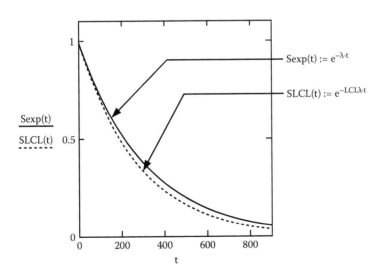

FIGURE 3.9
Exponential continuous survival function and LCL control actuator complete TTF data.

The continuous hazard function, $h_{exp}(t)$, for the control actuator is expressed as

$$h_{exp}(t) = \frac{f_{exp}(t)}{1 - F_{exp}(t)} = \frac{f_{exp}(t)}{S_{exp}(t)} = \frac{\lambda e^{-\lambda t}}{e^{-\lambda t}} = \lambda = 3.258 \cdot 10^{-3} \qquad (3.19)$$

The continuous lower confidence limit of the hazard function, $h_{LCL}(t)$, for the control actuator is expressed as

$$h_{LCL}(t) = \lambda_{LCL} = 3.675 \cdot 10^{-3} \qquad (3.20)$$

The plots for the exponential continuous hazard function and its lower confidence limit are illustrated in Figure 3.10.

Exponential hazard functions are constants equal to the failure rate. The TTF data frequency distribution infers that failure rates are not constant but rather vary with time.

The mission reliability function, $R(\tau \mid t)$, for the control actuator is the conditional probability that the control actuator will survive the next mission, $S(t + \tau)$, of duration τ, given that it has survived to the start of the next mission, $S(t)$.

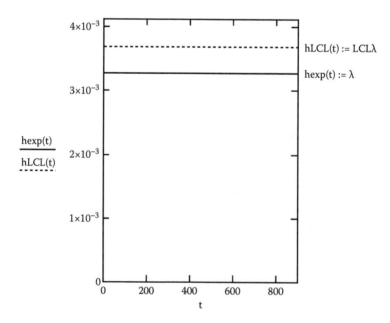

FIGURE 3.10
Exponential continuous hazard function and LCL control actuator complete TTF data.

The continuous mission reliability function, $R_{exp}(\tau|t)$, for the control actuator is expressed as

$$R_{exp}(\tau|t) = \frac{S_{exp}(t+\tau)}{S_{exp}(t)} = \frac{e^{-\lambda(t+\tau)}}{e^{-\lambda t}} = \frac{e^{-(\lambda t + \lambda \tau)}}{e^{-\lambda t}} = \frac{(e^{-\lambda t})(e^{-\lambda \tau})}{e^{-\lambda t}} = e^{-\lambda \tau} = e^{-0.003258(16)} = 0.949$$

(3.21)

The continuous lower confidence limit for the mission reliability function, $R_{LCL}(t)$, for the control actuator is expressed as

$$R_{LCL}(\tau|t) = e^{-\lambda_{LCL}\tau} = e^{-0.003675(16)} = 0.943$$

(3.22)

The plots for the exponential continuous mission reliability function and its lower confidence limit are illustrated in Figure 3.11.

Many texts and papers use the reliability expression, $R(t)$, for $1 - F(t)$, but this causes confusion. The use of $R(t)$ to represent both the life function, $S(t)$, and the mission reliability function, $R(\tau|t)$, which is often stated incorrectly as $R(t)$, should be stopped.

Spreadsheet computations for the parameters of the exponential failure model MTTF (θ) or failure rate (λ) do not distinguish between complete or censored data. The ToT is used to calculate MTTF. However, the calculation of the lower confidence limit treats all data as censored and uses the χ^2 sampling statistic rather than the standard normal z-statistic or the Student's t-statistic.

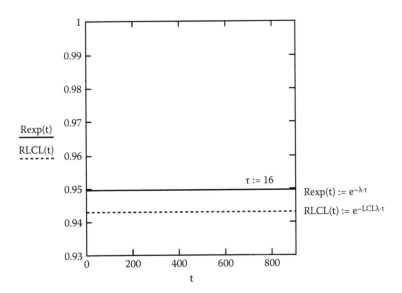

FIGURE 3.11
Exponential continuous mission reliability function and LCL control actuator complete TTF data.

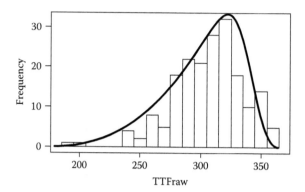

FIGURE 3.12
Complete data histogram and continuous curve fit.

Exponential Distribution: Minitab

Statistical software takes raw data and computes the descriptive statistics and exponential reliability distributions with far less work than using a spreadsheet requires.

Complete Data

The raw TTF data for the control actuator are entered into the Minitab worksheet. The first step in data analysis is to plot the frequency distribution in a histogram to observe the shape of the sample data. The histogram is plotted in Minitab, as shown in Figure 3.12.

The shape of the sample data is the same as that from the spreadsheet approach. Fitting a continuous curve over the histogram emphasizes the non-symmetrical distribution with a negative skew—obviously not an exponential distributed shape. The next step is to calculate the sample data descriptive statistics using the Minitab routine, as shown in Table 3.4.

The parameter of the exponential distribution is the failure rate ($\lambda = 0.00326 =$ 3.26E-03), which is calculated as the inverse of the mean time to failure. Minitab descriptive statistics provide the information to plot a box diagram for the TTF data provided in Figure 3.13.

TABLE 3.4

Minitab Descriptive Statistics: Control Actuator Complete Data

Variable	N	Mean	Median	Tr Mean	SD	SE Mean
TTF raw	189	306.96	310.90	308.16	28.72	2.09
Variable	Minimum	Maximum	Q1	Q3		
TTF raw	192.30	363.20	290.75	325.05		

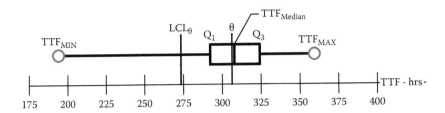

FIGURE 3.13
Box plot control actuator TTF data.

The lower confidence limit for the mean time to failure is calculated from the sum of all ToT, the sum of all TTF in the historical sample ($T = 58,015.26$) and the χ^2 statistic for the confidence levels, and $2r + 2$ degrees of freedom, where $r \equiv$ total number of observations, 189,

$$\mu_{LCL} = \frac{2T}{\chi^2_{(\alpha,2r+2)}} = \frac{2(58015.26)}{\chi^2_{(0.05,380)}} = 272.1 \tag{3.23}$$

and is plotted on the box plot scale. The lower confidence limit for the failure rate is calculated from the lower confidence limit of the mean time to failure as

$$\lambda_{LCL} = \frac{1}{\mu_{LCL}} = \frac{1}{272.01} = 3.68 \cdot 10^{-3} \tag{3.24}$$

The expressions for the reliability functions $f(t)$, $F(t)$, $S(t)$, $h(t)$, and $R(\tau|t)$, characterized by the point estimate of the mean time to failure (PEM) and calculated from complete TTF data, are shown in Table 3.5.

Time-Censored Data without Replacement

The time-censored data treatment is observed from a controlled experiment. Twenty test control actuators are placed on test in field systems for 270

TABLE 3.5

Exponential Reliability Function: Control Actuator Complete Data

	Point Estimate of the Mean	Lower Confidence Limit
$f(t)$	$3.26 \cdot 10^{-3} e^{-3.26 \cdot 10^{-3} t}$	$3.68 \cdot 10^{-3} e^{-3.68 \cdot 10^{-3} t}$
$F(t)$	$1 - e^{-3.26 \cdot 10^{-3} t}$	$1 - e^{-3.68 \cdot 10^{-3} t}$
$S(t)$	$e^{-3.26 \cdot 10^{-3} t}$	$e^{-3.68 \cdot 10^{-3} t}$
$h(t)$	3.26×10^{-3}	3.68×10^{-3}
$R(\tau|t)$	$e^{-3.26 \cdot 10^{-3} \tau}$	$e^{-3.68 \cdot 10^{-3} \tau}$

TABLE 3.6

Control Actuator
Time-Censored Data
without Replacement
Test Data

TTF
192.311
202.463
237.253
240.233
240.735
242.952
247.544
253.927
257.786
259.092
261.571
262.771
263.318
263.464
264.125
264.222
267.540
269.158
270.000
270.000

operating hours. The time to failure is entered in Minitab including the time on test for the control actuators that did not fail, as shown in Table 3.6.

Small sample experiments do not provide sufficient information to plot a meaningful frequency distribution, unlike complete data. The objective is to use the small sample size to gather sufficient information to fit the parameters of a probability distribution. Note that the data include both time to failure and censored data. It is a mistake to treat censored data as failures at the censor time. The analysis routine used by Minitab to fit the mean of exponential distribution is not possible in a spreadsheet. The Minitab exponential distribution analysis uses the median ranks least squares method and time censoring at 270 h to fit the parameter of the exponential distribution, as presented in Table 3.7.

The censoring information shows that 18 times to failure and two censored times on test were analyzed. "Scale" parameter is the Minitab term for the measure of central tendency. The mean TTF ($\theta = 217$ h) and the LCL of the mean TTF ($\theta_{LCL} = 154$ h) are calculated by Minitab. The failure rate

TABLE 3.7

Minitab Distribution Analysis: Control Actuator Time-Censored Data without Replacement

Distribution Analysis: TTF			
Variable: TTF			
Censoring Information	**Count**		
Uncensored value	18		
Right censored value	2		
Type 1 (time) censored at 270.0000			

Estimation Method: Least Squares—Failure Time (X) on Rank (Y)
Distribution: Exponential

Parameter	Estimate	Standard Error	95.0% Normal Bound Lower
Shape	1.00000		
Scale	217.36	45.19	154.41

Characteristics of Distribution:

	Estimate	Standard Error	95.0% Normal Bound Lower
Mean (MTTF)	217.3624	45.1857	154.4130
Standard deviation	217.3624	45.1857	154.4130
Median	150.6642	31.3203	107.0310
First quartile (Q1)	62.5313	12.9991	44.4219
Third quartile (Q3)	301.3283	62.6407	214.0619
Interquartile range (IQR)	238.7970	49.6416	169.6401

($\lambda = 4.61\text{E-}03$) and the LCL ($\lambda_{LCL} = 6.50\text{E-}03$) are calculated from the respective means. The characteristics of the distribution provide the values for the first quartile, median, and third quartile time to failure that are needed to construct the control actuator box plot for time-censored data without replacement, as shown in Figure 3.14.

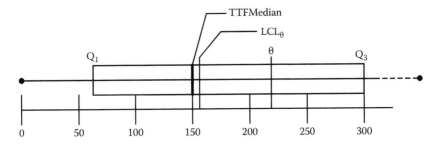

FIGURE 3.14
Control actuator box plot time-censored data without replacement.

TABLE 3.8

Control Actuator Reliability Functions Time-Censored Data without Replacement

	Point Estimate of the Mean	Lower Confidence Limit
$f(t)$	$4.61 \cdot 10^{-3} e^{-4.61 \cdot 10^{-3} t}$	$6.5 \cdot 10^{-3} e^{-6.5 \cdot 10^{-3} t}$
$F(t)$	$1 - e^{-4.61 \cdot 10^{-3} t}$	$1 - e^{-6.5 \cdot 10^{-3} t}$
$S(t)$	$e^{-4.61 \cdot 10^{-3} t}$	$e^{-6.5 \cdot 10^{-3} t}$
$h(t)$	4.61×10^{-3}	6.5×10^{-3}
$R(\tau\|t)$	$e^{-4.61 \cdot 10^{-3} \tau}$	$e^{-6.5 \cdot 10^{-3} \tau}$

Notice that the first quartile is a value less than the minimum sample observation and the third quartile is greater than the time-censored value of 270 h. Minitab infers the interquartile range from the data. One may infer that the minimum TTF is zero, but the maximum TTF is unknown. The small sample size and censoring cause a wide measure of dispersion. The expressions for the time-censored TTF data for the reliability functions $f(t)$, $F(t)$, $S(t)$, $h(t)$, and $R(\tau\|t)$ are shown in Table 3.8.

Time-Censored Data with Replacement

The time-censored data treatment is observed from a controlled experiment. Twenty test fixtures to test control actuators are placed on test in field systems for 400 operating hours. Failed control actuators are removed and replaced with new control actuators until the test is stopped at 400 h. Time to failure for failed parts and the time on test for the surviving parts are entered in Minitab. A censor column is entered to identify the control actuators that failed (F) and the control actuators that were operating at completion of the test duration (P) (see Table 3.9).

The Minitab exponential distribution analysis using the median ranks least squares method and time censoring at 400 h to fit the parameter of the exponential distribution is presented in Table 3.10. The censoring information shows that 22 control actuators failed and 18 were operating at 400 h. The mean TTF is estimated at 395 h with an LCL of 274 h. The interquartile range is 434 h from Q1 = 114 h and Q3 = 547 h. The failure rates are 2.53E-03 and 3.65E-03, respectively.

The box plot is shown in Figure 3.15. The box plot shows a wider interquartile range, largely due to the time-censored data from replaced control actuators that were still operating at 400 h but had logged low times on test.

The expressions for the time-censored TTF data for the reliability functions $f(t)$, $F(t)$, $S(t)$, $h(t)$, and $R(\tau\|t)$ are shown in Table 3.11.

TABLE 3.9

Control Actuator Time-Censored
Data with Replacement Test Data

TTF	Censor
192.311	F
202.486	F
237.253	F
240 263	F
240.725	F
242.952	F
247.544	F
253.927	F
257 788	F
259.092	F
261.571	F
262.771	F
263.318	F
263 464	F
264.125	F
264.222	F
267.840	F
268.158	F
270 631	F
271.410	F
207.689	F
197.514	F
162.747	P
159.737	P
159.275	P
157.048	P
152.456	P
146.073	P
142.212	P
140.908	P
138.429	P
137.229	P
136.682	P
136.536	P
135.875	P
135.778	P
132.160	P
131.842	P
129.369	P
128.590	P

TABLE 3.10

Minitab Distribution Analysis: Control Actuator Time-Censored Data with Replacement

Distribution Analysis: TTF			
Variable: TTF			
Censoring Information	**Count**		
Uncensored value	18		
Right censored value	22		
Censoring value: Cen = F			

Estimation Method: Least Squares—Failure Time (X) on Rank (Y)
Distribution: Exponential

Parameter	Estimate	Standard Error	95.0% Normal Bound Lower
Shape	1.00000		
Scale	394.76	87.69	273.94

Characteristics of Distribution:

	Estimate	Standard Error	95.0% Normal Bound Lower
Mean (MTTF)	394.7645	87.6924	273.9386
Standard deviation	394.7645	87.6924	273.9386
Median	273.6299	60.7838	189.8797
First quartile (Q1)	113.5667	25.2275	78.8072
Third quartile (Q3)	547.2598	121.5675	379.7595
Interquartile range (TQR)	433.6931	96.3400	300.9523

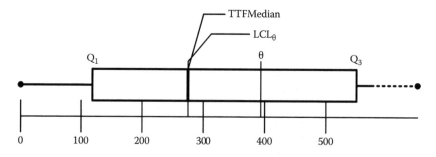

FIGURE 3.15
Control actuator box plot time-censored data with replacement.

TABLE 3.11

Control Actuator Reliability Functions Time-Censored Data with Replacement

	Point Estimate of the Mean	Lower Confidence Limit
$f(t)$	$2.53 \cdot 10^{-3} e^{-2.53 \cdot 10^{-3} t}$	$3.65 \cdot 10^{-3} e^{-3.65 \cdot 10^{-3} t}$
$F(t)$	$1 - e^{-2.53 \cdot 10^{-3} t}$	$1 - e^{-3.65 \cdot 10^{-3} t}$
$S(t)$	$e^{-2.53 \cdot 10^{-3} t}$	$e^{-3.65 \cdot 10^{-3} t}$
$h(t)$	2.53×10^{-3}	3.65×10^{-3}
$R(\tau\|t)$	$e^{-2.53 \cdot 10^{-3} \tau}$	$e^{-3.65 \cdot 10^{-3} \tau}$

Failure-Censored Data

Failure-censored data treatment applies when field failures are observed over a specified time until the nth failure is observed. The nth failure is used as the censor value. A spreadsheet approach to fit the mean of the exponential is not possible. The example for control actuators places 14 test articles on test until all have failed (see Table 3.12).

The example data are used as failure-censored data by entering the number of failures in the failure-censored block. The distribution analysis is tabulated in Table 3.13. Minitab censoring information treats the nth failure as the censored data point. The characteristics of the distribution provide the

TABLE 3.12
Control Actuator
Failure-Censored
Test Data

TTF
261.571
262.771
263.318
263.464
264.125
264.222
267.840
268.158
270.631
271.410
273.256
275.223
275.818
275.887

TABLE 3.13

Minitab Distribution Analysis: Control Actuator Failure Censored

Distribution Analysis: TTF			
Variable: TTF			
Censoring Information	**Count**		
Uncensored value	13		
Right censored value	1		
Type 2 (failure) censored at 14			

Estimation Method: Least Squares—Failure Time (X) on Rank (Y)
Distribution: Exponential

Parameter	Estimate	Standard Error	95.0% Normal Bound Lower
Shape	1.00000		
Scale	206.89	48.55	140.64

Characteristics of Distribution:

	Estimate	Standard Error	95.0% Normal Bound Lower
Mean (MTTF)	206.8903	48.5456	140.6443
Standard deviation	206.8903	48.5456	140.6443
Median	143.4055	33.6492	97.4872
First quartile (Q1)	59.5186	13.9657	40.4608
Third quartile (Q3)	266.3109	67.2984	194.9744
Interquartile range (TQR)	227.2923	53.3328	154.5135

values for the minimum, first quartile, median, third quartile, and maximum time to failure that are needed to construct the control actuator box plot for failure-censored data, as shown in Figure 3.16.

The expressions for the failure-censored TTF data for the reliability functions $f(t)$, $F(t)$, $S(t)$, $h(t)$, and $R(\tau|t)$ are shown in the Table 3.14.

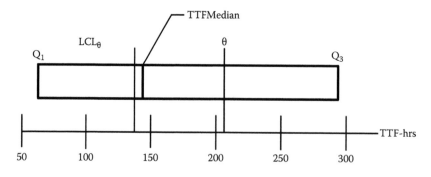

FIGURE 3.16
Control actuator box plot, failure censored.

TABLE 3.14

Control Actuator Reliability Data, Failure Censored

	Point Estimate of the Mean	Lower Confidence Limit
$f(t)$	$4.83 \cdot 10^{-3} e^{-4.83 \cdot 10^{-3} t}$	$7.1 \cdot 10^{-3} e^{-7.1 \cdot 10^{-3} t}$
$F(t)$	$1 - e^{-4.83 \cdot 10^{-3} t}$	$1 - e^{-7.1 \cdot 10^{-3} t}$
$S(t)$	$e^{-4.83 \cdot 10^{-3} t}$	$e^{-7.1 \cdot 10^{-3} t}$
$h(t)$	4.83×10^{-3}	7.1×10^{-3}
$R(\tau \mid t)$	$e^{-4.83 \cdot 10^{-3} \tau}$	$e^{-7.1 \cdot 10^{-3} \tau}$

Failure-Censored to nth Failure Data

This is the special case for failure-censored experimental analysis, but does not have relevance for fitting the parameters of exponential, or any other, reliability model. It is expressly applied to verify the expected likelihood that material and part failures meet a minimum criterion. The mean time to failure from failure analysis and empirical investigation fits a Poisson probability distribution:

$$P(x) = \frac{(\lambda t)^x e^{-(\lambda t)}}{x!} \tag{3.25}$$

where $x \equiv$ number of failures that take on values 1, 2, ..., n. Given $\theta = 307$, from complete data, the failure rate (λ) is the inverse of θ. The Poisson distribution applies queuing theory to evaluate the time likelihood to the first failure ($x = 1$) and the second failure ($x = 2$), as shown in Figure 3.17. Tests to first, second, ..., nth failure are run consecutively

Exponential Distribution: MathCAD

Engineering software has fewer functional routines but is more likely to be available for reliability failure analysis and math modeling. Data can be entered manually (a really bad idea) or by importing from MS Excel or Minitab spreadsheets. Manual entry just adds the opportunity for error. Data acquired from math modeling or empirical analysis are checked out for validity and errors. Text and program tables are pasted into MathCAD arrays, as shown in Figure 3.18. Data arrays can be condensed to save space in MathCAD worksheets and reports, as shown in the figure. The data are not affected by this.

The TTF data for the control actuator are entered into MathCAD by selecting the "insert" list, followed by the "data" list, followed by the "table" entry. The table is given the variable name "TTF" and the data are pasted in the table using the "paste table" command. MathCAD allows an intuitive approach to solve for and plot the point estimators for the reliability parameters: MTTF (θ) and λ. MathCAD paste functions perform many statistical operations. For example, the mean of the exponential probability distribution is calculated

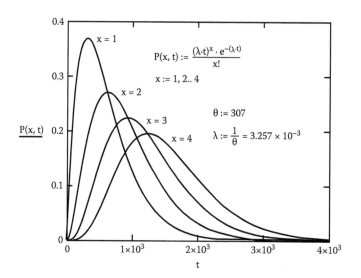

FIGURE 3.17
Poisson plots for *n*th failure.

by using the "mean(v)" paste function, where "v" is the vector of data, TTF. The point estimate for λ is calculated by writing the equation as one would write it on paper:

$$\lambda = \frac{1}{\theta} \tag{3.26}$$

The solution for the failure rate can be solved by writing "$\lambda =$" below the equation; a shortcut is to type "=" following the end of the equation, as follows:

$$\lambda = \frac{1}{\theta} = 3.258E - 03$$

The 95% lower confidence limit for the MTTF (LCLθ) is calculated using the ξ^2 sampling distribution:

$$LCL\theta \le \frac{2T}{\chi_{(\alpha,2r+2)}} \tag{3.27}$$

where
T is the total ToT
α is the level of significance; the complement of the confidence level ($\alpha = 1 - 95\% = 0.05$)
$2r + 2 = 2(27) + 2 = 59$ degrees of freedom, where r is the total number of failures

The lower 95% confidence limit for the failure rate (LCLλ) is the inverse of LCLθ. The lower confidence limits for the hazard function, $h_{LCL}(t)$, and survival

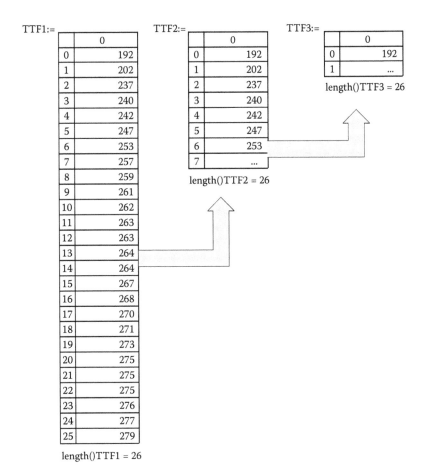

FIGURE 3.18
Array space options.

function, $S_{LCL}(t)$, are calculated using LCLλ. The lower confidence limit of the reliability expression, $R_{LCL}(\tau|t)$, is calculated using lower confidence level of the survival function and the mission duration (τ). Plots for the exponential reliability functions and the corresponding lower confidence limits are identical to those developed using the spreadsheet and will not be repeated here.

Weibull Distribution Approach

The failure model for the pump vane is presented using the three software tools. The FTA shows that the pump vane failure mode is the result of four failure modes: vane jams, vane fractures, vane corrodes, and vane spring fails. All four failure modes result from failure mechanisms that manifest wear out; therefore, the exponential probability distribution does not apply. The TTF data will be used to fit a Weibull failure model for the vane fractures failure mode. Methods for fitting a Weibull failure model by spreadsheet, Minitab, and MathCAD approaches using complete data from historical records and censored data from field and laboratory experiment are presented. The vane pump is scheduled to operate for a mission duration (τ) of 24 h.

Spreadsheet Approach

The spreadsheet approach implements the median ranks regression to characterize the parameters of the Weibull distribution. The median ranks regression is performed by the following steps that convert the cdf, $F(t)$, to a linear equation of the form

$$Y = b_o + b_1 X \tag{3.28}$$

where
Y is the dependent variable
b_o is the y-intercept
X is the independent variable
b_1 is the slope

TTF data of any shape—skewed, symmetrical, or exponential—will fit a Weibull pdf, $f(t)$. The parameters of the Weibull are shape (β) and scale or location (η). The general expression for the Weibull[4] pdf is

$$f(t) = \frac{\beta}{\eta} \left(\frac{t}{\eta} \right)^{\beta-1} e^{-\left(\frac{t}{\eta}\right)^{\beta}} \tag{3.29}$$

where t is the independent variable time in hours, operating hours, cycles, cold starts, etc. The Weibull cdf, $F(t)$, is derived from the indefinite integral of the pdf, $f(t)$, as follows:

$$F(t) = \int_0^t \left[\frac{\beta}{\eta} \left(\frac{t}{\eta} \right)^{\beta-1} e^{-\left(\frac{t}{\eta}\right)^{\beta}} \right] dt = 1 - e^{-\left(\frac{t}{\eta}\right)^{\beta}} \tag{3.30}$$

Conversion of the cdf to a linear form is performed by the following steps:

1. Start with the cdf and isolate the exponential function to the right side of the equation:

$$F(t) = 1 - e^{-\left(\frac{t}{\eta}\right)^{\beta}} \tag{3.31}$$

$$1 - F(t) = e^{-\left(\frac{t}{\eta}\right)^{\beta}}$$

(3.32)

2. Substitute $S(t)$ for $1 - F(t)$:

$$S(t) = e^{-\left(\frac{t}{\eta}\right)^{\beta}}$$

(3.33)

or

$$S(t) = \frac{1}{e^{\left(\frac{t}{\eta}\right)^{\beta}}}$$

(3.34)

3. Invert both sides of the equation:

$$\frac{1}{S(t)} = e^{\left(\frac{t}{\eta}\right)^{\beta}}$$

(3.35)

4. Take the natural logarithm of both sides of the equation to reduce the exponential function to an nth order equation:

$$\ln\left(\frac{1}{S(t)}\right) = \left(\frac{t}{\eta}\right)^{\beta}$$

(3.36)

5. Take the natural logarithm of both sides of the equation to reduce the nth order equation to a linear equation:

$$\ln\left[\ln\left(\frac{1}{S(t)}\right)\right] = \beta \ln(t) - \beta \ln(\eta)$$

(3.37)

The Weibull cdf becomes a linear equation where the dependent variable (Y) is defined as

$$Y = \ln\left[\ln\left(\frac{1}{S(t)}\right)\right]$$

(3.38)

the independent variable (X) is defined as

$$X = \beta \ln(t)$$

(3.39)

the slope of the equation (b_1) is defined as

$$b_1 = \beta$$

(3.40)

and the y-intercept (b_0) is defined as

$$b_o = -\beta \ln(\eta)$$

(3.41)

The linear equation is now expressed in terms of X and Y as

$$Y = b_0 + b_1 X \tag{3.42}$$

The parameters of the median ranks regression, b_0 and b_1, are used to fit the parameters of the Weibull distribution, η and β, as follows:

$$\eta = e^{-\left(\frac{b_0}{b_1}\right)} \tag{3.43}$$

This is the characteristic life, and

$$\beta = b_1 \tag{3.44}$$

is the shape parameter.

Complete Data

The spreadsheet approach applies to complete data (uncensored), time-censored without replacement data, time-censored with replacement data, and failure-censored data. Historical data for the vane pump are acquired and tabulated in rank order in MS Excel, including the TTF descriptive statistics, frequency distribution table, and histogram, as shown in Figure 3.19. (It is absolutely necessary to rank order the data to characterize the parameters of the Weibull distribution in both spreadsheet and engineering software applications.)

The frequency distribution shows a continuous distribution with a negative skew, with a mean and median of 448.7 and 461 h, respectively. The kurtosis ($2.27 > 1$) describes a peak higher than the normal distribution; skewness (-1.55) confirms the negative, left skew shape.

The spreadsheet approach to fit TTF data to parameters of the Weibull pdf requires sorting the raw TTF data, column B, in rank order from TTF_{MIN} to TTF_{MAX} and assigning an index number from one to n, column A (see Figure 3.20).

The independent variable, $X = \ln(\text{TTF}_i)$, is computed in column C. Computing the dependent variable, $Y = \ln[\ln(1/S(t))]$, column D, is done in three steps:

1. Compute the median rank estimator for $F(t)$, column E, given by Bartlett's median rank as follows:

$$\hat{F}(t) = \frac{i - 0.3}{n + 0.4} \tag{3.45}$$

2. Compute $S(t) = 1 - F(t)$, column F.
3. Compute Y using $S(t)$.

NOTE: The calculations for $F(\text{TTF}_i)$ and $S(\text{TTF}_i)$ are placed to the left of the column for $\ln[\ln(1/S(\text{TTF}_i))]$ to allow columns for X and y to be next to each other for ease of plotting.

TTF	Pump TTF Descriptive Statistics		TTF	Frequency
309	Mean	448.74	325	1
342	Standard Error	7.54	350	1
361	Median	461.00	375	1
381	Mode	483.00	400	2
399	Standard Deviation	44.62	425	2
414	Sample Variance	1990.55	450	7
424	Kurtosis	2.27	475	9
431	Skewness	-1.55	500	12
436	Range	186		
437	Minimum	309		
438	Maximum	495		
440	Sum	15706		
441	Count	35		
449	Confidence Level (95.0%)	-12.8		
452				
457				
459				
461				
466				
467				
471				
473				
473				
476				
476				
477				
481				
483				
483				
488				
490				
491				
492				
493				
495				

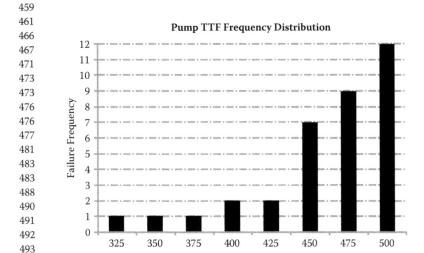

FIGURE 3.19
Vane pump vane fracture failure mode complete data.

Excel plots a scatter plot of X and Y. The least squares trend line operation is selected in Excel for the scatter plot and includes the equation of the line and its coefficient of determination, r^2. The equation of the line will be in terms of $Y = b_0 - b_1 X$.

The coefficient of determination, $r^2 = 0.9288$, provides information on the strength of the relationship between the independent variable, time, and the dependent variable, $F(t)$—the cumulative probability of failure for the failure mechanism under investigation. The coefficient of determination at 0.9288

i	TTF	Ln(TTF)	Ln[Ln(1/S(t))]	F(t)	S(t)
1	309	5.733	−3.913	0.020	0.980
2	342	5.835	−3.012	0.048	0.952
3	361	5.889	−2.534	0.076	0.924
4	381	5.943	−2.204	0.105	0.895
5	399	5.989	−1.949	0.133	0.867
6	414	6.026	−1.740	0.161	0.839
7	424	6.050	−1.562	0.189	0.811
8	431	6.066	−1.405	0.218	0.782
9	436	6.078	−1.266	0.246	0.754
10	437	6.080	−1.139	0.274	0.726
11	438	6.082	−1.022	0.302	0.698
12	440	6.087	−0.913	0.331	0.669
13	441	6.089	−0.811	0.359	0.641
14	449	6.107	−0.715	0.387	0.613
15	452	6.114	−0.623	0.415	0.585
16	457	6.125	−0.534	0.444	0.556
17	459	6.129	−0.449	0.472	0.528
18	461	6.133	−0.367	0.500	0.500
19	466	6.144	−0.286	0.528	0.472
20	467	6.146	−0.207	0.556	0.444
21	471	6.155	−0.129	0.585	0.415
22	473	6.159	−0.052	0.613	0.387
23	473	6.159	0.025	0.641	0.359
24	476	6.165	0.102	0.669	0.331
25	476	6.165	0.179	0.698	0.302
26	477	6.168	0.258	0.726	0.274
27	481	6.176	0.339	0.754	0.246
28	483	6.180	0.422	0.782	0.218
29	483	6.180	0.510	0.811	0.189
30	488	6.190	0.602	0.839	0.161
31	490	6.194	0.703	0.867	0.133
32	491	6.196	0.815	0.895	0.105
33	492	6.198	0.945	0.924	0.076
34	493	6.201	1.111	0.952	0.048
35	495	6.205	1.367	0.980	0.020

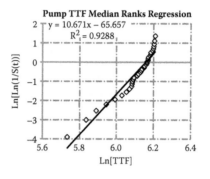

Pump TTF Median Ranks Regression

$y = 10.671x - 65.657$
$R^2 = 0.9288$

Ln[Ln(1/S(t))]

Ln[TTF]

$y_0 =$	−65.66	$v =$	34
$\beta =$	10.67	$sd =$	44.62
$\eta =$	470.32	$n =$	35
		$sem =$	7.54
Σ TTF $=$	15706 t	$crit =$	1.69
$r =$	35	$\eta LCL =$	457.57
$\theta =$	448.74		
$TTF_{MED} =$	461.00		

FIGURE 3.20
Spreadsheet approach to fit Weibull distribution complete data.

suggests that the pump vane fracture failure mode is a strong predictor of part failure.

The coefficient of determination ranges from $0 \leq r^2 \leq 1$ and states that a unit change of time causes $r^2 100\%$ change in the cumulative probability of failure; the corollary is that $(1 - r^2)100\%$ of the change in the cumulative probability of failure is due to something else.

For example, consider investigation of a tire valve stem failure mode resulting from ambient solar radiation and thermal shock acting on the tire material properties. Historical TTF data for tire failure due to valve stem failures are fit to a Weibull failure model with an r^2 of 9%. This suggests that the valve stem failure mode is not a statistically significant failure mechanism and that 91% of the cumulative failure of tires is caused by other failure mechanisms and modes.

Contrast tire stem failure to tire tread wear-out failure mode resulting from operational force loads acting on the tire material properties. Historical TTF data for tire failure due to tread wear-out failures are fit to a Weibull failure model with an r^2 of 86%. This suggests that the tire tread wear-out failure mode is a statistically significant failure mechanism and that only 14% of the cumulative failure of tires is caused by other failure mechanisms and modes.

The Weibull failure model is completed for statistically significant failure modes by characterization of the Weibull parameters, β and η, from the parameters of the equation of the line, $Y = b_0 + b_1 X$, as shown:

$$\beta = b_1 = 10.67 \tag{3.46}$$

$$\eta = e^{-\left(\frac{b_0}{b_1}\right)} = e^{-\left(\frac{-65.66}{10.67}\right)} = e^{6.154} = 470.32 \tag{3.47}$$

The Weibull probability distribution for failure, $f_w(t)$, is expressed as

$$f_w(t) = \frac{\beta}{\eta}\left(\frac{t}{\eta}\right)^{\beta-1} e^{-\left(\frac{t}{\eta}\right)^{\beta}} = 3.264 \times 10^{-28} t^{9.67} e^{-\left(\frac{t}{470.32}\right)^{10.67}} \tag{3.48}$$

The Weibull survival function, $S_w(t)$, is expressed as

$$S_w(t) = e^{-\left(\frac{t}{\eta}\right)^{\beta}} = e^{-\left(\frac{t}{470.32}\right)^{10.67}} \tag{3.49}$$

The Weibull reliability function, $R_w(\tau|t)$, is expressed as

$$R_w(\tau|t) = \frac{e^{-\left(\frac{t+\tau}{\eta}\right)^{\beta}}}{e^{-\left(\frac{t}{\eta}\right)^{\beta}}} = \frac{e^{-\left(\frac{t+24}{470.32}\right)^{10.67}}}{e^{-\left(\frac{t}{470.32}\right)^{10.67}}} \tag{3.50}$$

The Weibull hazard function, $h_w(t)$, is expressed as

$$h_w(t) = \frac{\beta}{\eta}\left(\frac{t}{\eta}\right)^{\beta-1} = 3.264 \times 10^{-28} t^{9.67} \tag{3.51}$$

The graphic illustration of the reliability functions is presented in Figure 3.21.

Time-Censored Data without Replacement

An experiment for pump failure for vane fracture is designed. Fourteen pumps are put in service at a field site under operational conditions of use for 500 operating hours. No replacement pumps are available. The times to failure for 12 pumps are entered in the spreadsheet in rank order from lowest to highest TTF and the two pumps that did not fail are entered at 500 h each.

Time-censored without replacement data influence the computation of the cumulative probability of failure estimator, $F(t)$. The cdf estimator is computed

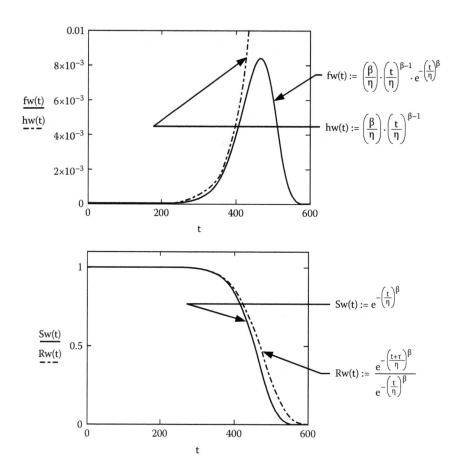

FIGURE 3.21
Vane fracture model reliability functions complete data.

for TTF only, but time-censored data include ToT data for parts that did not fail. The rank-ordered ToT places all the unfailed parts at the bottom of column B. The index number includes the unfailed parts in column A.

The independent variable, $X = \ln(TTF_i)$, is computed in column C for all rows of data. Computing the dependent variable, $Y = \ln[\ln(1/S(t))]$, column D, is done in three steps:

1. Compute the median rank estimator for $F(t)$, column E, for the failed parts only, given by Bartlett's median rank as follows:

$$\hat{F}(t) = \frac{i-0.3}{n+0.4}$$

(3.52)

where n is the last index number in the array of ToT data.

i	ToT	Ln(ToT)	Ln[Ln(1/S(t))]	F(t)	S(t)
1	361	5.889	−2.999	0.049	0.951
2	438	6.082	−2.074	0.118	0.882
3	441	6.089	−1.572	0.188	0.813
4	449	6.107	−1.214	0.257	0.743
5	452	6.114	−0.929	0.326	0.674
6	461	6.133	−0.685	0.396	0.604
7	466	6.144	−0.468	0.465	0.535
8	471	6.155	−0.268	0.535	0.465
9	476	6.165	−0.076	0.604	0.396
10	490	6.194	0.113	0.674	0.326
11	492	6.198	0.307	0.743	0.257
12	492	6.198	0.515	0.813	0.188
13	500	6.215			
14	500	6.215			

Tmax = 500-hrs

Pump TTF Median Ranks Regression

$y = 11.626x - 71.958$
$R^2 = 0.8797$

Ln[TTF]-Time Censored w/o Replacement

y_0 =	−71.96	v =	13
β =	11.63	sd =	36.40
η =	487.55	n =	14
		sem =	9.73
ΣToT =	6489	tcrit =	1.77
r =	12	ηLCL =	470.33
θ =	540.75		

FIGURE 3.22
Spreadsheet approach to fit Weibull failure model time-censored data without replacement.

2. Compute the estimator for $S(t) = 1 - F(t)$, column F, for the failed parts only.

3. Compute Y using $S(t)$ for the failed parts only.

The TTF data with two ToT-censored values are tabulated in Figure 3.22, including the computations to fit the median ranks regression, the fitted median ranks regression plot, and the characterization of the Weibull parameters.

The coefficient of determination at 0.8797 suggests that the pump vane fracture failure mode is a strong predictor of part failure. The Weibull failure model is completed for statistically significant failure modes by characterization of the Weibull parameters, β and η, from the parameters of the equation of the line, $Y = b_0 + b_1 X$, as shown:

$$\beta = b_1 = 11.63 \tag{3.53}$$

$$\eta = e^{-\left(\frac{b_0}{b_1}\right)} = e^{-\left(\frac{-71.96}{11.63}\right)} = e^{6.187} = 487.55 \tag{3.54}$$

The Weibull probability distribution for failure, $f_w(t)$, is expressed as

$$f_w(t) = \frac{\beta}{\eta}\left(\frac{t}{\eta}\right)^{\beta-1} e^{-\left(\frac{t}{\eta}\right)^{\beta}} = 6.547 \times 10^{-31} t^{10.63} e^{-\left(\frac{t}{487.55}\right)^{11.63}} \tag{3.55}$$

The Weibull survival function, $S_w(t)$, is expressed as

$$S_w(t) = e^{-\left(\frac{t}{\eta}\right)^{\beta}} = e^{-\left(\frac{t}{487.55}\right)^{11.63}}$$

(3.56)

The Weibull reliability function, $R_w(\tau|t)$, is expressed as

$$R_w(t) = \frac{e^{-\left(\frac{t+\tau}{\eta}\right)^{\beta}}}{e^{-\left(\frac{t}{\eta}\right)^{\beta}}} = \frac{e^{-\left(\frac{t+24}{487.55}\right)^{11.63}}}{e^{-\left(\frac{t}{487.55}\right)^{11.63}}}$$

(3.57)

The Weibull hazard function, $h_w(t)$, is expressed as

$$h_w(t) = \frac{\beta}{\eta}\left(\frac{t}{\eta}\right)^{\beta-1} = 6.547 \times 10^{-31} t^{10.63}$$

(3.58)

The graphic illustration of the reliability functions is presented in Figure 3.23.

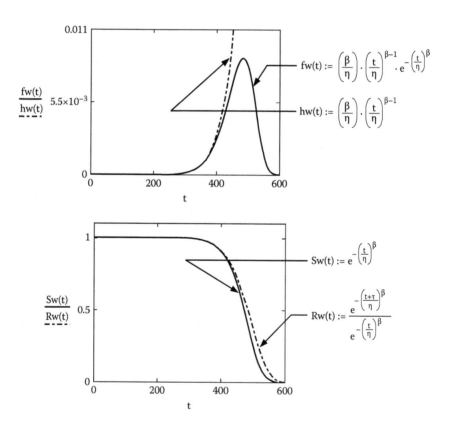

FIGURE 3.23
Vane fracture model reliability functions time censored without replacement.

Time-Censored Data with Replacement Data

An experiment for pump failure for vane fracture is designed. Fourteen pumps are put in service at a field site under operational conditions of use for 500 operating hours. Fourteen replacement pumps are available. The times to failure for 14 pumps are entered in the spreadsheet in rank order from lowest to highest TTF and the times on test for the 14 pumps that did not fail are entered at the time test accrued at 500 h (see Figure 3.24).

Time-censored with replacement data also influence the computation of the cumulative probability of failure estimator, $F(t)$. The cdf estimator is computed for TTF only, but time-censored data include ToT data for parts that did not fail. The rank ordered data mix TTF and unfailed ToT parts, now in column C. A censor entry is introduced as a new column B that marks a TTF entry with "F" and a ToT entry with "C." The index number includes the TTF and ToT data in column A.

The independent variable, $X = \ln(TTF_i)$, is computed in column D for all rows of data. Computing the dependent variable, $Y = \ln[\ln(1/S(t))]$, column E, is done in three steps:

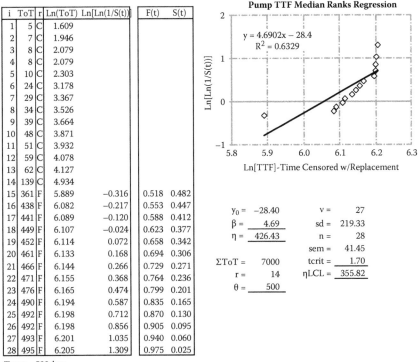

i	ToT	r	Ln(ToT)	Ln[Ln(1/S(t))]	F(t)	S(t)
1	5	C	1.609			
2	7	C	1.946			
3	8	C	2.079			
4	8	C	2.079			
5	10	C	2.303			
6	24	C	3.178			
7	29	C	3.367			
8	34	C	3.526			
9	39	C	3.664			
10	48	C	3.871			
11	51	C	3.932			
12	59	C	4.078			
13	62	C	4.127			
14	139	C	4.934			
15	361	F	5.889	−0.316	0.518	0.482
16	438	F	6.082	−0.217	0.553	0.447
17	441	F	6.089	−0.120	0.588	0.412
18	449	F	6.107	−0.024	0.623	0.377
19	452	F	6.114	0.072	0.658	0.342
20	461	F	6.133	0.168	0.694	0.306
21	466	F	6.144	0.266	0.729	0.271
22	471	F	6.155	0.368	0.764	0.236
23	476	F	6.165	0.474	0.799	0.201
24	490	F	6.194	0.587	0.835	0.165
25	492	F	6.198	0.712	0.870	0.130
26	492	F	6.198	0.856	0.905	0.095
27	493	F	6.201	1.035	0.940	0.060
28	495	F	6.205	1.309	0.975	0.025

Tmax = 500-hrs

Pump TTF Median Ranks Regression

$y = 4.6902x − 28.4$
$R^2 = 0.6329$

Ln[Ln(1/S(t))]

Ln[TTF]-Time Censored w/Replacement

$y_0 =$	−28.40	$v =$	27
$\beta =$	4.69	$sd =$	219.33
$\eta =$	426.43	$n =$	28
		$sem =$	41.45
$\Sigma ToT =$	7000	$tcrit =$	1.70
$r =$	14	$\eta LCL =$	355.82
$\theta =$	500		

FIGURE 3.24
Spreadsheet approach to fit Weibull failure model time censored with replacement.

1. Compute the median rank estimator for $F(t)$, column F, for the failed parts only, based on the censor entry, given by Bartlett's median rank, as follows:

$$\hat{F}(t) = \frac{i - 0.3}{n + 0.4} \qquad (3.59)$$

where n is the last index number in the array of all data.

2. Compute the estimator for $S(t) = 1 - F(t)$, column G, for the failed parts only.

3. Compute Y using $S(t)$ for the failed parts only.

The coefficient of determination at 0.6329 suggests that the pump vane fracture failure mode is a good predictor of part failure but that other failure modes are probably present. The Weibull failure model is completed for statistically significant failure modes by characterization of the Weibull parameters, β and η, from the parameters of the equation of the line, $Y = b_0 + b_1 X$, as shown:

$$\beta = b_1 = 4.69 \qquad (3.60)$$

$$\eta = e^{-\left(\frac{b_0}{b_1}\right)} = e^{-\left(\frac{-28.4}{4.69}\right)} = e^{5.97} = 426.43 \qquad (3.61)$$

The Weibull probability distribution for failure, $f_w(t)$, is expressed as

$$f_w(t) = \frac{\beta}{\eta}\left(\frac{t}{\eta}\right)^{\beta-1} e^{-\left(\frac{t}{\eta}\right)^{\beta}} = 2.174 \times 10^{-12} t^{3.69} e^{-\left(\frac{t}{426.43}\right)^{4.69}} \qquad (3.62)$$

The Weibull survival function, $S_w(t)$, is expressed as

$$S_w(t) = e^{-\left(\frac{t}{\eta}\right)^{\beta}} = e^{-\left(\frac{t}{426.43}\right)^{4.69}} \qquad (3.63)$$

The Weibull reliability function, $R_w(\tau|t)$, is expressed as

$$R_w(t) = \frac{e^{-\left(\frac{t+\tau}{\eta}\right)^{\beta}}}{e^{-\left(\frac{t}{\eta}\right)^{\beta}}} = \frac{e^{-\left(\frac{t+24}{426.43}\right)^{4.69}}}{e^{-\left(\frac{t}{426.43}\right)^{4.69}}} \qquad (3.64)$$

The Weibull hazard function, $h_w(t)$, is expressed as

$$h_w(t) = \frac{\beta}{\eta}\left(\frac{t}{\eta}\right)^{\beta-1} = 2.174 \times 10^{-12} t^{3.69} \qquad (3.65)$$

The graphic illustration of the reliability functions is presented in Figure 3.25.

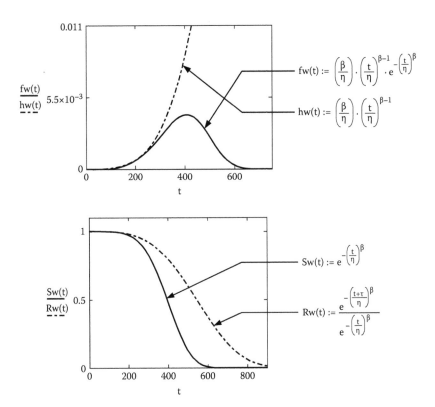

FIGURE 3.25
Reliability functions time censored with replacement.

Failure-Censored Data Approach

An experiment for pump failure for vane fracture is designed. Twelve pumps are put in service at a field site under operational conditions of use until all pumps fail. The time to failure for the 12 pumps is entered in the spreadsheet in rank order from lowest to highest TTF. Failure-censored data also influence the computation of the cumulative probability of failure estimator, $F(t)$. The cdf estimator is computed for all TTF data.

The independent variable, $X = \ln(TTF_i)$, is computed in column C for all rows of data. Computing the dependent variable, $Y = \ln[\ln(1/S(t))]$, column D, is done in three steps:

1. Compute the median rank estimator for $F(t)$, column E, given by Bartlett's median rank, as follows:

$$\hat{F}(t) = \frac{i - 0.3}{n + 0.4} \tag{3.66}$$

i	ToT	Ln(ToT)	Ln[Ln(1/S(t))]	F(t)	S(t)
1	361	5.889	−2.845	0.056	0.944
2	438	6.082	−1.914	0.137	0.863
3	441	6.089	−1.404	0.218	0.782
4	449	6.107	−1.037	0.298	0.702
5	452	6.114	−0.741	0.379	0.621
6	461	6.133	−0.485	0.460	0.540
7	466	6.144	−0.252	0.540	0.460
8	471	6.155	−0.030	0.621	0.379
9	476	6.165	0.190	0.702	0.298
10	490	6.194	0.422	0.782	0.218
11	492	6.198	0.687	0.863	0.137
12	492	6.198	1.056	0.944	0.056

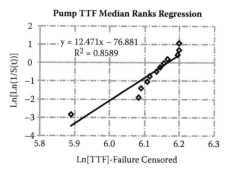

$y_0 =$	−76.88		$v =$	11
$\beta =$	12.47		$sd =$	35.82
$\eta =$	475.89		$n =$	12
			$sem =$	10.34
$\Sigma ToT =$	5489		$tcrit =$	1.80
$r =$	12		$\eta LCL =$	457.33
$\theta =$	457.42			

FIGURE 3.26
Spreadsheet solution to fit Weibull failure model, failure censored.

where n is the last index number in the array of all data.
2. Compute the estimator for $S(t) = 1 − F(t)$, column F.
3. Compute Y using $S(t)$.

The spreadsheet computations are summarized in Figure 3.26.

The coefficient of determination at 0.8589 suggests that the pump vane fracture failure mode is a strong predictor of part failure. The Weibull failure model is completed for statistically significant failure modes by characterization of the Weibull parameters, β and η, from the parameters of the equation of the line, $Y = b_0 + b_1 X$, as shown:

$$\beta = b_1 = 4.69 \tag{3.67}$$

$$\eta = e^{-\left(\frac{b_0}{b_1}\right)} = e^{-\left(\frac{-76.88}{12.47}\right)} = e^{6.165} = 475.89 \tag{3.68}$$

The Weibull probability distribution for failure, $f_w(t)$, is expressed as

$$f_w(t) = \frac{\beta}{\eta}\left(\frac{t}{\eta}\right)^{\beta-1} e^{-\left(\frac{t}{\eta}\right)^{\beta}} = 5.106 \times 10^{-33} t^{11.47} e^{-\left(\frac{t}{475.89}\right)^{12.47}} \tag{3.69}$$

The Weibull survival function, $S_w(t)$, is expressed as

$$S_w(t) = e^{-\left(\frac{t}{\eta}\right)^{\beta}} = e^{-\left(\frac{t}{475.89}\right)^{12.47}} \tag{3.70}$$

The Weibull reliability function, $R_w(\tau|t)$, is expressed as

$$R_w(t) = \frac{e^{-\left(\frac{t+\tau}{\eta}\right)^{\beta}}}{e^{-\left(\frac{t}{\eta}\right)^{\beta}}} = \frac{e^{-\left(\frac{t+24}{475.89}\right)^{12.47}}}{e^{-\left(\frac{t}{475.89}\right)^{12.47}}} \tag{3.71}$$

The Weibull hazard function, $h_w(t)$, is expressed as

$$h_w(t) = \frac{\beta}{\eta}\left(\frac{t}{\eta}\right)^{\beta-1} = 5.106 \times 10^{-33} t^{11.47} \tag{3.72}$$

The graphic illustration of the reliability functions is presented in Figure 3.27. The comparative linear plots for the range, mean, median, and characteristic life of the data analyzed in the spreadsheet approach are summarized in Figure 3.28.

The value for the mean TTF for complete data passes the common sense test, considering that it was calculated from a large sample size from field experience. The controlled experiments with small sample sizes yield mean

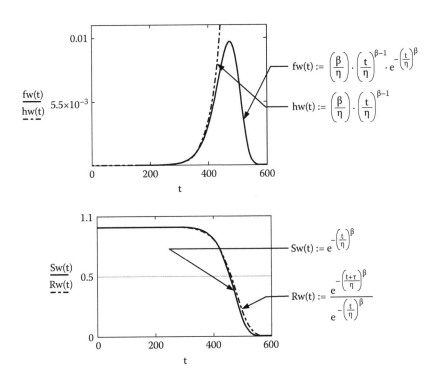

FIGURE 3.27
Reliability functions Weibull failure model failure censored.

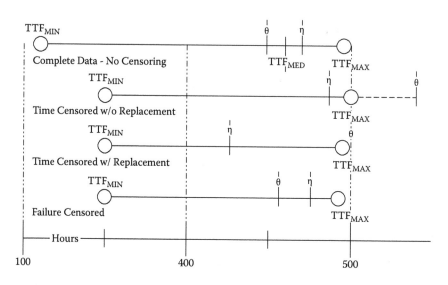

FIGURE 3.28
Comparative line plots Weibull failure models.

TTF values that are widely varied. The characteristic life has less variation between the sample methods. Time-censored with replacement data provides the most conservative estimator. All methods suggest that the TTF is greater than 300 h for the population and greater than 400 h for the measure of central tendency.

Weibull Distribution: Minitab

Statistical software programs like Minitab perform comprehensive TTF data analysis and parameter estimation in one step that the spreadsheet and MathCAD programs perform in step-by-step procedures. The TTF data are entered manually or pasted in the worksheet that has a spreadsheet format, or they can be imported from spreadsheet or database files. The pull-down menus provide the following options.

The parametric distribution analysis asks for the column containing the TTF data. The censor menu uses complete data—uncensored as the default, or censor information. The censor column is selected to identify the TTF rows that are censored (C) or failed (F). Time censor inputs the duration of the test and censors TTF values that equal the censor time. Failure censor inputs the number of failures that limit the test duration.

The estimate menu uses the maximum likelihood estimator (MLE) as the default method for parameter estimation. The least squares regression is the alternative method and the one used in this text. Confidence level uses 95% as the default with two-sided confidence intervals. Parameter estimation

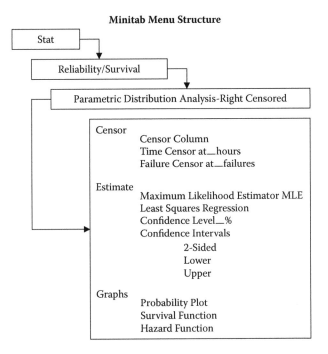

FIGURE 3.29
Minitab parameter estimation pull-down menu structure.

uses the lower confidence interval. Graphs plot the probability plot showing the fit of the data to the distribution and the survival and hazard functions. The plots are of limited utility and do not allow formatting. The plots do not transfer very well to other documents (e.g., Word, PowerPoint) and are not included here (see Figure 3.29).

The output information from the Minitab parameter estimation routine provides superb information: the censoring logic, parameter point estimates with the 95% lower confidence limit, and the characteristics of the distribution, as shown for complete and censored data.

Complete Data

The results from a complete data distribution analysis are provided in Table 3.15. The parameter estimates characterize the point estimates and lower confidence limits for the shape, β, and scale (location), η, parameters. The parameter estimates' confidence limits are calculated step by step by the engineer in spreadsheet and MathCAD programs. The characteristics of the distribution characterize the shape of the sample data and are provided in comparable box plots at the end of the Minitab discussion.

TABLE 3.15

Minitab Distribution Analysis: Weibull Complete Data

Distribution Analysis: TTF			
Variable: TTF			
Censoring Information	Count		
Uncensored value	35		

Estimation Method: Least Squares—Failure Time (X) on Rank (Y)
Distribution: Weibull1

Parameter	Estimate	Standard Error	95.0% Normal Bound Lower
Shape	11.610	2.568	8.098
Scale	468.336	7.243	456.572

Characteristics of Distribution:

	Estimate	Standard Error	95.0% Normal Bound Lower
Mean (MTTF)	448.2839	7.8233	435.5987
Standard deviation	46.7067	9.4397	33.4969
Median	453.8189	7.6333	441.4354
First quartile (Q1)	420.7967	11.7620	401.888C
Third quartile (Q3)	481.6639	8.C814	468.5529
Interquartile range (IQR)	60.8671	12.9703	42.8707

Time-Censored Data without Replacement

The TTF data are entered in the Minitab worksheet and rank ordered from minimum to maximum. The time-censored instruction is selected and the time, 500 h, is entered. All TTF data for unfailed parts ($k = 2$) will be entered at 500 h and will be censored by the distribution analysis routine. The distribution analysis is presented in Table 3.16.

Time-Censored Data with Replacement

The TTF data are entered in the Minitab worksheet and rank ordered from minimum to maximum. A censor column is added that identifies censored test articles (C) and failed test articles (F). The time-censored instruction is selected and the time (500 h) is entered. All TTF data for unfailed parts ($k = 14$) will be entered at the time accrued at the completion of the test and will be censored by the distribution analysis routine. The distribution analysis is presented in Table 3.17.

Failure-Censored Data

The TTF data are entered in the Minitab worksheet and rank ordered from minimum to maximum. The failure-censored instruction is selected and the

TABLE 3.16

Minitab Distribution Analysis: Weibull Time-Censored without Replacement

Distribution Analysis: TTF			
Variable: TTF			
Censoring Information	**Count**		
Uncensored value	14		
Type 2 (failure) censored at 500			
Estimation Method: Least Squares—Failure Time (X) on Rank (Y)			
Distribution: Weibull			
		Standard	95.0% Normal
Parameter	**Estimate**	**Error**	**Bound Lower**
Shape	15.466	4.861	9.223
Scale	479.006	8.757	464.816
Characteristics of Distribution:			
		Standard	95.0% Normal
	Estimate	**Error**	**Bound Lower**
Mean (MTTF)	462.9999	9.7383	447.2558
Standard deviation	36.7686	10.7363	22.7452
Median	467.7882	9.3682	452.6299
First quartile (Q1)	441.9324	14.0717	419.3822
Third quartile (Q3)	489.2299	9.3873	474.0303
Interquartile range (IQR)	47.2975	14.4264	28.6386

limiting number of failures ($k = 12$) is entered and will be censored by the distribution analysis routine. The distribution analysis is presented in Table 3.18.

Weibull Distribution: MathCAD Approach

Engineering software like MathCAD will fit the parameters of the Weibull reliability functions by performing the individual steps for the median ranks regression.

Complete Data

Complete data from the previous examples are used to illustrate fitting the Weibull parameters using MathCAD in the following step-by-step approach:

1. The pump's raw TTF data are entered as a table array, sorted from minimum to maximum, sort(TTF), and a corresponding index array is calculated as $n = 1, 2 ..., 35$ (see Figure 3.30).
2. The sample size (n) is expressed as the last number in the index[5] array, $n := index_{34} = 35$. Note that MathCAD arrays label the first row as 0 and the last row as $n - 1$; therefore, the last index number (35) is in row 34.

TABLE 3.17

Minitab Distribution Analysis: Weibull Failure-Censored Model with Replacement

Distribution Analysis: TTF
Variable: TTF

Censoring Information	Count
Uncensored value	14
Right censored value	14

Censoring value: censor = C
Estimation Method: Least Squares—Failure Time (X) on Rank (Y)
Distribution: Weibull

Parameter	Estimate	Standard Error	95.0% Normal Bound Lower
Shape	15.926	5.308	9.205
Scale	477.683	8.506	463.895

Characteristics of Distribution:

	Estimate	Standard Error	95.0% Normal Bound Lower
Mean (MTTF)	462.1324	9.4679	446.8185
Standard deviation	35.6809	11.0897	21.4000
Median	466.8154	9.0731	452.1275
First quartile (Q1)	441.7381	13.9914	419.3135
Third quartile (Q3)	467.5812	9.2793	472.5545
Interquartile range (IQR)	45.8431	14.8681	26.8900

3. The independent variable X is calculated as the natural logarithm of the sorted TTF table array (see Figure 3.31):

$$X = \ln(\text{TTF}) \tag{3.73}$$

4. The matrix estimator of the cdf, $F(\text{TTF})$, is calculated by Bartlett's median rank, where $i = $ index (see Figure 3.32):

$$F(\text{TTF}) = \frac{\text{index} - 0.3}{n + 0.4} \tag{3.74}$$

5. The survival function matrix, $S(\text{TTF})$, is calculated as $1 - F(\text{TTF})$ (see Figure 3.33).

6. The independent variable matrix Y is calculated as the natural logarithm of the natural logarithm of the inverse of the survival function matrix (see Figure 3.34).

$$Y = \ln\left[\ln\left(\frac{1}{S(\text{TTF})}\right)\right] \tag{3.75}$$

TABLE 3.18

Minitab Distribution Analysis: Weibull Failure Model, Failure Censored

Distribution Analysis: TTF

Variable: TTF

Censoring Information	Count
Uncensored value	11
Right censored value	1

Type 2 (failure) censored at 12

Estimation Method: Least Squares—Failure Time (X) on Rank (Y)

Distribution: Weibull

Parameter	Estimate	Standard Error	95.0% Normal Bound Lower
Shape	14.506	5.213	8.032
Scale	475.58	10.66	458.37

Characteristics of Distribution:

	Estimate	Standard Error	95.0% Normal Bound Lower
Mean (MTTF)	458.7635	10.9073	441.1688
Standard deviation	38.7425	13.0403	22.2713
Median	463.7182	10.6596	446.5120
First quartile (Q1)	436.4417	15.5125	411.6574
Third quartile (Q3)	466.4141	12.0935	466.9233
Interquartile range (IQR)	49.9724	17.5977	28.0009

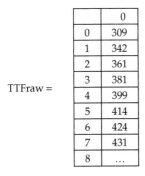

FIGURE 3.30
MathCAD TTF data array and index.

	0
0	5.733
1	5.835
2	5.839
3	...

$X =$

FIGURE 3.31
FD independent variable array: X.

	0
0	0.02
1	0.048
2	0.076
3	...

$F(TTF) =$

FIGURE 3.32
FD Bartlett's median ranks estimator for $F(t)$.

	0
0	0.98
1	0.952
2	0.924
3	...

$S(TTF) =$

FIGURE 3.33
FD survival function array: S(TTF).

	0
0	−3.913
1	−3.012
2	−2.534
3	...

$Y =$

FIGURE 3.34
FD independent variable array: Y.

7. MathCAD calculates the y-intercept (b_0) using the "intercept(X,Y)" operation, and the slope (b_1) using the "slope(X,Y)" operation:

$$b_0 = \text{intercept}(X,Y) = -65.657 \qquad (3.76)$$

$$b_1 = \text{slope}(X,Y) = 10.671 \qquad (3.77)$$

8. The point estimates for the characteristic life (η) and the shape parameter (β) are calculated from the y-intercept and slope:

$$\eta = e^{-\frac{b_0}{b_1}} = 470.186 \qquad (3.78)$$

$$\beta = b_1 = 10.671 \qquad (3.79)$$

9. The equation for the median ranks regression, MRR(X), is written in the form of a line:

$$\text{MRR}(X) = b_0 + b_1 X \qquad (3.80)$$

10. The mean for the X and Y matrices is calculated using the "mean(X)" and "mean(Y)" operations:

$$\text{Mean}(X) = 6.101 \qquad (3.81)$$

$$\text{Mean}(Y) = -0.556 \qquad (3.82)$$

11. The coefficient of correlation (ρ) is calculated using the "corr(X,Y)" operation. The coefficient of determination (r^2) is calculated as ρ^2:

$$\rho = \text{corr}(X,Y) = 0.964 \qquad (3.83)$$

$$r^2 = \text{corr}(X,Y)^2 = 0.929 \qquad (3.84)$$

12. The scatter plot and least squares regression, MRR(X), are plotted including the intersection of the point estimates of the means of X and Y (see Figure 3.35).

The two-sided confidence interval for the median ranks regression is calculated and plotted to evaluate the goodness of fit of the data by the following steps:

1. The sums of squares matrix (SXX, for X), Sxx(X), is calculated as the sum of the square of the difference between X and mean(X), $\Sigma Sxx(X)$:

$$Sxx(X) = (X - \text{mean}(X))^2 \qquad (3.85)$$

$$SXX = \Sigma Sxx(x) = 0.403 \qquad (3.86)$$

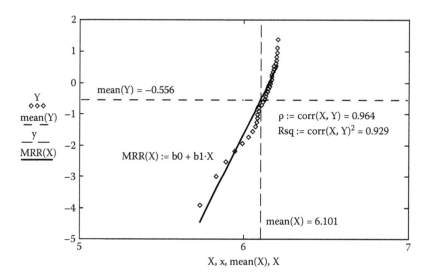

FIGURE 3.35
Median ranks regression scatter plot and trend line.

2. The matrix of the predicted value of Y (Y_{pred}) is the equation of the MRR(X).
3. The sums of squares matrix (*SSE*) for the residual of Y, *Res*(Y), is calculated as the sum of the square of the difference between Y and Y_{pred}, $\Sigma Res(Y)$:

$$Y_{pred} = b_0 + b_1 X \tag{3.87}$$

$$Res(Y) = (Y - Y_{pred})^2 \tag{3.88}$$

$$SSE = SRes(Y) = 3.513 \tag{3.89}$$

4. The variance of the regression (VAR) is calculated as the sums of squares of the residuals, error, divided by the degrees of freedom, $index_{34} - 1$:

$$VAR = \frac{SSE}{index_{34} - 1} = 0.103 \tag{3.90}$$

5. The standard error of the regression matrix, $se(X)$, is calculated by

$$se(X) = \sqrt{VAR\left[\left(\frac{1}{n}\right) + \left(\frac{Sxx(X)}{SXX}\right)\right]} \tag{3.91}$$

6. The lower and upper values of the Student's *t*-sampling distribution (tL and tU) are calculated using the MathCAD operation—$qt(\alpha/2,v)$ and $qt(1 - \alpha/2,v)$, respectively, where $v = n - 2 = index_{34} - 2$:

$$tL = qt\left(\frac{\alpha}{2}, \quad index_{34} - 2\right) = -2.035$$

$$tU = qt\left(1 - \frac{\alpha}{2}, \quad index_{34} - 2\right) = 2.035$$

where $C = 95\%$ and $\alpha = 0.05$.

7. The lower confidence limit factor, CLI, defines the lower confidence boundary of the median ranks regression:

$$CLI(X) = tLse(X) \tag{3.92}$$

The upper confidence limit factor, CUI, defines the upper confidence boundary of the median ranks regression:

$$CUI(X) = tUse(X) \tag{3.93}$$

8. The lower confidence interval matrix is calculated as

$$CIL(X) = MRR(X) + CLI(X) \tag{3.94}$$

The upper confidence interval matrix is calculated as

$$CIU(X) = MRR(X) + CUI(X) \tag{3.95}$$

9. The median ranks regression, $MRR(X)$, and the confidence intervals, $CIL(X)$ and $CIU(X)$, are plotted as shown in Figure 3.36.

Evaluation of the median ranks regression is based on the width of the confidence interval where tight intervals are preferred over wide intervals. The tighter the interval is, the better the fit is.

The point estimates of the Weibull parameters are applied to solve for the reliability functions with plots. The corresponding exponential reliability functions are included to show the lack of applicability. This exercise graphically demonstrates the error that occurs by improper use of the exponential failure model that is too commonly used. Nonengineers can see the differences and acknowledge the error (see Figure 3.37).

The Weibull pdf, $f_w(t)$, fits the sample frequency distribution where the exponential clearly does not, $f_e(t)$. The mean of the exponential (θ) is clearly higher than the Weibull characteristic life, η (see Figure 3.38).

The differences in the survival functions for the exponential, $S_e(t)$, and the Weibull, $S_w(t)$, failure models have logistics ramifications. The consumption rate for the exponential would lead to overstock spare parts with the associated costs and wasted resources (see Figure 3.39).

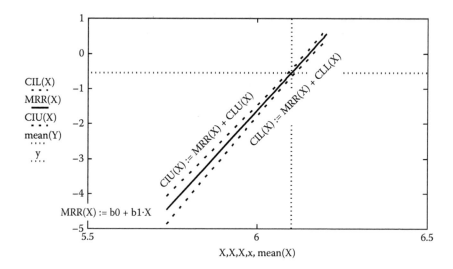

FIGURE 3.36
Median ranks regression with upper and lower confidence limits.

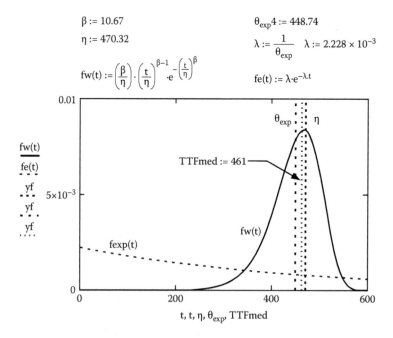

FIGURE 3.37
MathCAD Weibull comparative pdf complete data.

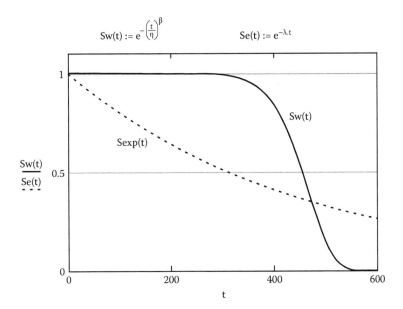

FIGURE 3.38
MathCAD Weibull comparative survival functions complete data.

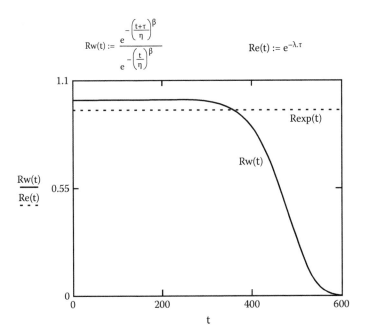

FIGURE 3.39
MathCAD Weibull comparative reliability function complete data.

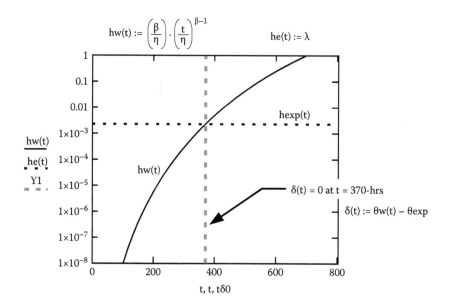

FIGURE 3.40
MathCAD comparative hazard functions complete data.

The exponential reliability forecasts constant probability of failure-free operation during missions that is lower in the early usage than the Weibull and does not reflect the decrease in reliability as the part wears through operational use (see Figure 3.40).

NOTE: The Weibull reliability is not 100% at the first mission, but very nearly so.

The exponential hazard function is constant and useless for maintenance planning and probabilistic risk assessment. The Weibull hazard function can forecast the age when a part reaches a risk threshold of failure. For example, if allowable operational or safety risk is defined as up to one failure/1,000 h, the part can operate up to approximately 390 h. A more precise replacement time is found by solving for $h_w(t) = 1,000$.

The lower confidence limits for the survival, reliability, and MTTF(t) functions are calculated for the pump vane complete data and plotted in Figure 3.41.

The survival and reliability functions describe a part design that can be expected to function very nearly failure free up to an operating life of 200 h. The MTTF(t) function describes a part that degrades to 1,000 h mean time to failure at approximately 390 h as expected from the hazard function. The MTTF(t) drops by an order of magnitude to 100 h mean time to failure in less than 50 additional operating hours. Concurrently, the survival and reliability functions drop by 50% in the same time interval.

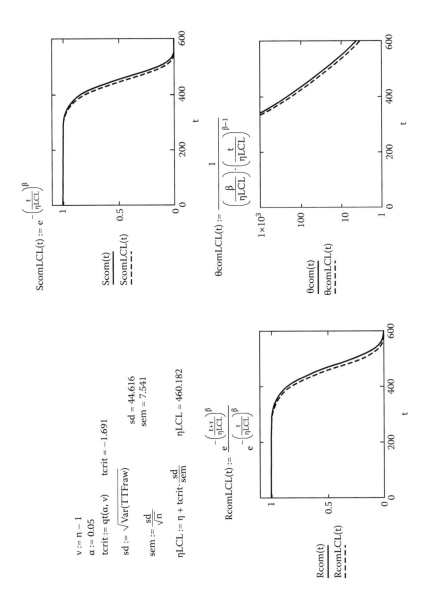

$$\text{ScomLCL}(t) := e^{-\left(\frac{t}{\eta LCL}\right)^{\beta}}$$

$$\frac{\text{Scom}(t)}{\text{ScomLCL}(t)}$$
- - - - -

$$\theta\text{comLCL}(t) := \cfrac{1}{\left(\frac{\beta}{\eta LCL}\right) \cdot \left(\frac{t}{\eta LCL}\right)^{\beta-1}}$$

$$\frac{\theta\text{com}(t)}{\theta\text{comLCL}(t)}$$
- - - - -

$v := n - 1$

$\alpha := 0.05$

$\text{tcrit} := \text{qt}(\alpha, v) \qquad \text{tcrit} = -1.691$

$\text{sd} := \sqrt{\text{Var(TTFraw)}} \qquad \text{sd} = 44.616$

$\qquad\qquad\qquad\qquad\qquad \text{sem} = 7.541$

$\text{sem} := \dfrac{\text{sd}}{\sqrt{n}}$

$\eta\text{LCL} := \eta + \text{tcrit} \cdot \text{sem} \qquad \eta\text{LCL} = 460.182$

$$\text{RcomLCL}(t) := \cfrac{e^{-\left(\frac{t+\tau}{\eta LCL}\right)^{\beta}}}{e^{-\left(\frac{t}{\eta LCL}\right)^{\beta}}}$$

$$\frac{\text{Rcom}(t)}{\text{RcomLCL}(t)}$$
- - - - -

FIGURE 3.41

MathCAD reliability functions with LCL complete data.

Time-Censored Data without Replacement

The distinction between characterization of the parameters of Weibull reliability functions for complete data and censored data is that the estimator for the Bartlett's median ranks estimator is applied only to the failed parts. The TTF array includes only the times to failure for the failed parts. The index array is assigned only to the TTF array only and not to all of the data. Therefore, for this example, the index array assigns rank only to the 12 failed parts: index = 1, 2, ..., 12. The difference between number of failures and the total parts on test is k. In this example, 14 pumps were put on test and $k = 2$ did not fail by the time-censored duration, 500 h. The number of parts on test, n, is equal to index plus k. The number of parts on test influences the value of F(TTF), where the denominator of Bartlett's median ranks is $n + 0.4$. The MathCAD parameter estimation, confidence intervals, and reliability parameters are all materially affected by this distinction. The TTF array and index are shown in Figure 3.42.

The parameters of the Weibull distribution are characterized by the following median ranks regression entirely in MathCAD. The parameters of the Weibull distribution are calculated using the median ranks regress, as shown in Figure 3.43. The index ranges from 1, 2, ..., 12 and the sample size, $n = 14$, for Bartlett's median ranks.

The estimators for the parameters of the Weibull are $\eta = 487.62$ and $\beta = 11.23$, with a coefficient of correlation, $\rho = 0.938$, and a coefficient of determination, $r_{sq} = 0.88$. The median ranks regression, MRR(X) = $b_0 + b_1 X$, is plotted in Figure 3.44.

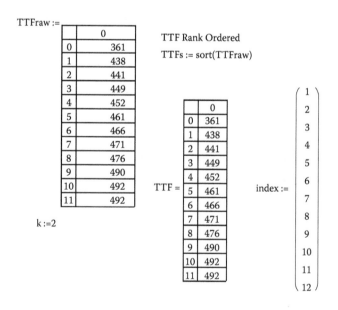

FIGURE 3.42
MathCAD approach to fit Weibull failure model time-censored data without replacement rank order and index.

$n := index_{11} + k$ $\qquad$ $n = 14$

$X := \ln(\text{TTF})$ $\qquad$ $F(\text{TTF}) := \dfrac{index - 0.3}{n + 0.4}$

$\qquad\qquad\qquad\qquad$ $S(\text{TTF}) := 1 - F(\text{TTF})$

$Y := \ln\left(\ln\left(\dfrac{1}{S(\text{TTF})} \right) \right)$

$b_0 := \text{intercept}(X, Y)$ $\qquad$ $b1 := \text{slope}(X, Y)$

$b_0 = -71.958$ $\qquad\qquad$ $b1 = 11.626$

$\eta := e^{\frac{b0}{b1}}$ $\qquad\qquad\qquad$ $\beta := b1$

$\eta = 487.618$ $\qquad\qquad$ $\beta = 11.626$ $\qquad\qquad$ $\rho := \text{corr} < X, Y)$

$\text{MRR}(X) := b0 + b1 \cdot X$ $\qquad$ $\text{mean}(X) = 6.123$ $\qquad$ $\rho = 0.938$

$\qquad\qquad\qquad\qquad\qquad$ $\text{mean}(Y) = -0.779$ $\qquad$ $\text{rsq} = \rho^2$

$\qquad\qquad\qquad\qquad\qquad\qquad\qquad\qquad\qquad$ $\text{rsq} = 0.88$

FIGURE 3.43
MathCAD approach to fit Weibull failure model time-censored data without replacement Weibull parameter estimators.

The confidence limits of the median ranks regression are calculated and plotted in Figure 3.45. The degrees of freedom for the variance are a number of failed parts, n = 12, minus 1; n − 1 = 11.

The confidence limits provide a visual indication of the adequacy of the Weibull failure model. The wider the confidence limits are, the less descriptive the model is. Ideally, the confidence limits will be very narrow about the median ranks regression line. The confidence limits widen for time that differs from the mean time (see Figure 3.46).

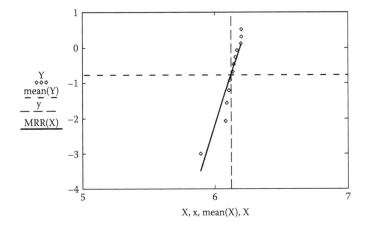

FIGURE 3.44
MathCAD approach to fit Weibull failure model time-censored data without replacement median ranks regression.

$Sxx(X) := (X - mean(X))^2$ $Y\ pted := b0 + b1 \cdot X$
$SXX := \sum Sxx(X)$ $Res(Y) := (Y\text{-}Y\ pted)^2$
$SXX = 0.078$ $SSE := \sum Res(Y)$
$SSE = 1.438$

$$VAR := \frac{SSE}{index_{13} - 1}$$

$C := 0.95$
$\alpha := 1 - C$ $VAR = 0.31$

$$se(X) := \sqrt{VAR \times \left[\left(\frac{1}{index_{13}}\right) + \frac{(X - mean\ (X))^2}{SXX}\right]}$$

$tL := qt\left(\frac{\alpha}{2}, index_{13} - 2\right)$ $tL = -2.228$

$tU := qt\left(1 - \frac{\alpha}{2}, index_{13} - 2\right)$ $tU = 2.228$

$CLL(X) := tL \cdot se(X)$ $CLU(X) := tU \cdot se(X)$
$CIL(X) := MRR(X) +$ $CIU(X) := MRR(X) +$
$\quad CLL(X)$ $\quad CLU(X)$

FIGURE 3.45
MathCAD approach to fit Weibull failure model time-censored data without replacement median ranks regression confidence interval.

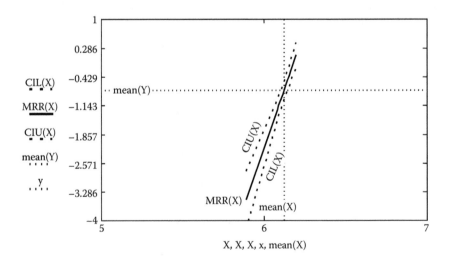

FIGURE 3.46
MathCAD approach to fit Weibull failure model time-censored data without replacement median ranks regression confidence interval plot.

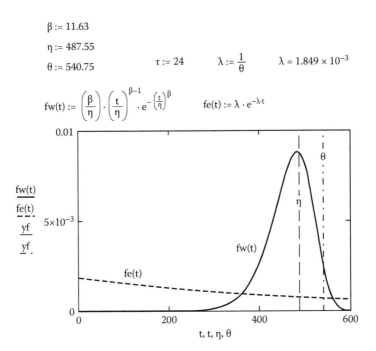

$\beta := 11.63$

$\eta := 487.55$

$\theta := 540.75$ $\qquad \tau := 24 \qquad \lambda := \dfrac{1}{\theta} \qquad \lambda = 1.849 \times 10^{-3}$

$$fw(t) := \left(\frac{\beta}{\eta}\right) \cdot \left(\frac{t}{\eta}\right)^{\beta-1} \cdot e^{-\left(\frac{t}{\eta}\right)^{\beta}} \qquad fe(t) := \lambda \cdot e^{-\lambda \cdot t}$$

FIGURE 3.47
MathCAD Weibull comparative pdf time-censored data without replacement.

The exponential mean time to failure and failure rate are calculated from the TTF data. Comparative plots of the exponential and Weibull pdf failure models confirm that the exponential probability distribution is not appropriate (see Figure 3.47).

The exponential pdf of failure, $f_e(t)$, suggests that failures occur throughout the time interval, $0 \leq t \leq 225$ h; yet, the Weibull pdf of failure suggests that failures commence at $t = 225$ h. The comparative exponential and Weibull survival functions are plotted in Figure 3.48.

The comparative exponential and Weibull mission reliability functions are plotted in Figure 3.49. The comparative mission reliability for a 24-h duration at $t = 0$ shows that the part is far more reliable than the exponential distribution infers. Only after 300 h does the Weibull mission reliability fall below the exponential mission reliability. This is just what one would expect from a part that exhibits wear out or fatigue. By the time the Weibull reliability drops to near zero, at 600 h, the exponential mission reliability is still at 95.7%.

The comparative Weibull and exponential hazard functions plot is the most important characteristic of the part for design and sustainment. The exponential approach understates the part and, by extension, the system reliability. Design decisions that have an impact on cost and performance are

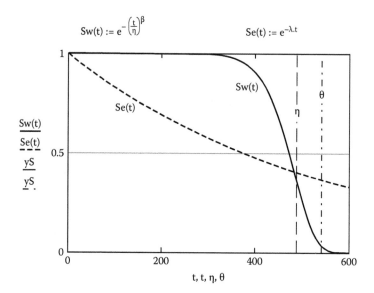

FIGURE 3.48
MathCAD Weibull comparative survival functions time-censored data without replacement.

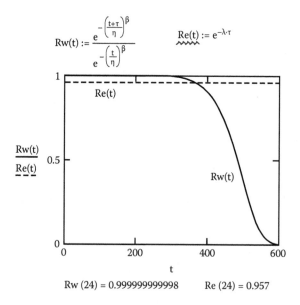

FIGURE 3.49
MathCAD Weibull comparative reliability functions time-censored data without replacement.

influenced by the estimate for the failure rate. Logistical support calculates spare parts inventory using the failure rate as a consumption rate and is more likely to overstock spare parts. The excess spare parts risk degradation in stockpile and the life-cycle cost is unnecessarily high. Reorder quantities are reduced as part failures do not meet expectations, only to increase as the system ages. Then spare parts are in short supply, emergency spares are ordered at premium costs, and system downtime awaiting spare parts causes lost opportunity costs.

Maintenance resources are determined from predicted consumption rates that become idle until the system ages to the point that the organization finds itself playing keep up. Organizations that seek to implement proactive maintenance to replace parts prior to failure are not able to use the exponential failure rate as a factor. The Weibull approach describes the actual failure behavior of the part and allows trend analysis to determine the time in use that will reach a threshold risk of failure that can be used to schedule proactive maintenance actions. Spare parts inventory can be controlled by better understanding the failure behavior of the part and its impact on the system. Note the comparative hazard functions shown in Figure 3.50.

The hazard function example illustrates the inapplicability of the exponential distribution in probabilistic risk assessment. An organization designates a risk threshold of one failure/100,000 h. The exponential failure model would reject use of the part; the Weibull failure model suggests that the part can be used for up to 200 h before replacement for meeting the risk threshold. Weibull characterization of the lower confidence limits for the Weibull reliability functions is calculated and plotted in Figure 3.51. The time-based

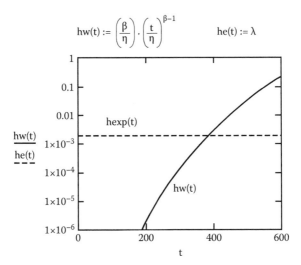

$$hw(t) := \left(\frac{\beta}{\eta}\right) \cdot \left(\frac{t}{\eta}\right)^{\beta-1} \qquad he(t) := \lambda$$

FIGURE 3.50
MathCAD Weibull comparative hazard functions time-censored data without replacement.

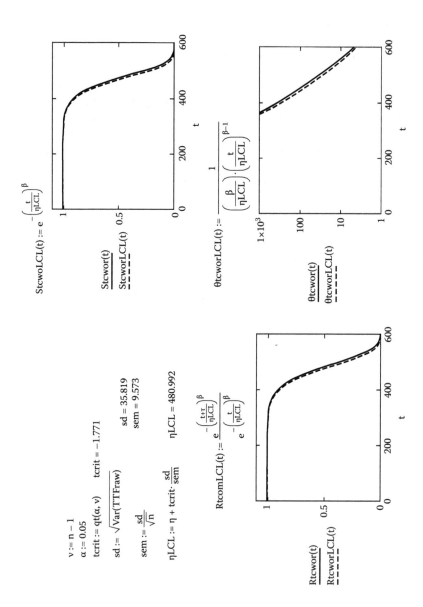

FIGURE 3.51
MathCAD reliability functions with LCL time-censored data without replacement.

mean time to failure, $\theta(t)$, and its lower confidence limit, $\theta_{LCL}(t)$, show how a part ages over time. Prior to 300 h, the expected MTTF is greater than 1,000 h. The MTBF is an alternative metric to the hazard function for risk.

Time-Censored Data with Replacement

The time-censored TTF table array includes only TTF for failed parts; times on test for unfailed parts are not entered. In the example in Figure 3.52, 28 pumps were put on test. Failed pumps were replaced with new pumps. Fourteen pumps were functioning at the conclusion of the test duration; $k = 14$ at 500 h. The number of pumps on test, n, is equal to index_{13} plus k.

The parameters of the Weibull distribution are calculated using the median ranks regress, as shown in Figure 3.53. The index ranges from 1, 2, ..., 14 and the sample size, $n = 28$, for Bartlett's median ranks.

The estimators for the parameters of the Weibull are $\eta = 528.22$ and $\beta = 10.86$, with a coefficient of correlation, $\rho = 0.951$, and a coefficient of determination, $r_{sq} = 0.91$. The median ranks regression is plotted in Figure 3.54.

The confidence limits of the median ranks regression are calculated in Figure 3.55. Note that the variance degrees of freedom are found by

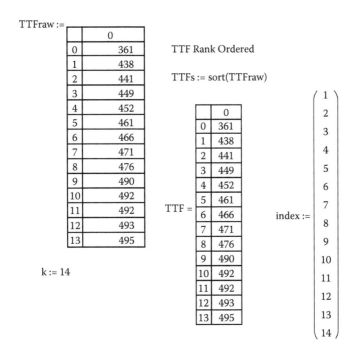

FIGURE 3.52
MathCAD approach to fit Weibull failure model time-censored data with replacement rank order and index.

$n := \text{index}_{13} + k$ $\qquad$ $n = 28$

$X := \ln(\text{TTF})$ $\qquad$ $F(\text{TTF}) := \dfrac{\text{index} - 0.3}{n + 0.4}$

$S(\text{TTF}) := 1 - F(\text{TTF})$

$Y := \ln\left(\ln\left(\dfrac{1}{S(\text{TTF})}\right)\right)$

$b_0 := \text{intercept}(X, Y)$ $\qquad$ $bl := \text{slope}(X, Y)$
$b_0 = -68.106$ $\qquad$ $bl = 11.626$

$\eta := e^{\frac{b0}{bl}}$ $\qquad$ $\beta := bl$

$\eta = 528.223$ $\qquad$ $\beta = 10.863$ $\qquad$ $\rho := \text{corr}(X, Y)$

$\text{MRR}(X) := b0 + bl \cdot X$ $\qquad\qquad\qquad\qquad$ $\rho = 0.951$

$\qquad\qquad\qquad\qquad$ $\text{mean}(X) = 6.134$ $\qquad$ $\text{rsq} = \text{corr}(X, Y)2$

$\qquad\qquad\qquad\qquad$ $\text{mean}(Y) = -1.473$ $\qquad$ $\text{rsq} = 0.905$

FIGURE 3.53
MathCAD approach to fit Weibull failure model time-censored data with replacement rank Weibull parameter estimators.

subtracting 1 from the number of failed pumps, $\text{index}_{13} = 14$. Confidence limits are a function of sample size. Small sample size will yield wider confidence limits than complete data, as shown in the median ranks regression confidence limits in Figure 3.56. The exponential MTTF, θ, is included with the parameters of the Weibull to characterize the probability density functions of the respective failure models, as shown in Figure 3.57.

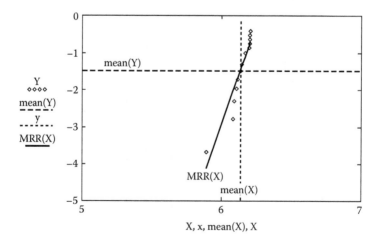

FIGURE 3.54
MathCAD approach to fit Weibull failure model time-censored data with replacement median ranks regression.

$Sxx(X): = (X - mean(X))^2$

$SXX := \sum Sxx(X)$

$SXX = 0.089$

$Y\ pted:=b0 + b1 \cdot X$

$Res(Y):=(Y-Y\ pted)^2$

$SSE:= \sum Res(Y)$

$SSE = 1.105$

$$VAR := \frac{SSE}{index_{13} - 1}$$

$C:= 0.95$

$\alpha:=1 - C$

$VAR = 0.085$

$$se(X) := \sqrt{VAR \times \left[\left(\frac{1}{index_{13}} \right) + \frac{(X - mean\ (X))^2}{SXX} \right]}$$

$tL := qt\left(\frac{\alpha}{2}, index_{13} - 2 \right)$ $tL = -2.179$

$tU := qt\left(1-\frac{\alpha}{2}, index_{13} - 2 \right)$ $tU = 2.179$

$CLL(X):=tL \cdot se(X)$ $CLU(X) := tU \cdot se(X)$

$CIL(X):= MRR(X) + CLL(X)$ $CIU(X):= MRR(X) + CLU(X)$

FIGURE 3.55
MathCAD approach to fit Weibull failure model time-censored data with replacement median ranks regression confidence interval.

The comparative survival functions for the exponential probability distribution and the Weibull time-censored with replacement data are shown in Figure 3.58. One can visualize that the exponential survival function suggests that 37% of the pumps will survive to the mean time to failure ($\theta = 500$); the

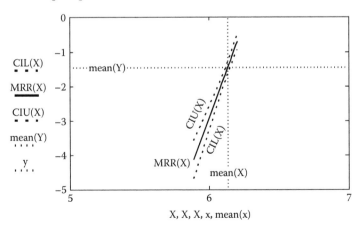

FIGURE 3.56
MathCAD approach to fit Weibull failure model time-censored data with replacement median ranks regression confidence interval plot.

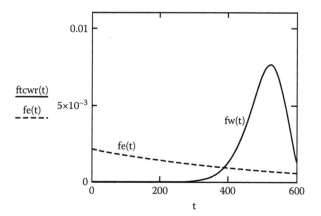

FIGURE 3.57
MathCAD Weibull comparative pdf time-censored data with replacement.

Weibull survival function suggests that only about 10% will survive to 500 h. The exponential unrealistically assumes longer life for a larger proportion of parts than the Weibull. In that regard, the Weibull is more conservative.

The comparative exponential and Weibull mission reliability functions are shown in Figure 3.59. The Weibull distribution suggests the more realistic scenario that a part becomes less reliable as it ages; the exponential distribution assumes that the mission reliability does not change with age.

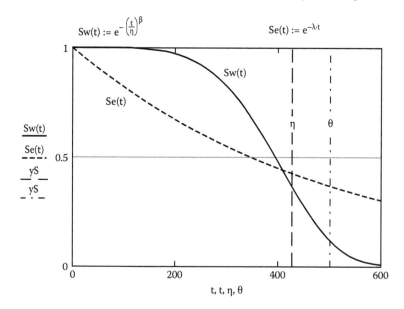

FIGURE 3.58
MathCAD comparative survival functions time-censored data with replacement.

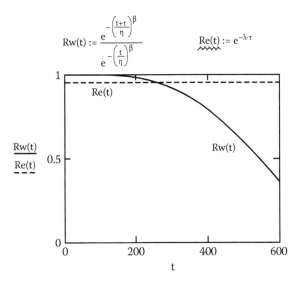

FIGURE 3.59
MathCAD Weibull comparative reliability functions time-censored data with replacement.

The comparative exponential and Weibull hazard functions in Figure 3.60 show that the Weibull distribution predicts that the part will not increase to the point where it equals the exponential failure rate until after over 200 h of use.

The reliability functions' lower confidence limits are plotted in Figure 3.61 along with evaluations at the point estimates of the parameters for the Weibull.

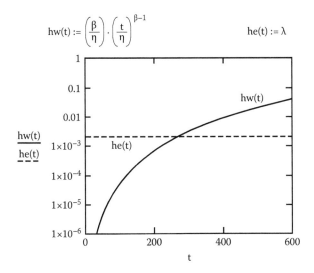

FIGURE 3.60
MathCAD Weibull comparative hazard functions time-censored data with replacement.

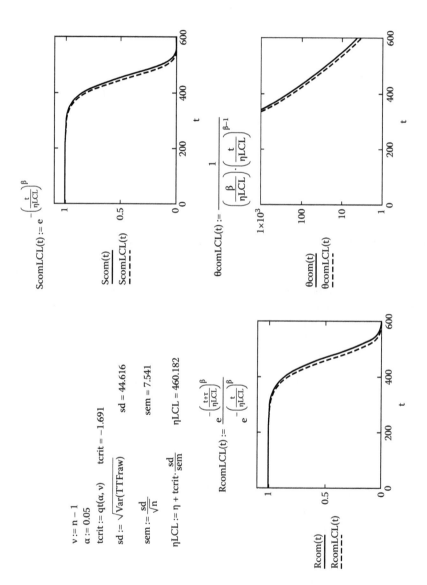

FIGURE 3.61
MathCAD reliability functions with LCL time-censored data with replacement.

Failure-Censored Data

Failure-censored experiments end when the last part fails. The TTF for all part TTF is entered in the worksheet. In the example in Figure 3.62, 12 parts were run to failure. There is no time on test for unfailed parts, $k = 0$, and $n = $ index$_{11}$. The median ranks regression is computed in the same manner for complete data. There is no distinction between the index and the number of observations for Bartlett's median ranks (see Figure 3.63).

The estimators for the parameters of the Weibull are $\eta = 475.79$ and $\beta = 12.47$, with a coefficient of correlation, $\rho = 0.927$, and a coefficient of determination, $r_{sq} = 0.859$. The median ranks regression is plotted in Figure 3.64. The confidence limits are calculated in Weibull as shown in Figure 3.65. The median ranks regression, MRR(X), is plotted with its upper and lower confidence limits in Figure 3.66. The comparative exponential and Weibull pdf failure models are plotted in Figure 3.67. The comparative exponential and Weibull survival functions are plotted in Figure 3.68. The comparative exponential and Weibull reliability functions are plotted in Figure 3.69. The comparative exponential and Weibull hazard functions are plotted in Figure 3.70. The Weibull reliability functions are plotted for the point estimates of the characteristic life and shape parameters and for the lower confidence limit of the characteristic life in Figure 3.71.

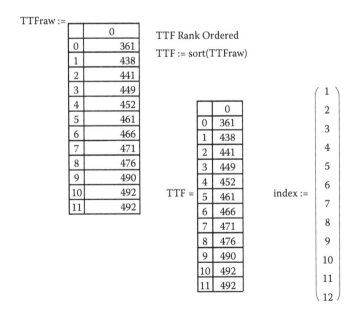

FIGURE 3.62
MathCAD approach to fit Weibull failure model failure-censored data rank order and index.

$n := \text{index}_{11}$ $n = 12$

$X := \ln(\text{TTF})$ $F(\text{TTF}) := \dfrac{\text{index} - 0.3}{n + 0.4}$

$S(\text{TTF}) := 1 - F(\text{TTF})$

$Y := \ln\left(\ln\left(\dfrac{1}{S(\text{TTF})}\right)\right)$

$b_0 := \text{intercept}(X,Y)$ $bl := \text{slope}(X,Y)$
$b_0 = -76.881$ $bl = 11.626$

$\eta := e^{\frac{b0}{bl}}$ $\beta := bl$

$\eta = 475.791$ $\beta = 12.471$ $\rho := \text{corr}(X,Y)$
$MRR(X) := b0 + bl \cdot X$ $\rho = 0.927$
                    $rsq = \text{corr}(X,Y)^2$
            $\text{mean}(X) = 6.123$ $rsq = 0.859$
            $\text{mean}(Y) = -0.53$

FIGURE 3.63
MathCAD approach to fit Weibull failure model failure-censored data Weibull parameter estimators.

Pump Failure Math Model

The pump failure math model comprises the failure math models of each failure mode: vane, control actuator, rotor, and bearing failure. The failure math model parameters' estimators are tabulated in Table 3.19.

The failure math model for the control actuator is the exponential distribution from the first example. The failure math models for the rotor and bearing

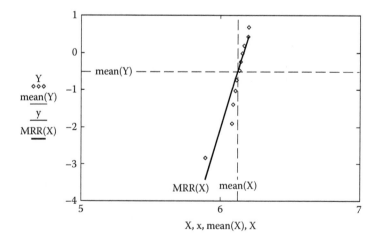

FIGURE 3.64
MathCAD approach to fit Weibull failure model failure-censored data median ranks regression.

$Sxx(X) := (X - mean(X))^2$ $Y\ pted := b0 + b1 \cdot X$

$SXX := \Sigma Sxx(X)$ $Res(Y) := (Y - Y\ pted)^2$

$SXX = 0.078$ $SSE := \Sigma Res(Y)$

$SSE = 1.988$

$$VAR := \frac{SSE}{index_{13} - 1}$$

$C := 0.95$

$\alpha := 1 - C$ $VAR = 0.181$

$$se(X) := \sqrt{VAR \times \left[\left(\frac{1}{index_{13}}\right) + \frac{(X - mean\ (X))^2}{SXX}\right]}$$

$tL := qt\left(\frac{\alpha}{2}, index_{13} - 2\right)$ $tL = -2.228$

$tU := qt\left(1 - \frac{\alpha}{2}, index_{13} - 2\right)$ $tU = 2.228$

$CLL(X) := tL \cdot se(X)$ $CLU(X) := tU \cdot se(X)$

$CIL(X) := MRR(X) + CLL(X)$ $CIU(X) := MRR(X) + CLU(X)$

FIGURE 3.65
MathCAD approach to fit Weibull failure model failure-censored data median ranks regression confidence intervals.

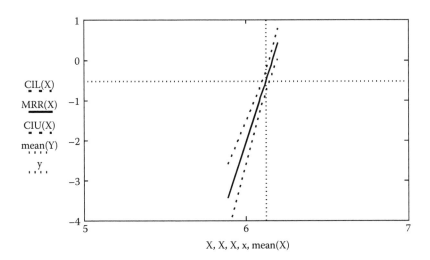

FIGURE 3.66
MathCAD approach to fit Weibull failure model failure-censored data median ranks regression plot.

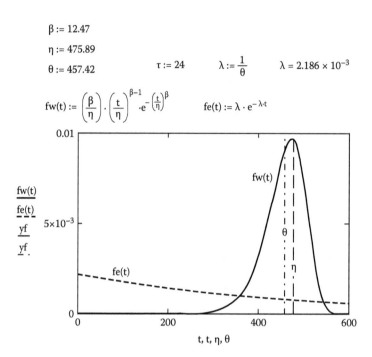

FIGURE 3.67
MathCAD Weibull comparative pdf failure-censored data.

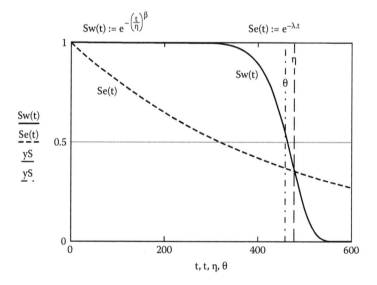

FIGURE 3.68
MathCAD Weibull comparative survival functions failure-censored data.

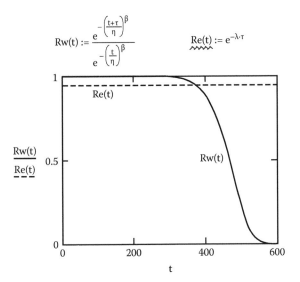

FIGURE 3.69
MathCAD Weibull comparative reliability functions failure-censored data.

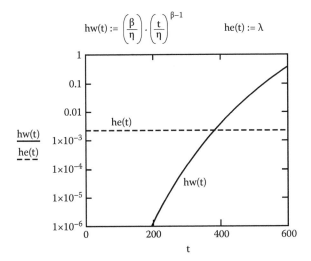

FIGURE 3.70
MathCAD Weibull comparative hazard functions failure-censored data.

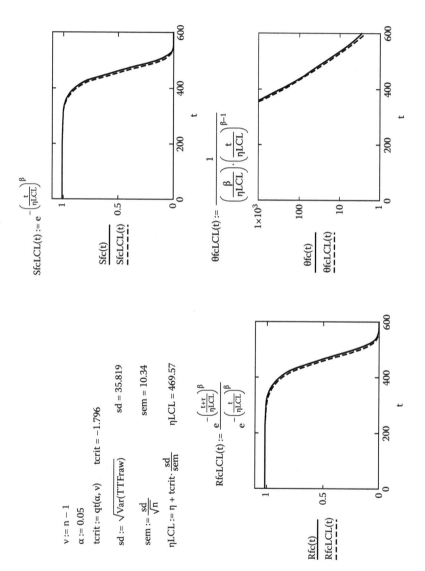

TABLE 3.19

Vane Pump Failure Model Parameters

Failure Mode	Failure Mode	MTTF (θ)		Characteristic Life (η)		Shape Parameter (β)
		Point Estimate	Lower Confidence	Point Estimate	Lower Confidence	Point Estimate
Vane	See Table 3.20					
Control actuator	Exponential	25.36	18.39			
Rotor	Weibull			967	849	7.65
Bearing	Weibull			542	481	6.95

are provided using the Weibull distribution. The pump vane failure model is composed of the failure models of each failure mode: vane jams, vane fractures, van rotor fails, and vane spring fails. The Weibull vane fracture failure math model from the time-censored with replacement experiments is used with $\eta = 525$ h, $\eta_{LCL} = 519$, and $\beta = 10.86$. The failure model parameters' estimators are tabulated in Table 3.20.

The vane failure model is expressed as the survival, reliability, and hazard functions. The vane survival function, $S_{vane}(t)$, is the product of the four failure modes' survival functions, expressed as

$$S_{vane}(t) = S_{jam}(t) \cdot S_{fract}(t) \cdot S_{corr}(t) \cdot S_{spr}(t) \tag{3.96}$$

The lower confidence limit of the vane survival function, $S_{vane}LC(t)$, is the product of the four failure modes' survival functions evaluated at the lower confidence limits of the characteristic life (see Figure 3.72), expressed as

$$S_{vane}LC(t) = S_{jam}LC(t) \cdot S_{fract}LC(t) \cdot S_{corr}LC(t) \cdot S_{spr}LC(t) \tag{3.97}$$

TABLE 3.20

Vane Failure Model Parameters

Failure Mode	Failure Mode	MTTF (θ)		Characteristic Life (η)		Shape Parameter (β)
		Point Estimate	Lower Confidence	Point Estimate	Lower Confidence	Point Estimate
Vane jam	Weibull			1031	997	12.32
Vane fracture	Weibull			528	519	10.86
Vane corrosion	Weibull			967	849	7.65
Vane spring	Weibull			542	481	6.95

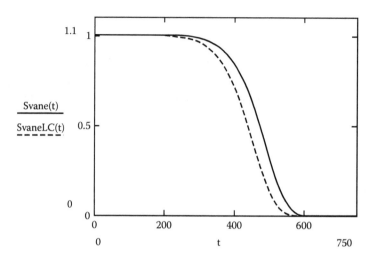

FIGURE 3.72
Vane survival function and MTBF.

The mean time to failure for the vane, θ_{vane} (464.23 h), is the indefinite integral of the vane survival function, $S_{vane}(t)$:

$$\theta = \int_0^{Tmax} S(t)dt \tag{3.98}$$

The lower confidence limit for the mean time to failure for the vane, $\theta_{vane}LC$ (430.54), is the indefinite integral of the vane survival function, $S_{vane}LC(t)$. In both integrals, the definite integral from 0 to 750 h is used to minimize the burden on the computer. MathCAD can get bogged down with true indefinite integral operation and will freeze up for more than five factors in the survival function. One can incrementally increase the upper limit of integration until there is no change in the solution.

The mean time to failure for a population of parts that have aged to a specified time can be evaluated by setting the lower limit of integration to the achieved time in service (τ). For example, at 240 h time in service, the mean time to failure for the vane $\theta_{vane}\tau$ (224.34 h) becomes the integral from 240 to 750 h.

The estimate for MTBF is a valuable population metric for logistical support analysis but holds little utility to design and safety engineering analysis. The hazard function, $h(t)$, found as the inverse of the MTBF estimates the risk of failure of the part for the failure mechanism and mode that can be compared to the functional requirements and environmental conditions of use. Given a risk threshold of one failure/100 operating hours, we find that the part remains within acceptable limits at 10 missions, where h (240 h) is approximately 0.004 failure/hour.

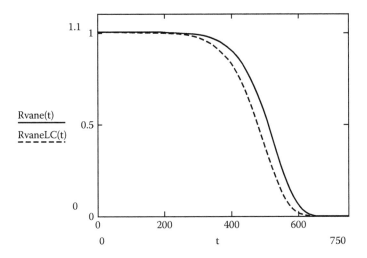

FIGURE 3.73
Vane reliability function.

The vane mission reliability function, $R_{vane}(\tau|t)$, and its lower confidence limits are equal to the product of the failure modes' mission reliability functions, evaluated at mission duration, $\tau = 24$ h, as shown in Figure 3.73.[6] Reliability plots like this one are commonplace in failure analysis reports and presentation and should prompt the engineer to demand, "So what?!" There are two reasons to characterize and plot the reliability function with its lower confidence limit:

- Compare the reliability of the part to its allocated reliability specification; the allocated specification should be plotted with the part plot. A decision must be made concerning whether the design achieves the allocation if its plot is below the lower confidence limit.
- Compare two or more design alternatives to the allocated specification and select the best reliability choice.

The vane pump survival function shown in Figure 3.74, $S_{pump}(t)$, is the product of the survival functions for the four parts: control actuator, vane, rotor, and bearing. Note that the control actuator is expressed as an exponential distribution, and the other three parts are expressed by the Weibull distribution. Also note that the vane survival function is expressed symbolically as $S_{vane}(t)$, which is located on the MathCAD worksheet and not repeated in the equation.

The pump survival function is dominated by the exponential factor. The failure behaviors of the mechanical and structural pump components are masked by the control actuator. The pump mean time to failure, θ_{pump} (25.35 h), is the integral of the survival function of the pump; the lower confidence limit of the pump survival function, $\theta_{pump}LC$ (18.39 h), is the integral of the lower

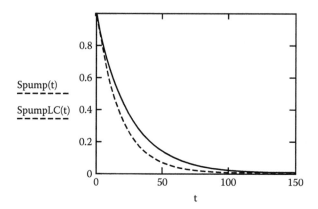

FIGURE 3.74
Pump survival function and MTBF with control actuator component.

confidence limit of the pump survival function. The mean time to failure for a population of parts that have aged to $\tau = 48$ h is evaluated by setting the lower limit of integration to the achieved time in service, τ: θ_{pump} $\tau = 3.8$ h.

The pump mission reliability function, $R_{\text{pump}}(t)$, is the product of the mission reliability functions of the four parts, as shown in the algorithms and plots in Figure 3.75. The reliability plot shows a constant value to the time where the control actuator is no longer a contributor factor, and then the failure behavior of the mechanical and structural components shows the decreasing reliability.

The failure analysis for the mechanical and structural components provides little information for comparative analysis between the components

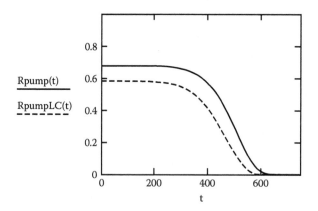

FIGURE 3.75
Pump reliability function with control actuator component.

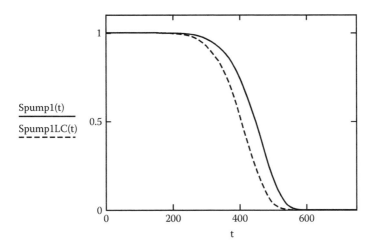

FIGURE 3.76
Pump survival function and MTBF without control actuator component.

and their respective allocations or for comparative analysis between design alternatives. This problem is resolved by removing the exponentially distributed component, as shown in Figure 3.76.

The pump survival function and MTBF for the mechanical and structural components are better understood by excluding the control actuator component, as is the reliability function, shown in Figure 3.77.

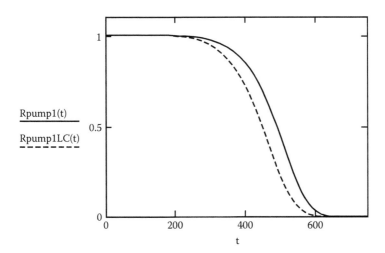

FIGURE 3.77
Pump reliability function without control actuator component.

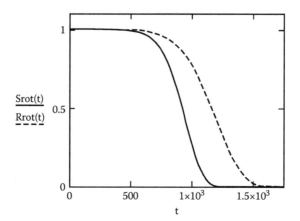

FIGURE 3.78
Pump rotor survival and reliability functions.

Though the pump allocated reliability specification is not included on the reliability plot, let us assume that the pump does not achieve the allocation. The answer is found by seeking to understand the "weak" component. The reliability of a part is influenced by the lowest component reliability. Determination of an acceptable pump is achieved by understanding whether the vane, rotor, or bearing is the limiting component.

Calculating the survival and reliability functions and plots for rotor and bearing components and comparing each to the allocated reliability specification inform us how the pump can be specified to achieve the allocation, as shown in Figures 3.78 and 3.79.

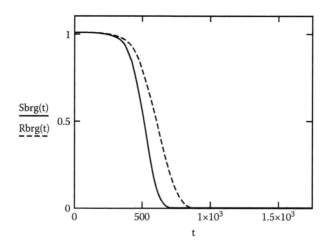

FIGURE 3.79
Pump bearing survival and reliability functions.

This analytical approach shows that the bearing is clearly the limited component. If we assume that the vane and rotor components are acceptable, then we can seek a pump that has a bearing component that is capable of withstanding the failure mechanism.

Triangular Distribution

Data are not always available to develop a failure model; yet, a part on the critical items list demands failure analysis. Part failure that is logically Weibull distributed due to wear-out failure modes can be subjectively modeled with the triangular probability distribution. The parameters of the triangular failure distribution, minimum, mode, and maximum time to failure can be subjectively estimated by an Adelphi survey.

An Adelphi survey is performed by contacting three or more subject matter experts (SMEs), who use their experience with the design and field performance of the part to estimate the parameters. Their responses are used to characterize the worst-case TTF, the lowest estimate of the minimum TTF; the most likely TTF, the average of their estimates for the mode; and the best-case TTF, the lowest estimate of the maximum TTF. The guidelines of using the lowest estimates for the worst and best cases introduce conservatism in the triangular failure model that mimics the appropriate confidence limits. This logic is illustrated next.

Adelphi Survey

Let us assume that the pump includes an electric power supply that transmits rotation to the pump shaft using a pulley and single V-belt design. The failure behavior of the elastomeric V-belt under operating and ambient conditions of use is not known, but it is deemed an operational critical item. Three V-belt SMEs are queried after informing them of the V-belt specifications, operational torque, and the conditions of use. The results of their subjective analysis of the time to failure are summarized in Table 3.21.

TABLE 3.21

Adelphi Survey for Triangular Distribution Parameters

V-Belt TTF	Minimum	Mode	Maximum
SME1	0	70	110
SME2	25	75	125
SME3	15	80	100
Minimum estimate	0		100
Mean estimate		75	

The results of the Adelphi survey can be evaluated in two ways:

- Worst case: the minimum estimates for the minimum, mode, and maximum are selected: $t_{MIN} = 0$, $t_{MODE} = 70$, and $t_{MAX} = 100$.
- Modified worst case: the mean of the mode is used with the minimum estimates for the minimum and maximum: $t_{MIN} = 0$, $t_{MODE} = 75$, and $t_{MAX} = 100$.

The mean of the triangular distribution is expressed as

$$\frac{t_{MIN} + t_{MODE} + t_{MAX}}{3} = \frac{0 + 75 + 100}{3} = 58.33 \tag{3.99}$$

The probability density function for time to failure, $f(t)$, is expressed as

$$f(t) = \begin{cases} \dfrac{2(t - t_{MIN})}{(t_{MAX} - t_{MIN})(t_{MODE} - t_{MIN})} & if \quad t_{MIN} \le t \le t_{MODE} \\[2ex] \dfrac{2(t_{MAX} - t)}{(t_{MAX} - t_{MIN})(t_{MAX} - t_{MODE})} & if \quad t_{MODE} < t \le t_{MODE} \end{cases} \tag{3.100}$$

The cumulative probability density function for time to failure, $F(t)$, is expressed as

$$F(t) = \begin{cases} \dfrac{(t - t_{MIN})^2}{(t_{MAX} - t_{MIN})(t_{MODE} - t_{MIN})} & if \quad t_{MIN} \le t \le t_{MODE} \\[2ex] \dfrac{(t_{MAX} - t)^2}{(t_{MAX} - t_{MIN})(t_{MAX} - t_{MODE})} & if \quad t_{MODE} < t \le t_{MAX} \end{cases} \tag{3.101}$$

The survival function is expressed as

$$S(t) = 1 - F(t) \tag{3.102}$$

The hazard function is expressed as

$$h(t) = \frac{f(t)}{S(t)} \tag{3.103}$$

The triangular distribution is not suited to spreadsheet or most statistical software programs. Engineering programs like MathCAD are able to express and calculate the triangular distribution.

Triangular Distribution: MathCAD Approach

The information from the Adelphi survey is entered into a MathCAD worksheet. The frequency distribution of part failure, $f(t)$, is expressed, evaluated,

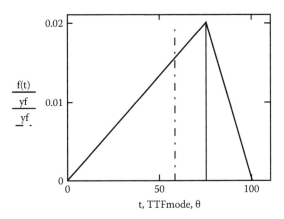

FIGURE 3.80
MathCAD triangular failure model pdf.

and plotted in Figure 3.80. The mean time to failure (TTF$_{mean}$ = 58.33 h) is shown as the dot-dashed line. The cumulative frequency distribution of part failure, $F(t)$, and the survival function, $S(t)$, are expressed, evaluated, and plotted in Figure 3.81.

The survival and cumulative failure functions intersect at the value of the mean. The instantaneous part failure rate, the hazard function, $h(t)$, is expressed, evaluated, and plotted in Figure 3.82. The hazard function shows that risk of failure is low up to 50 h and then increases rapidly. One may infer that part replacement at 50 h will significantly reduce the risk of unscheduled

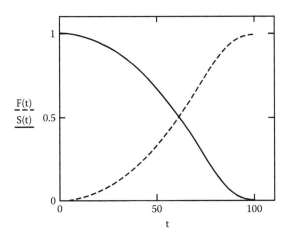

FIGURE 3.81
MathCAD triangular failure model cdf.

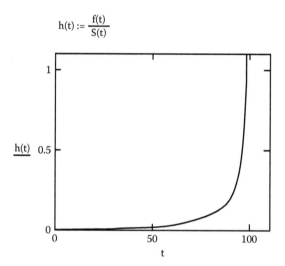

FIGURE 3.82
MathCAD triangular failure model hazard function.

failure during scheduled operations. The mission reliability function, $R(\tau|t)$, for $\tau = 10$ h, and the survival function are expressed, evaluated, and plotted in Figure 3.83. One may surmise that the information provided can be used to compare two or more design options but provides little help in planning sustainability.

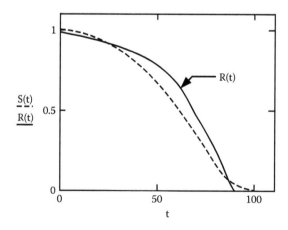

FIGURE 3.83
MathCAD triangular failure model reliability function.

Notes

1. I assume that the engineer has an understanding of basic probability and statistics. Probability and statistical methods are applied to model failure and influence design.
2. The pure statistician words the inference differently, stating that insufficient information exists to reject the null hypothesis when | test statistic < critical statistic |.
3. Time to failure and time to replacement define the time to a part failed state and differ only in the logistical distinction that purists use the former for repairable parts and the latter for unrepairable parts. In the same context, "mean time between failure" is used for repairable parts and "mean time to failure" is used for unrepairable parts.
4. The expression provided here is the two-parameter Weibull that assumes that failures will occur randomly for all $t > 0$. The three-parameter Weibull includes a threshold value that assumes that no failures occur prior to that time. The two-parameter Weibull is the best distribution for materials and parts that experience wear out to failure.
5. The index array numbers the rows from 0 to $n - 1$; therefore, the value of the index array at row number 0 is the first rank, 1, and the value of the index array at the last row, $n - 1$, is last rank, n.
6. MathCAD limits the function expression to one independent variable for this calculation, $R(t)$.

4

Part Maintainability and Availability

Bias for action, and stay close to the customer.

Tom Peters, *In Search of Excellence*

Introduction

Maintainability is defined as the probability that a part will be restored to full functionality and it is measured by mean time to repair (MTTR). Characterization of the MTTR, the upper confidence limit (UCL) of the MTTR, C% UCL MTTR, and the C% UCL of the sample time to repair (TTR) is the goal. Maintainability is stated as an MTTR of 1.5 h with a 95% UCL of the mean of 1.75 h and a 95% UCL of the TTR of 2.25 h. The confidence level is the risk of accepting a maintenance event that takes longer than the estimated TTR. There is no risk associated with overstating TTR.

Maintenance actions associated with characterization for TTR of a failed part include:

- system fault detection: realization that the system is in a down state or a degraded mode of operations
- part fault detection: diagnostic actions that identify the cause for the system down state drilling down the system through the design configuration hierarchy to the failed part
- part fault isolation: specification of the boundary conditions of the failed part and interface components to determine the extent of part removal required

System downing events and part failure detection and isolation are directly correlated to failure modes:

- prerepair administrative downtime: planning and documentation of resource allocation and scheduling of the maintenance action
- prerepair logistics downtime: implementation of the preparations for the maintenance action, including moving the system to the maintenance facility or moving the maintenance resources to the system,

cleaning the failed part and interfaces, performing safety controls, acquisition of tools and spare parts, etc.

- part replacement: removal of the failed part and installation of the replacement part

Part replacement is directly correlated to failure mechanisms and complexity of the design configuration:

- part functionality checkout: verification of the effectiveness of the maintenance action, reiteration of the system fault detection, and subsequent steps for ineffective outcome
- postrepair logistics downtime: replenishment of fluids and interfaces removed during part replacement, cleanup and recovery of the failed part, and disposal of interface components
- postrepair administrative downtime: documentation of the failure event, root cause failure analysis, corrective action, and direct labor and materials costs
- return to service: actions required to place the system in full functionality

Design analysis can characterize only MTTR for part replacement. Characterization of MTTR for the other maintenance actions is performed by systems engineering and operations and management sustainment engineering. The utility of this analysis is maintenance planning:

- Budgeting for labor, specialty tools, and facility needs to perform this maintenance action uses the UCL of the mean TTF. Given past history, 25 of these maintenance actions can be expected to occur in the next reporting period, requiring two mechanics; then (25)(2) (0.796) = 39.8, or 40 direct mechanic labor hours will be budgeted.
- A single TTR will take less than 0.965 h, or 58 min, at the 95% confidence level. This is a good estimate to give a plant foreman for a repair task; there is little complaint when the work is completed in less time than estimated.

Both examples address dealing with management risk. The first case is budget risk; the second is operating risk.

Part Mean Time to Repair

The design analysis goal is to characterize the point estimates and upper confidence limits for the sample TTR and MTTR. Recall that the mean of a distribution is the location of the measure of central tendency of the data, but its utility in maintenance planning is not useful. The confidence limits for a sample and its mean are far more meaningful. The 95% UCL of the sample time to repair allows the inference that a maintenance action will take up to the determined time. The point estimate of the MTTR only claims that some repair actions will take less and some more than the mean TTR. For example, a maintenance planner determines from historical TTR data that a maintenance action to remove and replace the rear tire on a truck has a point estimate of 1.25 h and a 95% UCL of 1.75 h. The maintenance planner infers, correctly, that removal and replacement of the rear tire will take less than 1.75 h, with 95% confidence, and bases resource scheduling on the UCL.

The 95% upper confidence limit of the mean TTR is an effective planning tool for time and resource budgeting for all maintenance actions for a given part over a designated reporting period. The maintenance operations manager, in the preceding example, used the same TTR data to calculate the 95% upper confidence limit of the mean TTR for tire removal and replacement. The 95% UCL of the 1.25 h MTTR is found to be 1.33 h. The annual budget forecast prognosticates that an expected 200 tire removals and replacements will consume no more than 267 man-hours.

The prevailing school of thought is to characterize the parameters for MTTR for the same parts that are on the critical items list (CIL) for which failure math models are conducted. This is not always the case. Failure analysis is focused on parts that pose failure consequences; maintenance analysis is focused on parts that pose failure and economic consequences. Failure analysis leads to mitigation of failure consequences through design. Ideally, catastrophic failures are eliminated and operational failures are designed to degraded modes. Maintenance analysis is focused on costs and frequency of part failure, methods of inspection to measure wear-out condition indicators, commonality of interface and fastener components, and ease of access to failed parts.

Just as not all failure consequences are equal, not all repair maintenance actions are equal. The order of preference for acquisition of TTR data is suggested to be the following:

1. Historical data for any and all parts that engineering judgment determines a need for maintenance analysis where available and demands a few days and small costs of data acquisition and analysis.

2. An Adelphi survey for any and all parts that engineering judgment determines need maintenance analysis and for which the costs of the maintenance actions are deemed low and the TTR is small (such parts have limited influence on system availability) demands a few days and small costs of data acquisition and analysis.

3. Maintenance experiments for any and all parts that engineering judgment determines need maintenance analysis and for which

the costs of the maintenance actions are deemed high and the TTR is large (such parts have huge influence on system availability) demand weeks of test planning and performance and high costs for data acquisition and analysis.

Time-to-repair data can be acquired from historical records for parts with field use or similar parts that may differ functionally but are very similar in removal and replacement maintenance actions. Time-to-repair data for parts lacking historical data can be acquired empirically through maintenance experiments or by an Adelphi survey. Time to repair is a lower-is-best population that is positively skewed and fits a lognormal probability distribution.

Maintenance Experiment

Maintenance experiments are designed to acquire representative TTR data under expected conditions of use and are performed by skilled employees. Satisfying the representative requirement demands following these guidelines:

- The test article must be installed on a comparable system or on a test fixture designed to replicate the actual system, including actual interface components and proximate parts or dummy components that limit access to the test article.
- Removal and replacement procedures must be written and validated through consultation with vendors, skilled maintenance personnel, and maintenance supervisors.
- Facility and tool requirements must be specified and the experiment must closely approximate the conditions of the maintenance action; the experiment must not be performed in a lab setting that is uncharacteristic of actual conditions.
- The experiment must be performed by appropriately skilled maintenance personnel—not fellow engineers.
- Learning curve trials must be run to reduce special causes of variability.
- Trials for record must be accurately timed in units of minutes or hours:minutes; restating all TTR in hours can be done in the spreadsheet.
- Each trial must be performed without comment by the test engineer.
- Each trial must be reviewed for uncharacteristic actions that influence time to repair.

The lognormal math model is used due to its ease in transforming skewed data to approximate the normal probability distribution so that the standard normal z-statistic, or Student's t-statistic, can be used to calculate the upper confidence limit. The best practice for the lognormal math model uses the natural logarithm (ln) in base e, ln(TTR), although the base-10 logarithm can also be used. The following steps are used to apply the lognormal distribution to TTR data:

1. Tabulate the TTR data from either historical or experimental sources. Convert the TTR data to hours when the TTR is measured in minutes.
2. Calculate the natural logarithm of the TTR data in hours, ln(TTR).
3. Calculate the mean and standard deviation of the logarithm of the TTR data.
4. Determine the confidence level (C%) for the upper confidence limit:
   a. Calculate the upper confidence limit of the sample: $TTR_{UCL} = MTTR + z_{C\%}s$, where $z_{C\%}$ is the z-statistic for 95% confidence and s is the standard deviation of the natural logarithms for the sample TTR.
   b. Calculate the upper confidence limit of the MTTR:

$$MTTR_{UCL} = MTTR + z_{C\%}\left(\frac{s}{\sqrt{n}}\right) \qquad (4.1)$$

where n is the sample size of the TTR data.

Excel Spreadsheet Approach

The spreadsheet approach can be used to calculate the parameters of the lognormal distribution with focus on the UCL for the sample and mean TTR:

1. Input raw TTR data in minutes (column 1).
2. Restate raw TTR data in hours (column 2).
3. Excel calculates descriptive statistics for TTR.
4. Evaluate the parameters and shape of the distribution from the descriptive statistics.
5. Excel calculates the frequency distribution table.
6. Use Excel to plot the frequency distribution histogram to confirm the shape of the distribution visually.
7. Calculate natural logarithms for TTR hours (column 3).
8. Excel calculates UCL of sample TTR in hours using spreadsheet paste function.

TABLE 4.1

MS Excel TTR Table: Complete Data (Minutes)

TTR Minutes (Raw)							
2000	2001	2002	2003	2004	2005	2006	2007
39	45	40	46	38	44	41	47
38	49	39	50	37	48	40	51
44	46	45	47	43	45	46	48
42	49	43	50	41	48	44	51
43	52	44	53	42	51	45	54
40	52	41	53	39	51	42	54
41	57	42	58	40	56	43	59
44	61	45	62	43	60	46	63

9. Calculate UCL of MTTR in ln(TTR).
10. Calculate UCL of MTTR using antilog of UCL of MTTR in ln(TTR).

Complete Data

An illustrative example for historical, or complete, TTR data is presented in Table 4.1. The TTR in minutes is restated in hours in Table 4.2. Descriptive statistics for TTR are calculated by MS Excel and provide values for the MTTR, standard error (se) TTR = standard deviation/sqrt(n), median, the 50th percentile TTR, standard deviation and variance, kurtosis, skewness, range, and the minimum and maximum TTR (Table 4.3).

The distribution for repair time is located at 47 min, 0.78 h and has a minor kurtosis that is very nearly the same as the normal distribution, and a positive skew, which is not characteristic of the normal distribution. MS Excel creates the frequency distribution table and plots the histogram (see Figure 4.1). The histogram illustrates a significant shape for a skew to the right—the classic condition for use of the lognormal transformation of the data.

TABLE 4.2

MS Excel TTR Table: Complete Data (Hours)

TTR Hours (Raw)							
0.650	0.750	0.667	0.767	0.633	0.733	0.683	0.783
0.633	0.817	0.650	0.833	0.617	0.800	0.667	0.850
0733	0.767	0.750	0.783	0.717	0.750	0.767	0.800
0.700	0.817	0.717	0.833	0.683	0.800	0.733	0.850
0.717	0.867	0.733	0.883	0.700	0.850	0.750	0.900
0.667	0.867	0.683	0.883	0.650	0.850	0.700	0.900
0.683	0.950	0.700	0.967	0.667	0.933	0.717	0.983
0.733	1.017	0.750	1.033	0.717	1.000	0.767	1.050

TABLE 4.3

TTR Descriptive Statistics

	TTRmin	TTRhrs
Mean	46.875	0.781
Standard error	0.810	0.014
Median	45	0.75
Standard deviation	6.482	0.108
Sample variance	42.016	0.012
Kurtosis	−0.127	−0.127
Skewness	0.762	0.762
Range	26	0.433
Minimum	37	0.617
Maximum	63	1.050
Sum	3000	50
Count	64	64

The box plot is another way to see the shape of the data. The first and third quartiles (25th and 75th percentiles, respectively) are calculated in MS Excel by the paste function, "QUARTILE(array,k)," where the array is the TTR data and $k \equiv$ the quartile (i.e., 1 and 3, respectively):

$$QUARTILE(Row3:Row66, 1) = 42 \text{ min}$$

$$QUARTILE(Row3:Row66, 3) = 51 \text{ min}$$

The interquartile range (IQR) is the difference between the third and first quartile values and equals 9 h. Descriptive statistics provides TTRmin = 37 min and TTRmax = 63 min; the range is 26 min (Figure 4.2).

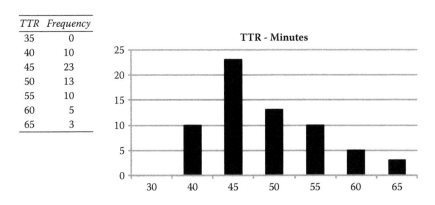

TTR	Frequency
35	0
40	10
45	23
50	13
55	10
60	5
65	3

FIGURE 4.1
TTR frequency distribution table and histogram: minutes.

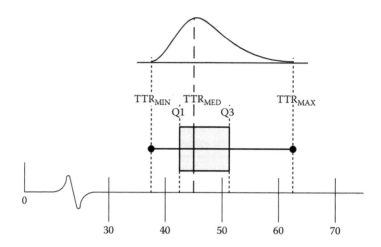

FIGURE 4.2
MS Excel TTF complete data histogram and box plot.

We observe that the sample TTR data are positively skewed; therefore, the data cannot be evaluated using the normal probability distribution and the lognormal math model is selected. The natural logarithms of the TTR hours are calculated as well as calculation of the mean and standard deviation for the natural logarithms for TTR (see Table 4.4).

MS Excel calculates the descriptive statistics for the ln(TTR) data, as shown in Table 4.5. The only values needed are the mean, standard error, and standard deviation. The 95% UCL for the sample TTR data is calculated in Excel using the paste function:

$$= loginv(C\%, ln(mean), ln(stdev))$$

which becomes

$$= loginv(0.95, -0.256, 0.134) = 0.965 \text{ h}$$

TABLE 4.4

Natural Logarithms for TTF Complete Data

Natural Logarithm (TTRhrs)							
−0.431	−0.288	−0.405	−0.266	−0.457	−0.310	−0.381	−0.244
−0.457	−0.203	−0.431	−0.182	−0.483	−0.223	−0.405	−0.163
−0.310	−0.266	−0.288	−0.244	−0.333	−0.288	−0.266	−0.223
−0.357	−0.203	−0.333	−0.182	−0.381	−0.223	−0.310	−0.163
−0.333	−0.143	−0.310	−0.124	−0.357	−0.163	−0.288	−0.105
−0.405	−0.143	−0.381	−0.124	−0.431	−0.163	−0.357	−0.105
−0.381	−0.051	−0.357	−0.034	−0.405	−0.069	−0.333	−0.017
−0.310	0.017	−0.288	0.033	−0.333	0.000	−0.266	0.049

TABLE 4.5

Descriptive Statistics for Natural Logarithm (TTR) Data

	Natural Logarithm (TTRhrs)
Mean	−0.256
Standard error	0.017
Standard deviation	0-134
C% =	0.95
zC% =	1.645
sem =	0.027
Natural logarithm UCL MTTR =	−0.228
UCL MTTR =	0.796 h
Natural logarithm UCL sample =	−0.036
UCL sample =	0.965 h

The UCL of the mean for TTR is calculated from the lognormal TTR summary statistics as

$$\text{MTTR}_{\text{UCL}} = \text{MTTR} + z_{\text{UCL}}\left(\frac{s}{\sqrt{n}}\right) = -0.256 + 1.645\left(\frac{0.134}{\sqrt{64}}\right) = -0.228 \qquad (4.2)$$

The UCL of the mean TTR is found by taking the antilog of −0.228 that equals 0.796 h. The conclusion from the complete data for time to repair is that MTTR = 0.781 h with a UCL of the mean of 0.796 h and that the UCL for TTR is 0.965 h at the 95% confidence level.

Empirical Data

Typically, TTR data are acquired from maintenance experiments and the sample sizes are far smaller than historical, complete data. Observe an experiment for the same part from the preceding section with 16 sample trials (see Table 4.6). Time and budget constraints allow an experiment of 16 trials. The raw data are converted from minutes to hours. MS Excel calculates the summary statistics and the frequency distribution.

We note from the descriptive statistics that 50% of all TTR fall between 0.733 and 0.854 h, the interquartile range, first and third quartile, with a median of 0.783 h and a mean of 0.799 h. The fastest TTR is 0.650 h and the slowest is 0.95 h. The frequency distribution is plotted as a histogram and box plot (see Figure 4.3).

The distribution of TTF data is positively skewed so the lognormal distribution will be used to analyze the data. The natural logarithms of the TTR hours are calculated as well as calculations of the mean and standard deviation for the natural logarithms for TTR (see Figure 4.4).

TABLE 4.6

Empirical TTR Data (Minutes and Hours): Summary Statistics and Frequency Distribution

TTR Minutes (Raw)		TTR Hours (Raw)		Summary Statistics		Frequency Distribution	
						TTR	f(TTR)
57	46	0.950	0.767	Mean	0.799	0.60	0
43	44	0.717	0.733	Standard deviation	0.087	0.70	1
51	45	0.850	0.750	Range	0.300	0.80	8
51	51	0.850	0.850	Minimum	0.650	0.90	5
48	39	0.800	0.650	Maximum	0.950	1.00	2
52	53	0.867	0.883	First quartile	0.733	1.10	0
56	44	0.933	0.733	Median	0.783		
43	44	0.717	0.733	Third quartile	0.854		
						Σf(TTR)	16

The sample 95% UCL for TTR is calculated in Excel as

$$= \text{loginv}(C\%,\ln(\text{mean}),\ln(\text{stdev}))$$

which becomes

$$= \text{loginv}(0.95,-0.230,0.108) = 0.949 \text{ h}$$

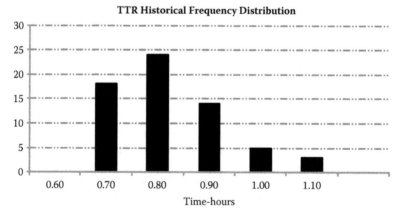

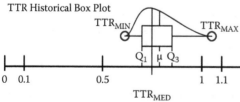

FIGURE 4.3
MS Excel empirical TTR data histogram and box plot.

Ln(TTR-hrs)	
−0 05 −0.27	Mean −0.230
−0.33 −0.31	Standard Deviation 0.108
−0.16 −0.29	
−0.16 −0.16	
−0.22 −0.43	TTR at 95% Confidence Level
−0.14 −0.12	C = 0.95
−0.07 −0.31	TTR95 = 0.949 hrs
−0.33 −0.31	

FIGURE 4.4
Excel ln(TTR) empirical data.

The Excel solution is stated in hours, and the functional arguments are stated in natural logarithms. The UCL of the mean for TTR is calculated from the lognormal TTR summary statistics as

$$\text{MTTR}_{\text{UCL}} = \text{MTTR} + z_{\text{UCL}}\left(\frac{s}{\sqrt{n}}\right) = -0.230 + 1.645\left(\frac{0.108}{\sqrt{16}}\right) = -0.186 \qquad (4.3)$$

The UCL of MTTR is found by taking the antilog of −0.186 that equals 0.830 h. The conclusion from the empirical data for time to repair is that MTTR = 0.799 h with a UCL of the mean of 0.830 h and that the UCL for the sample TTR is 0.949 h at the 95% confidence level.

Minitab Approach

Statistical software performs many of the calculations directly from the data, whereas spreadsheets require step-by-step calculations using the data. Statistical software also provides the capability to characterize the lognormal and the Weibull distributions for TTR. The statistical program approach requires the following step-by-step procedure to characterize the UCL for the sample and mean TTR:

1. Input TTR raw data in minutes (column 1).
2. Restate TTR data in hours (compute TTR/60) (column 2).
3. Minitab calculates descriptive statistics for TTR hours in the worksheet.
4. Minitab calculates natural logarithms for TTR hours in the worksheet.
5. Minitab calculates confidence limits for the parameters of the lognormal distribution of the sample data.

6. Calculate the UCL of the sample TTR using the Minitab output for the sample mean and the sample standard deviation and the value of the sampling critical statistic, $z_{C\%}$ or $t_{C\%}$.

7. Calculate UCL of the sample in hours using the antilog of the UCL for the MTTR.

8. Calculate UCL of the mean using the Minitab output for the sample mean and the upper confidence limit.

9. Calculate UCL of the MTTR in hours using the antilog of the UCL for the MTTR.

Complete Data Lognormal Distribution

The TTF data are entered in Minitab and the reliability/survival routines are used to fit the data to a probability distribution (see Table 4.7). Minitab treats the data as "failure time"; however, the use of this routine to fit time-to-repair data is valid. Minitab fits the data to the lognormal distribution with the least squares estimation methods where time to repair is the independent variable. The MTTR is read directly from the characteristics of the distribution to be 0.76 h with a 95% UCL of the mean equal to 0.78 h. The sample 95% upper confidence limit is solved using the equation for the UCL of the sample, as shown:

$$\text{TTR}_{\text{UCL}} = \text{MTTR} + z_{\text{UCL}}\sigma = 0.7602 + 1.645(0.0914) = 0.910 \qquad (4.4)$$

TABLE 4.7

Minitab TTR Lognormal Distribution Analysis Complete Data

Distribution Analysis: Complete			
Censoring Information			**Count**
Uncensored value			64
Estimation method: least squares—failure time (X) on rank (Y)			

Distribution: Lognormal Base e			
Parameter	**Estimate**	**Standard Error**	**95.0% Normal Bound Upper**
Location	−0.28135	0.01498	−0.25672
Scale	0.11980	0.01050	0.13839

Characteristics of Distribution:			
	Estimate	**Standard Error**	**95.0% Normal Bound Upper**
Mean (HTTF)	0.7602	0.01142	0.7792
Standard deviation	0.0914	0.00830	0.1061
Median	0.7548	0.01130	0.7736
First quartile (Q1)	0.6962	0.01153	0.7154
Third quartile (Q3)	0.8183	0.01356	0.8409
Interquartile range (IQR)	0.1221	0.01088	0.1414

Complete Data—Weibull

The Weibull distribution can be used to characterize skewed distributions and is therefore an accepted method to characterize time-to-repair data. The complete data in Minitab are used to fit the Weibull, as shown in the Minitab distribution analysis in Table 4.8.

The Weibull characteristics of TTR directly calculate an MTTR of 0.7795 h with a 95% UCL of the MTTR of 0.7998 h. The sample 95% upper confidence limit is solved using the equation for the UCL of the sample, as shown:

$$TTR_{UCL} = MTTR + z_{UCL}\sigma = 0.7795 + 1.645(0.0934) = 0.933 \qquad (4.5)$$

The estimates for MTTR and the UCL differ from the lognormal characterizations because the lognormal distribution is used to transform the skewed TTR data to a normal distribution for ease of drawing inferences, and the Weibull directly fits the skewed TTR data. This is illustrated graphically in the plot shown in Figure 4.5. The lognormal transform of the actual distribution of TTR, $f(TTR)$, introduces error in the estimates for calculation of the UCL of the mean and sample; the Weibull fits the actual distribution of TTR.

TABLE 4.8

Minitab TTR Weibull Distribution Analysis Complete Data

Distribution Analysis: TTRhrs
Variable: TTRhrs

Censoring Information			**Count**
Uncensored value			64
Estimation method: least squares—failure time (X) on rank (Y)			

Distribution: Weibull

Parameter	**Estimate**	**Standard Error**	**95.0% Normal Bound Upper**
Shape	10.0406	0.5909	11.0611
Scale	0.81924	0.01134	0.83811

Characteristics of Distribution:

	Estimate	**Standard Error**	**95.0% Normal Bound Upper**
Mean (MTTF)	0.7795	0.01218	0.7998
Standard deviation	0.09343	0.004202	0.1006
Median	0.7899	0.01214	0.8101
First quartile (Q1)	0.7236	0.01412	0.7472
Third quartile (Q3)	0.8463	0.01069	0.8641
Interquartile range (IQR)	0.1227	0.005900	0.1328

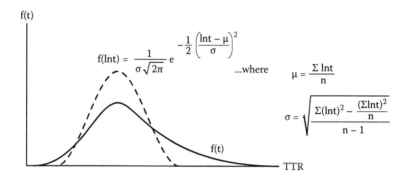

FIGURE 4.5
Lognormal transformation of skewed TTR data.

Empirical Data—Lognormal

A maintainability experiment records 16 sample TTR for the same maintenance action. Minitab is applied to characterize the parameters of the lognormal distribution from the preceding TTR data (see Table 4.9). The TTF data are treated as right censored to the last (16th) event. The lognormal characteristics of TTR directly calculate an MTTR of 0.8065 h with a 95% UCL of the MTTR of 0.844 h. The sample 95% upper confidence limit is solved using the equation for the UCL of the sample, as shown:

$$\text{TTR}_{\text{UCL}} = \text{MTTR} + z_{\text{UCL}}\sigma = 0.8065 + 1.645(0.08705) = 0.950 \qquad (4.6)$$

Empirical Data—Weibull

The empirical TTR data are fit by Minitab to a Weibull distribution as shown in the distribution analysis in Table 4.10. The TTF data are treated as right censored to the last (16th) event. The Weibull characteristics of TTR directly calculate an MTTR of 0.8052 h with a 95% UCL of the MTTR of 0.841 h. The sample 95% upper confidence limit is solved using the equation for the UCL of the sample, as shown:

$$\text{TTR}_{\text{UCL}} = \text{MTTR} + z_{\text{UCL}}\sigma = 0.8052 + 1.645(0.0826) = 0.9411 \qquad (4.7)$$

MathCAD Approach

Engineering statistical software provides the capability to characterize the UCL of the MTTR and the sample. The Weibull distribution can be used to characterize TTR because a Weibull fit will closely approximate the lognormal distribution. The Weibull distribution is presented alongside the

TABLE 4.9

Minitab TTR Lognormal Distribution Analysis Empirical Data

Distribution Analysis: TTRemp
Variable: TTRemp

Censoring Information	Count
Uncensored value	15
Right censored value	1
Censoring value: censor = 0	
Estimation method: least squares—failure time (X) on rank (Y)	

Distribution: Lognormal base e

Parameter	Estimate	Standard Error	95.0% Normal Bound Upper
Location	−0.22089	0.02744	−0.17575
Scale	0.10763	0.02020	0.14656

Characteristics of Distribution:

	Estimate	Standard Error	95.0% Normal Bound Upper
Mean (MTTF)	0.8065	0.02229	0.8440
Standard deviation	0.08705	0.01691	0.1198
Median	0.8018	0.02200	0.8388
First quartile (Q1)	0.7457	0.02240	0.7834
Third quartile (Q3)	0.8622	0.02693	0.9076
Interquartile range (TQR)	0.1165	0.02230	0.1596

lognormal distribution with comparative values for the 95% UCL for the sample TTR and MTTR. MathCAD performs the lognormal transform for TTF data by steps.

Complete Data—Weibull

1. Input TTR data array in minutes, TTRraw, using the paste table function from a spreadsheet (Figure 4.6).
2. Restate TTR data array in hours, TTR (Figure 4.7).
3. Plot the frequency distribution of the TTR data as a histogram (Figure 4.8).
4. The cumulative Weibull distribution for the TTR data is calculated using the median ranks regression, as shown in Figure 4.9. The data are stacked into a single column array. The sample size is calculated using the MathCAD function "length(array)." The data are rank ordered from minimum to maximum using the MathCAD function "sort(array)." The index array is calculated from 1...*n*. The independent variable, *X*, is calculated by taking the natural logarithm of

TABLE 4.10

Minitab TTR Weibull Distribution Analysis Empirical Data

Distribution Analysis: TTRemp
Variable: TTRemp

Censoring information	Count
Uncensored value	15
Right censored value	1

Censoring value: censor = 0
Estimation method: least squares—failure time (*X*) on rank (*Y*)

Distribution: Weibull

Parameter	Estimate	Standard Error	95.0% Normal Bound Upper
Shape	11.832	2.028	15.684
Scale	0.84070	0.01936	0.87316

Characteristics of Distribution:

	Estimate	Standard Error	95.0% Normal Bound Upper
Mean (MTTF)	0.8052	0.02128	0.8410
Standard deviation	0.08260	0.01224	0.1054
Median	0.8151	0.02098	0.8503
First quartile (Q1)	0.7567	0.02631	0.8012
Third quartile (Q3)	0.8642	0.01855	0.8953
Interquartile range (IQR)	0.1076	0.01684	0.1392

the rank-ordered TTR data, "ln(Tsort)." The cumulative distribution estimator, *F*, is calculated using Bartlett's median ranks:

$$F = \frac{index - 0.3}{n + 0.4} \tag{4.8}$$

TTRraw :=

	0	1	2	3	4	5	6	7
0	39	45	40	46	38	44	41	47
1	38	49	39	50	37	48	40	51
2	44	46	45	47	43	45	46	48
3	42	49	43	50	41	48	44	51
4	43	52	44	53	42	51	45	54
5	40	52	41	53	39	51	42	54
6	41	57	42	58	40	56	43	59
7	44	61	45	62	43	60	46	63

FIGURE 4.6
MathCAD TTR complete data: minutes.

$$\overline{TTR} := \frac{TTRraw}{60}$$

	0	1	2	3	4	5	6	7
0	0.65	0.75	0.667	0.767	0.633	0.733	0.683	0.783
1	0.633	0.817	0.65	0.833	0.617	0.8	0.667	0.85
2	0.733	0.767	0.75	0.783	0.717	0.75	0.767	0.8
TTR = 3	0.7	0.817	0.717	0.833	0.683	0.8	0.733	0.85
4	0.717	0.867	0.733	0.883	0.7	0.85	0.75	0.9
5	0.667	0.867	0.683	0.883	0.65	0.85	0.7	0.9
6	0.683	0.95	0.7	0.967	0.667	0.933	0.717	0.983
7	0.733	1.017	0.75	1.033	0.717	1	0.767	1.05

FIGURE 4.7
MathCAD TTR complete data: hours.

The dependent variable (Y) is calculated from the cumulative distribution estimator:

$$Y = \ln\left[\ln\left(\frac{1}{1-F}\right)\right]$$
(4.9)

The y-intercept (y_0) is found using the MathCAD function "intercept(X,Y)." The slope (β) is found using the MathCAD

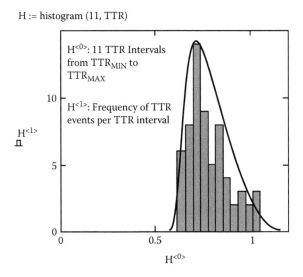

H := histogram (11, TTR)

$H^{<0>}$: 11 TTR Intervals from TTR_{MIN} to TTR_{MAX}

$H^{<1>}$: Frequency of TTR events per TTR interval

FIGURE 4.8
MathCAD TTR histogram: complete data.

$$T := \text{stack}\left(TTR^{<0>}, TTR^{<1>}, TTR^{<2>}, TTR^{<3>}, TTR^{<4>}, TTR^{<5>}, TTR^{<6>}, TTR^{<7>}\right)$$

$n := \text{length}(T)$ $\text{Tsort} := \text{sort}(T)$

$i := 1,2..n =$

ln(Tsort) =		0
	0	−0.483
	1	...

	1
	...

$X := \ln(\text{Tsort})$ $F := \dfrac{i - 0.3}{n + 0.4}$ $Y := \ln\left(\ln\left(\dfrac{1}{1 - F}\right)\right)$

$y0 := \text{intercept}(X,Y)$ $y0 = 1.621$

$\beta := \text{slope}(X,Y)$ $\beta := 8.541$ $\eta := e^{-\left(\frac{y0}{\beta}\right)}$ $\eta := 0.827$ $\longrightarrow$ $Fw(t) := 1 - e^{-\left(\frac{t}{\eta}\right)^{\beta}}$

FIGURE 4.9
MathCAD Weibull cumulative density function (cdf) characterization: complete data.

function "slope(X,Y)." The characteristic life (η) is found by solving for the antilog of (y_0/β):

$$\eta = e^{-\left(\frac{y0}{\beta}\right)} \tag{4.10}$$

The 95% UCL for the sample is found by entering values for TTR until the Weibull cumulative probability distribution, $Fw(t)$, is equal to C%. That procedure yields the 95% UCL of the MTTR of 0.9405 h. The 95% UCL of the MTTR is calculated from the mean and standard deviation of the Weibull TTR distribution. The Weibull mean is expressed as the characteristic life (η_w) times the gamma function of ($1 + 1/\beta$), as shown:

$$\mu_w = \eta\Gamma\left(1 + \frac{1}{\beta}\right) \tag{4.11}$$

Solving in MathCAD gives $\mu_w = 0.781$.

The Weibull standard deviation of the TTR is the square root of the Weibull variance (VAR$_w$):

$$VAR_w = \eta^2\Gamma\left(1 + \frac{2}{\beta}\right) - \mu_w^2 \tag{4.12}$$

Now, some engineers would view this equation as just a tad inconvenient—and it was before the engineering software and computers became commonplace. This calculation is the reason why the lognormal transform was so popular in the past and is still considered the best practice from intellectual inertia.

Solving in MathCAD gives $\sigma_w = 0.992$ h.
The 95% UCL of the MTTR, MTTR_{UCL}, can be calculated as

$$\text{MTTR}_{UCL} = \mu_w + z_{0.95}\left(\frac{\sigma_w}{\sqrt{n}}\right) \qquad (4.13)$$

$$\text{MTTR}_{UCL} = 0.781 + 1.645\left(\frac{0.992}{\sqrt{64}}\right) = 0.985 \qquad (4.14)$$

Complete Data Lognormal

MathCAD is applied to characterize the parameters of the lognormal distribution using the complete TTR data. The mean and standard deviation of the lognormal TTR array are calculated using the MathCAD functions "mean(array)" and "stdev(array)," respectively. The MathCAD expression for the cumulative normal distribution, $F_{\ln}(t)$, is "pnorm(t, μ, σ)," where $t \equiv \ln(t)$, $\mu \equiv$ mean of the $\ln(t)$ array, and $\sigma \equiv$ the standard deviation of the $\ln(t)$ array. The 95% UCL for the sample is found by entering values for TTR until the lognormal cumulative probability distribution, $F_{\ln}(t)$, is equal to C%. That procedure yields the 95% UCL of the MTTR of 0.965 h (see Figure 4.10). The 95% UCL of the lognormal MTTR (MTTR_{UCL}) is calculated as

$$\text{MTTR}_{UCL} = \mu_{\ln} + z_{0.95}\left(\frac{\sigma_{\ln}}{\sqrt{n}}\right) \qquad (4.15)$$

$$\text{MTTR}_{UCL} = -0.256 + 1.645\left(\frac{0.134}{\sqrt{64}}\right) = -0.228 \qquad (4.16)$$

The 95% UCL of the MTTR is the antilog of the 95% UCL of the lognormal MTTR, antilog $(-0.228) = 0.796$ h.

	0
$\ln(\text{Tsort}) =$ 0	−0.483
1	...

$\mu\ln := \text{mean}(\ln(\text{Tsort})) \quad \sigma\ln := \text{Stdev}(\ln(\text{Tsort}))$

$\mu\ln = -0.256 \qquad \sigma\ln = 0.134 \quad \Longrightarrow \quad \text{Fln}(t) := \text{pnorm}(\ln(t), \mu\ln, \sigma\ln)$

OR ──────────

$$\text{fln}(t) := \left(\frac{1}{\sigma\ln.\sqrt{2\cdot\pi}}\right)\cdot e^{-\left(\frac{1}{2}\right)\left(\frac{\ln(t)-\mu\ln}{\sigma\ln}\right)^2} \quad \Longrightarrow \quad \text{Fln}(t) := \left|\int\left(\frac{1}{\sigma\ln.\sqrt{2\cdot\pi}}\right)\cdot e^{-\left(\frac{1}{2}\right)\cdot\left(\frac{\ln(t)-\mu\ln}{\sigma\ln}\right)^2} dt\right.$$

FIGURE 4.10
MathCAD lognormal cdf characterization: complete data.

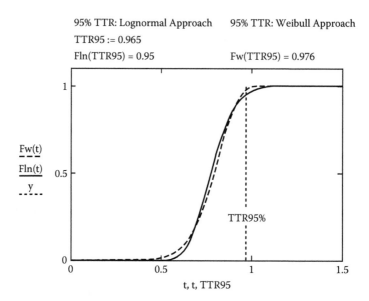

FIGURE 4.11
MathAD UCL calculation and plot: complete data.

The close relationship between the Weibull and the lognormal cumulative probability distributions is illustrated in the plot in Figure 4.11. It shows that the 95% UCL of the sample TTR is 0.965 h for the lognormal. The Weibull distribution provides a slightly more conservative value for the same TTR: a 97.6% confidence at 0.965 h.

Empirical Data

The same comparative approach for the lognormal and Weibull analysis for TTR is presented for the empirical data, as shown in Figure 4.12. The TTR data

TTRraw :=

	0	1
0	57	46
1	43	44
2	51	45
3	51	51
4	48	39
5	52	53
6	56	44
7	43	44

TTR :=

TTR=

	0	1
0	0.95	0.767
1	0.717	0.733
2	0.85	0.75
3	0.85	0.85
4	0.8	0.65
5	0.867	0.883
6	0.933	0.733
7	0.717	0.733

FIGURE 4.12
MathCAD TTR empirical data analysis.

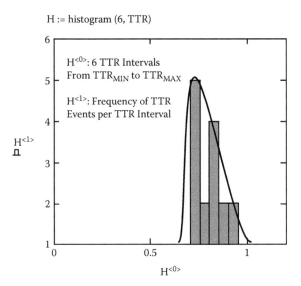

FIGURE 4.13
MathCAD TTR histogram: experimental data.

are entered in MathCAD in minutes and converted to hours. The histogram of the empirical data is plotted and shows a positive skew (Figure 4.13).

The TTR data are fitted to both a lognormal and a Weibull distribution and the cumulative probability distributions are expressed as $F_{ln}(t) = \text{plnorm}(t, \mu, \sigma)$ and

$$Fw(t) = 1 - e^{-\left(\frac{t}{\eta}\right)^{\beta}} \qquad (4.17)$$

respectively (see Figure 4.14).

The 95% UCL for the sample TTR is found by finding the value of TTR that solves the cumulative probability distributions for $F(t) = C\%$. The cumulative distributions are plotted in Figure 4.15 with the lognormal solution for the 95% UCL of the sample TTR.

The 95% UCL of the lognormal MTTR (MTTR_{UCL}) is calculated as

$$\text{MTTR}_{UCL} = \mu_{ln} + z_{0.95}\left(\frac{\sigma_{ln}}{\sqrt{n}}\right) \qquad (4.18)$$

$$\text{MTTR}_{UCL} = -0.23 + 1.645\left(\frac{0.108}{\sqrt{16}}\right) = -0.1856 \qquad (4.19)$$

The 95% UCL of the MTTR is the antilog of the 95% UCL of the lognormal MTTR, antilog $(-0.1865) = 0.83$ h.

$n := \text{length}(T)$ $\text{Tsort} := \text{sort}(T)$ $I := 1,2.. \, n=$

1
...

$\ln(\text{Tsort}) =$

	0
0	−0.431
1	...

$\mu\ln := \text{mean}(\ln(\text{Tsort}))$ $\sigma\ln := \text{Stdev}(\ln(\text{Tsort}))$

$\mu\ln = -0.23$ $\sigma\ln := 0.108$

$\text{Find}(t) := \text{plnorm}(t, \mu\ln, \sigma\ln)$

$X := \ln(\text{Tsort})$ $F := \dfrac{i - 0.3}{n + 0.4}$ $Y := \ln\left(\ln\left(\dfrac{1}{1-F}\right)\right)$
$y0 := \text{intercept}(X,Y)$ $y0 = 1.81$

$$-\left(\dfrac{y0}{\beta}\right)$$

$\beta := \text{slope}(X,Y)$ $\beta = 10.216$ $\eta := e$ $\eta = 0.838$

$Fw(t) := 1 - e^{\left(\frac{t}{\eta}\right)^{\beta}}$ $\rho = \text{corr}(X,Y)$ $\rho = 0.957$

FIGURE 4.14
MathCAD cdf characterization: experimental data.

The Weibull mean is expressed as the characteristic life (η_w) times the gamma function of $(1 + 1/\beta)$, as shown:

$$\mu_w = \eta\Gamma\left(1+\frac{1}{\beta}\right) \tag{4.20}$$

Solving in MathCAD gives $\mu_w = 0.798$ h.
The Weibull standard deviation of the TTR is the square root of the Weibull variance, VAR_w, expressed as

$$VAR_w = \eta^2\Gamma\left(1+\frac{2}{\beta}\right) - \mu_w^2 \tag{4.21}$$

Solving in MathCAD gives $\sigma_w = 0.094$ h.

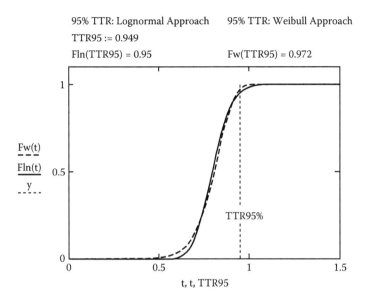

FIGURE 4.15
MathAD UCL calculation and plot: experimental data.

The Weibull 95% UCL of the MTTR ($MTTR_{UCL}$) is calculated as

$$MTTR_{UCL} = \mu_w + z_{0.95}\left(\frac{\sigma_w}{\sqrt{n}}\right); \tag{4.22}$$

$$MTTR_{UCL} = 0.798 + 1.645\left(\frac{0.094}{\sqrt{16}}\right) = 0.837 \tag{4.23}$$

Part and System Availability[1,2]

Part availability, A, is the probability that a part is in a full functional state when it is scheduled for use; it is determined by the reliability and maintainability of the part design. Assembly, subsystem, and system availability, A, is the probability that the design is not in a degraded mode functional state when it is scheduled for use and is a function of the availability of all subordinate design configuration levels. Availability is the ratio of reliability (uptime) to reliability plus maintainability (downtime) and can be expressed in general terms as

$$A = \frac{\text{Uptime}}{\text{Uptime} + \text{Downtime}} \tag{4.24}$$

Availability is expressed as a constant and is typically referred to as the part's or system's steady-state availability. Availability is calculated for four phases of system design and operation, each based on how uptime and downtime are measured:

1. Inherent availability: part to system design phase, predictive of operational performance
2. Instantaneous availability: part to system design phase, predictive of operational performance
3. Operational availability: system integration design and system sustainment phases, predictive of operational performance
4. Achieved availability: system sustainment phase, descriptive of operational performance

Inherent Availability

Inherent availability, A_i, is the idealized design measure of availability that is calculated by the design engineer.

Part Inherent Availability

The inherent availability of a part is calculated as the ratio of the part MTBF and the sum of part MTBF and part MTTR:

$$A_i = \frac{\text{MTBDE}}{\text{MTBDE} + \text{MTTR}} \tag{4.25}$$

Part A_i is a constant and assumes that the part provides the same availability over its useful life until it fails and is replaced.

System Inherent Availability

The inherent availability of a design, assembly, subsystem, and system is calculated as the ratio of the mean time between downing event (MTBDE) and the sum of MTBDE and MTTR$_{\text{design}}$:

$$A_i = \frac{\text{MTBF}}{\text{MTBF} + \text{MTTR}_{\text{design}}} \tag{4.26}$$

Design A_i is a constant and assumes that the design part provides the same availability over its useful life until its failure and is replaced. The estimator for design MTTR$_{\text{design}}$ is calculated as the weighted average of part mean time to repair (MTTR$_i$) using the respective exponential failure rates, λ_i, as the weighting factors:

$$\text{MTTR}_{\text{design}} = \frac{\sum_{i=1}^{n} \lambda_i \text{MTTR}_i}{\sum_{i=1}^{n} \lambda_i} \tag{4.27}$$

This approach to calculate design A_i applies to a system composed wholly of electronic and digital parts but does not correctly characterize structural and dynamic designs where parts experience wear out. The expression for design $\text{MTTR}_{\text{design}}$ for a structural and dynamic design must use the hazard function, $h_i(t)$, for the failure rate, λ_i, and becomes a function of time, $\text{MTTR}_{\text{dessign}}(t)$:

$$\text{MTTR}_{\text{design}}(t) = \frac{\sum_{i=1}^{n} h_i(t) \text{MTTR}_i}{\sum_{i=1}^{n} h_i(t)} \tag{4.28}$$

Instantaneous Availability

Instantaneous availability, A_{inst}, is a special case of inherent availability applied to single shot devices and to communications and digital networks that operate for very brief mission durations (τ). Instantaneous availability is a design measure of availability that is calculated by the design engineer. The parameters of instantaneous availability are the exponential distribution failure rate (λ) and the repair rate, MTTR^{-1}, μ:

$$A_{\text{inst}} = \frac{\mu}{\mu + \lambda} + \frac{\lambda}{\mu + \lambda} e^{-(\lambda + \mu)t} \tag{4.29}$$

The relationship between the instantaneous availability (A_{inst}) and the inherent availability (A_i) is illustrated in Figure 4.16. The instantaneous availability degrades to the inherent availability in a very brief time. The application of instantaneous availability is especially relevant for standby switch mechanisms in mechanical design, integration, and sustainment.

Operational Availability

Part Operational Availability

Operational availability, A_O, is both a system integration measure and a system sustainment measure of availability, and it is subject to interpretation and confusion by both system integrator and system sustainer. The parameters for part A_O are the MTBF, mean corrective maintenance time (M_{CT}),

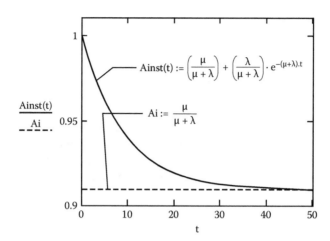

FIGURE 4.16
Instantaneous and inherent availability.

mean preventive maintenance time (M_{PT}), mean servicing time (M_{ST}), and mean administrative and logistics downtime (ALD). The system integrator estimates M_{CT} by analysis, proposes and estimates M_{PT} and M_{ST}, and estimates or guesses ALD. Mean corrective maintenance time is the mean time to repair.

Mean preventive maintenance actions are defined differently by organizations and the allocation of the time is inconsistent. Preventive maintenance actions range from little more than periodic inspections to time-determined overhaul, or they may be implemented through reliability-centered maintenance (RCM). Organizations that schedule preventive maintenance between missions may not charge the time as unavailable because the system is not scheduled to function. The logic states that a system is neither available nor unavailable if it is not scheduled to operate. Conversely, a system can only be available or unavailable during scheduled operations. Few sources acknowledge M_{ST} or assume it to be included in M_{PT}.

Servicing is equally ambiguous. It is not a maintenance action that is performed by replacing the part, but rather includes replenishing consumable materials (e.g., fuel, lubricants, coolant, etc.). Servicing time may also be treated like preventive maintenance time.

Operational availability of a part is calculated as the ratio of the part MTBF and the MTBF plus the sum of the part's M_{CT}, M_{PT} (inclusive of servicing time), and the mean administrative and logistical maintenance time:

$$A_O = \frac{\text{MTBF}}{\text{MTBF} + M_{CT} + M_{PT} + \text{ALD}} \tag{4.30}$$

The calculation of A_O by systems integrators differs from the calculation by system sustainers. System integrators use estimates for the parameters based on analysis and experiments to calculate a one-time, constant A_O to describe the system when it is new. System sustainers use historical performance periodically to calculate an ever-changing A_O that reflects updated parameters. The system sustainment characterizations for MTBF, M_{CT}, M_{PT}, and ALD are based on use of the part in actual conditions of use and will typically differ significantly from the predicted A_O by systems integration. More importantly, system sustainers see the parameters of A_O as management metrics. Mean time between failure is a metric to be increased, mean corrective maintenance time is a metric to be driven to zero, mean preventive maintenance time is a metric to be controlled, and administrative and logistics downtime is a metric to be minimized.

System Operational Availability

Operational availability (A_O) of an assembly, subsystem, and system as calculated by system integrators using the current best-practice approach is wrought with sources of error from the specious estimators for mean corrective maintenance time, mean preventive maintenance time, and mean administrative and logistics downtime:

$$Ao = \frac{MTBDE}{MTBDE + M_{CT} + M_{PT} + ALD} \qquad (4.31)$$

The systems integration estimator for design M_{CT} is calculated as the weighted average of $MTTR_i$, using the respective exponential failure rates (λ_i) as the weighting factors:

$$M_{CT_{design}} = \frac{\sum_{i=1}^{n} \lambda_i MTTR_i}{\sum_{i=1}^{n} \lambda_i} \qquad (4.32)$$

This applies to a system composed wholly of electronic and digital parts but does not correctly characterize structural and dynamic designs where parts experience wear out. The expression for design M_{CT} for a structural and dynamic design must use the hazard function, $h_i(t)$, for the failure rate (λ_i) and becomes a function of time, $M_{CT}(t)$:

$$M_{CT}(t) = \frac{\sum_{i=1}^{n} h_i(t) MTTR_i}{\sum_{i=1}^{n} h_i(t)} \qquad (4.33)$$

An accepted best-practice system integration predictive estimator for design M_{PT} is not defined in the literature. Two anecdotal approaches have been observed:

- A system is assigned periodic maintenance intervals (SMI_i), every 100, 500, and 2,000 operating hours, by the system integrator with prescribed lists of inspections, adjustments, and parts to be replaced for each maintenance interval. The time to perform each scheduled preventive maintenance action is estimated by analysis or experimentation. The design M_{PT} is calculated as the weighted average of the part M_{PT} per scheduled maintenance interval and the rate of scheduled maintenance intervals, SMI_i^{-1}, where the 100 h SMI has a rate of 1/100 h = 0.01 100-h SMIs/hour:

$$M_{PT_{design}} = \frac{\sum_{i=1}^{n} SMI_i^{-1} M_{PT_i}}{\sum_{i=1}^{n} SMI_i^{-1}} \tag{4.34}$$

- A Delphi survey is conducted to define scheduled maintenance intervals with prescribed lists of inspections, adjustments, and parts to be replaced for each maintenance interval and characterizes the median preventive maintenance time for each. Equation 4.11 also applies to this method.

Characterization of the mean administrative and logistics downtime by the system integrator is not possible; yet, the definition of A_O demands an estimator. No two organizations have the same administrative burden on maintenance actions; indeed, no two maintenance organizations within the same organization have the same administrative burden. Ideally, administrative burden should not exist; the better the maintenance manager is, the less the administrative burden will be.

Logistical downtime is equally unique to an organization. Spare parts inventory management, employment of skilled-trades employees, and investment in specialty tools, equipment, and facilities differ extensively. Use of contract maintenance services differs from using in-house maintenance organizations. Logistical downtime is a management factor that is lowest when management is effective. But the system integrator has no control over the system sustainer's management skills and experience.

Operational availability (A_O) of an assembly, subsystem, and system as calculated by system sustainers is a good predictor. The parameters are calculated from historical data that reflect the actual conditions of use experienced by the system at the organization:

$$A_O = \frac{MTBDE}{MTBDE + MMT + ALD} \tag{4.35}$$

Mean time between downing events is recorded by maintenance action and calculated by a simple average.

Mean maintenance time is the weighted average of mean corrective maintenance time and mean preventive maintenance time and their respective frequencies of occurrence, f_{CT} and f_{PT}:

$$\text{MMT} = \frac{f_{CT} M_{CT} + f_{PT} M_{PT}}{f_{CT} + f_{PT}} \tag{4.36}$$

Mean administrative and logistics downtime are recorded by maintenance action and calculated by a simple average.

Achieved Availability

Achieved availability, A_a, is a system sustainment metric; it is the least understood availability metric and is treated differently by various sources. Some view A_a as a predictive statistic, a redundant variant on A_O. This book treats A_a as the system sustainment measure of availability that is calculated by the reliability sustainment engineer to report the reliability experienced following an operational reporting period. It is used to compare actual availability, A_a, against the expected availability, A_O. Achieved availability is calculated as the ratio of the system MTBDE and the MTBDE plus the mean maintenance downtime (MDT):

$$A_a = \frac{\text{MTBDE}}{\text{MTBDE} + \text{MDT}} \tag{4.37}$$

Mean maintenance downtime is recorded by summing the means of all causes of maintenance downtime per maintenance event and calculated as

$$\text{MDT} = \text{MTTR} + \text{ALD} + M_{CT} + M_{PT} \tag{4.38}$$

Notes

1. Ireson, W. G., C. F. Coombs, Jr., and R. Y. Moss. 1996. *Handbook of reliability engineering and management*, 2nd ed. New York: McGraw–Hill.
2. O'Conner, P. D. T. 2002. *Practical reliability engineering*, 4th ed. New York: John Wiley & Sons.

5

Part Reliability Based on Stress-Strength Analysis

> This is the law of the Yukon, that only the strong shall thrive;
> That surely the weak shall perish, and only the fit survive.
>
> **Robert Service**

Introduction

Let us restate the core logic of reliability engineering: Failure mechanisms act on parts that cause part failure modes. Part failure modes cause part failure effects that extend to the next higher design configuration, the assembly. Assembly failure effects range from design and functional degraded mode to downing event. Part failure effects extend farther up to the end design configuration—the system—ranging from no effect, design and functional degraded mode, to system downing event. Part failure results in maintenance events, which reduce the availability of the system.

Part Stress

Failure mechanisms are the response of a material to stress. Rothbart defined stress as "the force per unit area acting on an elemental plane in the body."[1] Physical stresses act on a material as axial and transverse loads as described in various texts.[2,3] The three basic stresses are tension, shear, and compression loads. Torsion acts to elongate the body, shear acts to slice a body, and compression acts to shorten a body. A variant on shear is torsion that acts to twist the body.

Hudson defines the states of stress that act on the body[4]:

- Ultimate stress is the greatest stress that can be produced in a body before rupture occurs.
- Allowable stress or working stress is the intensity of stress that the material is designed to resist.

- Factor of safety is the ratio of the ultimate stress and the allowable stress.
- Yield stress is the intensity of stress beyond which the change in strain increases rapidly with little increase in stress.

Perry[5] expands on Hudson:

- Stress cycle is the smallest section of the stress-time function that is repeated periodically and identically.
- Nominal stress is calculated on the body by simple theory without taking into account the variation in stress conditions caused by geometric discontinuities such as hole, groves, fillets, etc.
- Maximum stress is the highest value of stress in the stress cycle; tensile stress is considered positive and compressive stress negative.
- Minimum stress is the lowest value of stress in the stress cycle; tensile stress is considered positive and compressive stress negative.

Part Failure

That stress is a failure mechanism is one of the two factors describing part failure; material strength is the other. Collins defines mechanical failure as "any change in the size, shape or material properties of a structure ... that renders it incapable of satisfactorily performing its intended function."[6] "Intended function" is the design requirement of the part—not the functionality of the next higher assembly or the system. The intended function of a fastener is to join two or more elements of a part, not the function of the assembly it constructs. Tensile elongation, shear, or embrittlement of the fastener changes the size, shape, and material properties of the bolt that render it incapable of performing its function. Likewise, the intended function of a shaft is to achieve a torque requirement, a ball in a bearing to reduce friction and bear a load, and a heat sink in a circuit assembly to conduct heat.

Superficially, failure appears to be intuitive: Part failure occurs when stress exceeds strength. Rigorous definition of failure is provided by stress theories. Collins[7] summarizes the most relevant stress theories as they apply to mechanical design:

- Rankine's maximum normal stress theory: Failure occurs when the applied normal stress load is equal to or greater than the maximum allowable normal stress based on empirical characterization of the part's maximum normal stress.

- Tresca–Guest's maximum shear stress theory: Failure occurs when the applied shear load is equal to or greater than the maximum allowable shear stress based on empirical characterization of the part's maximum shear stress.

- St. Venant's maximum normal strain theory: Failure occurs when the measured strain resulting from a stress load is equal to or greater than the maximum allowable strain based on empirical characterization of the part's maximum strain.

- Beltrami's total strain energy theory: Failure occurs when the total applied strain energy per unit volume is equal to or greater than the allowable total strain energy based on empirical characterization of the part's total strain energy.

- Huber, Von-Mises, and Henchy's distortion energy theory: Failure occurs when the applied distortion energy per unit volume is equal to or greater than the allowable distortion energy based on empirical characterization of the part's distortion energy.[8]

Emphasis on empirical characterization in each stress theory is relevant to design for reliability. Metrics for material strength or allowable stress in handbooks are point estimates of the mean stress empirically established by manufacturers, universities, and materials laboratories. Conditions of use for specific design requirements cannot be assumed to be met by handbook characterizations of stress. Handbook stress metrics provide the design engineer with a general order of magnitude that narrows the analysis to a manageable number of material options. Design for reliability demands that critical part materials require empirical investigation to understand the actual behavior of part failure that will be realized by the system in its specified conditions of use. The stress theories also provide understanding of the conditions of failure that can be the basis for a transition to reliability-centered maintenance in system sustainment.

Stress theory leads to the premise of this chapter:

> *Time does not cause part failure!*
> *Failure mechanisms cause part failure!*

Time-to-Failure Reliability Functions

Failure and reliability math models currently in use make simplifying assumptions that qualify the use of statistical distributions to fit time-to-failure data. The assumptions include the following:

- A part has a single dominant failure mechanism that is assumed to characterize the reliability of the part.
- A part operates at full functionality throughout the system mission at continuous, constant load.
- A part is not loaded when the system is not in operation.
- A system operates at full functionality throughout its useful life.
- System reliability and system capability are not correlated.

Statisticians are comfortable with the assumptions; engineers are not. Engineers know the following:

- A part has multiple failure mechanisms acting on it, and no one mechanism defines the reliability of the part.
- A part operates at different functional levels throughout the system mission and bears static and dynamic loads ranging from continuously to phased levels, each of which varies throughout the mission.
- A part may continue to be under load by physical and chemical stresses independently of system operations; indeed, a system can experience a downing event between missions.
- A system operates at degraded functional modes as parts wear out throughout the parts' useful lives.
- System reliability and system capability are correlated.

The comparative distinctions are illustrated respectively. System mechanical efficiency degrades over system use:

- A switch lever arm experiences bending, torsion, and corrosive failure mechanisms; one or two mechanisms may weaken material properties sufficiently that the third failure mechanism causes the failure mode–fracture.
- A car starter motor operates for a moment of time and then is idle and unloaded for the duration of a trip; the drive shaft operates continuously but at varying torque and rpm for stop-and-go and highway driving conditions of use.
- Battery posts experience corrosive loads during and between trips; mounting fasteners experience shear loads from the weight of the mounted assembly independently of system operation.
- Engine cylinder compression decreases as piston and cylinder wall surfaces wear.
- Redundant material handling process systems lose productivity as part failures take production equipment off-line; communication and network systems lose online capacity as heat stresses activate

system-protection mechanisms to prevent part failure; industrial lighting and phased array radar systems degrade lumens and sensitivity as individual light and radiation parts fail.

Interface parts introduce another failure issue that demands attention from design analysis: common parts used under different loads in the same system. Bolt fasteners illustrate such a part application:

- Use of a bolt to mount an assembly enclosure to a system structural member can be characterized by time-to-failure data. Time to failure (TTF) provides an acceptable estimation for reliability parameters for noncatastrophic, nonoperational, degraded mode parts when failure mechanisms act at static loads less than the yield strength over the useful life of the part.
- Use of the same fastener to seal a pressure vessel can be characterized only by understanding the behavior of the failure mechanisms acting on the bolt. Failure is not caused by time in service but rather by load levels that exceed the strength of the bolt.

Example TTF Reliability Functions for Hex Bolt

Fastening and joining are crucial to design and sustainment of structural and dynamic parts and higher design levels. Consider a simple hex head bolt used to fasten an assembly to a system structure (Figure 5.1). Historical data for hex bolt TTF are provided by maintenance records and are tabulated in MS Excel, as shown in Table 5.1. We view the TTF data as complete and uncensored, and assume the data are from a representative sample of the population of the useful life of the bolt exposed to the system's operating and ambient conditions of use.

System mission duration τ is 3–h, during which the bolt is loaded in shear and tension from vibration and the weight of the assembly. The exponential and Weibull reliability models are presented for comparative evaluation of the respective findings. Two Weibull reliability models are developed

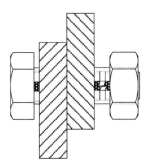

FIGURE 5.1
Hex head bolt.

TABLE 5.1

Hex Head Bolt Time-to-Failure Data

Historical Raw Data		
2005	2006	2007
39	119	37
122	63	53
47	115	139
55	129	49
131	56	41
55	135	126
56	61	140
127		50
TTF hours		118

for the TTF data: (1) a single failure mechanism model, and (2) two failure mechanisms, shear and tension.

Exponential Failure Distribution Approach

The exponential probability model is developed to characterize the hex bolt failure and reliability model parameters. It assumes that failure occurs from the instant the bolt is placed in use and continually thereafter due to applied loading as expected in the design requirements. The data table is stacked in MS Excel and imported to MathCAD (see Figure 5.2).[9] Note that the table is shown whole and collapsed. The collapsed view will be used throughout the remaining procedures to save space.

Expressions and calculations for the exponential bolt mean time to failure (μTTF) and failure rate (τ) are performed in MathCAD from the TTF data table, as shown in Figure 5.3.

The mean TTF is a population statistic that suggests that the useful life of all hex bolts is roughly 86 h. The constant failure rate is a population statistic that suggests that 1.2% of all hex bolts will fail during the next mission. The expressions for the exponential failure probability density function, $f_{exp}(t)$, survival function, $S_{exp}(t)$, and mission reliability function, $R_{exp}(\tau|t)$, are written in MathCAD, by subscript 'ept,' designating the exponential probability distribution, and the independent variable data source, TTF, as shown in Table 5.1. MathCAD does not allow the conditional statement "$\tau|t$," so we abbreviate with τ.

The exponential probability density function (pdf) for the bolt is

$$f_{exp}(t) = 0.012e^{-0.012t}$$

TTF:=

	0
0	39
1	122
2	47
3	55
4	131
5	55
6	56
7	127
8	119
9	63
10	115
11	129
12	56
13	135
14	61
15	37
16	53
17	139
18	49
19	41
20	126
21	140
22	50
23	118

TTF:=

	0
0	39
1	122
2	...

FIGURE 5.2
MathCAD array for time-to-failure data: hex bolt.

and, from the general equation for the exponential pdf,

$$f_{exp}(t) = \lambda e^{-\lambda t}$$ (5.1)

The survival function is

$$S_{exp}(t) = e^{-0.012t}$$

$\mu TTF := mean(TTF)$	$\mu TTF = 85.958$
$\lambda := \dfrac{1}{\mu TTF}$	$\lambda = 0.012$

FIGURE 5.3
Exponential mean TTF and failure rate: hex bolt.[10]

and, from the general equation for the survival function,

$$S_{exp}(t) = e^{-\lambda t} \tag{5.2}$$

The numerical solutions for the pdf and survival functions are variable over time, but the solution of the reliability expression for exponentially distributed failure is a constant. The mission reliability function for a mission ($\tau = 3$ h) is a constant equal to

$$R_{exp}(\tau|t) = e^{-(0.012)(3)} = 0.966 = 96.6\% \tag{5.3}$$

and, from the general equation for the mission reliability function,

$$R_{exp}(\tau|t) = e^{-\lambda \tau} \qquad e^{-\lambda \tau} = e^{-(0.012)(3)} = 0.966 \tag{5.4}$$

The mission reliability is a population statistic that suggests that 96.6% of all hex bolts will survive a mission of 3 h.

The exponential hazard function, $h(t)$, is written in MathCAD from its definition as the ratio of the pdf to the survival function, as shown next. The hazard function is constant and equal to the failure rate for exponentially distributed failure and is written in its simple form:

$$h_{exp}(t) = \frac{f_{exp}(t)}{S_{exp}(t)} = \frac{\lambda e^{-\lambda t}}{e^{-\lambda t}} = \lambda = 0.012 \tag{5.5}$$

The plots over time for the pdf, $f_{exp}(t)$, and the hazard function, $h_{exp}(t)$, are plotted in MathCAD, as shown in Figure 5.4.

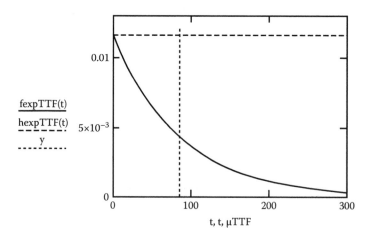

FIGURE 5.4
Plots for the pdf and hazard function: hex bolt.

The constantly decreasing pdf for hex bolt failure suggests that bolts will fail at a constant rate beginning during the first mission. The hazard function defines the constant failure rate over the hex bolt useful life. The relationship between the point estimate for the mean TTF and the distribution of failure illustrates how feeble the mean TTF is as a predictive tool for the expected reliability of the hex bolt. The exponential cumulative probability distribution, $F_{exp}(t)$, evaluated at the mean TTF is 63.2%:

$$F_{exp}(t = \mu TTF) = \int_0^{\mu TTF} f_{exp}(t)\,dt = 0.632 \tag{5.6}$$

Almost two-thirds of the hex bolts will fail prior to 86 h, the mean TTF. The plots over time for the survival function, $S_{exp}(t)$, and reliability function, $R_{exp}(t)$, are plotted in MathCAD, as shown in Figure 5.5.

The survival function is known as the life function for the hex bolt. It is a population statistic that "predicts" the steady decrease of survivors for a part as they age over time. Engineering judgment and field experience with fasteners suggest that bolt failure does not behave as described by the exponential probability distribution.

Single Failure Mechanism Weibull Model Approach

A Weibull reliability model is characterized on the assumption that a single, dominant failure mechanism is acting on the bolt to wear it out over its useful life. The TTF data are fit to the Weibull distribution using the median ranks regression method.[11] Median ranks regression transforms the cumulative

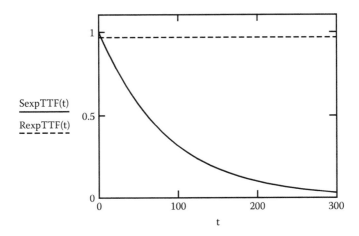

FIGURE 5.5
Plots for the survival and reliability functions: hex bolt.

probability distribution to a linear equation. The Weibull cumulative probability distribution is expressed as

$$F(t) = 1 - e^{-\left(\frac{t}{\eta}\right)^{\beta}}$$

(5.7)

The equation is restated as

$$1 - F(t) = e^{-\left(\frac{t}{\eta}\right)^{\beta}}$$

(5.8)

The exponential factor is shifted to the denominator to eliminate the minus sign:

$$1 - F(t) = \frac{1}{e^{\left(\frac{t}{\eta}\right)^{\beta}}}$$

(5.9)

The equation is restated as

$$\frac{1}{1 - F(t)} = e^{\left(\frac{t}{\eta}\right)^{\beta}}$$

(5.10)

The natural logarithm of both sides of the equation eliminates the exponential factor:

$$\ln\left(\frac{1}{1 - F(t)}\right) = \left(\frac{t}{\eta}\right)^{\beta}$$

(5.11)

The second natural logarithm yields a linear equation:

$$\ln\left[\ln\left(\frac{1}{1 - F(t)}\right)\right] = \beta \ln(t) - \beta \ln(\eta)$$

(5.12)

Let Y be defined as the dependent variable

$$Y = \ln\left[\ln\left(\frac{1}{1 - F(t)}\right)\right]$$

(5.13)

Let X be defined as the independent variable

$$X = \ln(t)$$

(5.14)

Let b_0 be defined as the y-intercept

$$b_0 = -\beta \ln(\eta)$$

(5.15)

$$\text{TTFsort} := \text{sort(TTF)} \quad \text{TTF sort}$$

	0
0	37
1	39
2	...

FIGURE 5.6
Sorted TTF data: hex bolt.

Let b_1 be defined as the slope of the equation

$$b_1 = \beta \tag{5.16}$$

The linear equation is expressed in terms of X and Y:

$$Y = b_0 + b_1 X \tag{5.17}$$

Median rank regression begins with rank ordering the raw data from minimum to maximum. MathCAD rank orders data tables using the "sort" command, as shown in Figure 5.6. A preferred approach is to rank order the data in a spreadsheet and import to a MathCAD array. The independent variable, $X = \ln(\text{TTFsort})$, is calculated in MathCAD in a column matrix, as shown in Figure 5.7. This step can also be performed in a spreadsheet and imported to a MathCAD array, thereby skipping the previous MathCAD step.

The column matrix is calculated as a range variable from $1 - n$ and is designated as "index" in MathCAD, as shown in Figure 5.8. This calculation can also be performed in a spreadsheet and imported to a MathCAD array.

The cumulative probability percent estimators are calculated in MathCAD in a column matrix designated as F, as shown in Figure 5.9. This is the y-axis for a probability paper used to evaluate sample data against the expected cumulative density function (cdf) for a selected probability distribution (exponential, Weibull, normal, etc.). The value for n is calculated using the "length(X)" command. The dependent variable,

$$Y = \ln\left[\ln\left(\frac{1}{1-F}\right)\right] \tag{5.18}$$

is calculated in a column matrix designated as Y.

$$X := \ln(\text{TTFsott})$$

	0
0	3.611
1	3.664
2	...

FIGURE 5.7
Independent variable, X: hex bolt.

index = 1.2 ... 24 =

1
2
...

FIGURE 5.8
Index: hex bolt.

The parameters of the median ranks regression, y-intercept (defined as b_0), and slope (defined as b_1) are calculated in MathCAD using the "intercept(X,Y)" and "slope(X,Y)" commands, as shown in Figure 5.10. The median rank regression is expressed in the linear equation form,

$$Y = b_0 + b_1 X$$

(5.19)

where Y is expressed by the "MRR(X)" command.

The median ranks regression is evaluated by the coefficients of correlation, "corr(X,Y)," and determination, R_{sq}, calculated as

$$R_{sq} = \text{corr}(X, Y)^2$$

(5.20)

The median ranks regression is plotted in Figure 5.10. The coefficient of determination, R_{sq}, suggests that the independent variable (TTF) treated as a metric for a single failure mechanism causes 83% of the change in Y, the cumulative probability of survival.[12] This is not an unusual magnitude for empirical data but it suggests further investigation of other failure mechanisms when the part under consideration has a critical failure effect. The plot of the data points alerts us to nonrandomness of the data; a clustering appears that suggests that the TTF data were drawn from two populations and strongly suggests that a single failure mechanism is an unacceptable assumption for failure analysis and reliability characterization.

$$F := \frac{\text{index} - 0.3}{\text{length}(X) + 0.4} \qquad F =$$

	0
0	0.029
1	0.07
2	...

$$Y := \ln\left[\ln\left(\frac{1}{1-F}\right)\right] \qquad Y =$$

	0
0	−3.537
1	−2.628
2	...

FIGURE 5.9
MathCAD cumulative distribution (F) and dependent variable (Y): hex bolt.

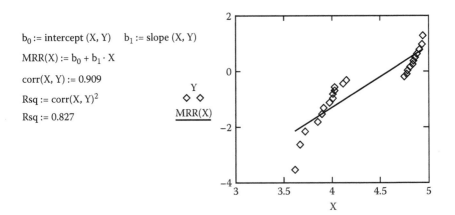

$b_0 := $ intercept (X, Y) $b_1 := $ slope (X, Y)

$MRR(X) := b_0 + b_1 \cdot X$

$corr(X, Y) := 0.909$

$Rsq := corr(X, Y)^2$

$Rsq := 0.827$

FIGURE 5.10
Median ranks regression and R_{sq}: hex bolt.

We continue with the characterization of the Weibull reliability model by calculating the Weibull parameters in MathCAD, as shown in Figure 5.11. The shape parameter, β, is equal to the slope of the median ranks regression line.[13] The characteristic life (η) is calculated as the antilog$_e$, or $\ln^{-1}$, of the ratio of the y-intercept (b_0) and the line slope (b_1):

$$\eta = \ln^{-1}\left(\frac{b_0}{b_1}\right) = e^{-\left(\frac{b_0}{b_1}\right)} \tag{5.21}$$

Recall that the y-intercept (b_0) was defined as

$$b_0 = -\beta \ln(\eta) \tag{5.22}$$

The equation is restated as

$$\ln(\eta) = -\frac{b_0}{\beta} = -\frac{b_0}{b_1} \tag{5.23}$$

Taking the antilog of both sides of the equation gives us

$$\ln^{-1}(\ln(\eta)) = \eta = \ln^{-1}\left(-\frac{b_0}{b_i}\right) = e^{-\left(\frac{b_0}{b_1}\right)} \tag{5.24}$$

$\beta := b_1$ $\beta = 2.201$

$\eta := e^{-\left(\frac{b_0}{\beta}\right)}$ $\eta = 9S.69S$

FIGURE 5.11
Single Weibull parameters: hex bolt.

The characteristic life suggests that the measure of central tendency for the hex bolt's useful life is 99 h—13 h more than the exponential mean TTF of 86 h. The Weibull expressions for the pdf, $f_w(t)$, of hex bolt failure; survival function, $S_w(t)$; and mission reliability function, $R_w(t)$, are written in MathCAD, as shown:

$$f_w(t) = \left(\frac{\beta}{\eta}\right)\left(\frac{t}{\eta}\right)^{\beta-1} e^{-\left(\frac{t}{\eta}\right)^{\beta}}$$

(5.25)

$$S_w(t) = e^{-\left(\frac{t}{\eta}\right)^{\beta}}$$

(5.26)

$$R_w(\tau \mid t) = \frac{e^{-\left(\frac{t+\tau}{\eta}\right)^{\beta}}}{e^{-\left(\frac{t}{\eta}\right)^{\beta}}}$$

(5.27)

MathCAD requires that the mission reliability expression be written for the independent variable, $R_w(t)$, rather than $R_w(\tau \mid t)$.

The Weibull reliability expression, $R_w(t)$, is evaluated over the continuous random variable, t. The single failure mechanism Weibull survival and reliability functions are plotted in Figure 5.12. The survival function suggests that the maximum lifetime of the bolt is $\text{TTF}_{\text{MAX}} = 250$ h. The calculation for bolt mission reliability for five selected system missions is shown on the survival and reliability functions' plots: first mission, where $t = 0$; three missions

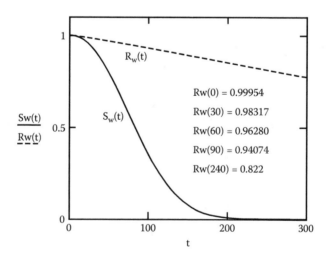

FIGURE 5.12
Single Weibull survival and reliability plots: hex bolt.

at 30 h intervals, $t = 30$, 60, and 90 h; and a mission near the maximum bolt lifetime, TTF$_{MAX}$, $t = 240$ h. Note how the mission reliability decreases over time as the hex bolt ages. This appears to make more sense than the constant reliability over time provided by the exponential probability distribution approach. The probability that a bolt will function without failure for a 3 h mission on the condition that it survives to 30 h is expected to be 98.3%, 96.3% at $t = 60$ h, 94.1% at 90 h, and 82.2% at 240 h.

The Weibull failure TTF pdf of failure, $f_w(t)$, and hazard function, $h_w(t)$, are plotted in Figure 5.13. The hazard function plot is repeated at a scale that provides more information about its behavior over time. The pdf suggests that the probability of bolt failure is near zero at TTF$_{MAX}$. The hazard function suggests that the bolt instantaneous failure rate is 0.068 failures/hour at TTF$_{MAX}$ and continues to increase beyond TTF$_{MAX}$.

The Weibull pdf and hazard functions are both equal to zero at $t = 0$, unlike the corresponding exponential pdf and hazard function, which are nonzero and equal at $t = 0$.

The complete expressions for the single failure mechanism, TTF-based reliability functions are summarized in Table 5.2. The cumulative hazard function suggests that 100% of the bolts will fail by 800 h.

Now is a good time to inquire whether the exponential and Weibull distribution using time-to-failure data and assuming a single failure mechanism provides an accurate characterization of the reliability functions of the bolt.

- We have observed that the two approaches yield different characterizations of failure and the reliability functions.
- Engineering judgment suggests that the Weibull approach is more realistic because it appears to model wear-out failure behavior.
- The Weibull instantaneous failure rate increases over time as the hex bolt ages, but engineering judgment tells us that the upper limit is not infinity. We can explore the behavior of the hazard function by calculating the cumulative hazard function. The Weibull cumulative hazard rate, $H_w(t)$, is calculated as the integral of the hazard function,

$$H_w(t) = \int h_w(t)\,dt \qquad (5.28)$$

and is plotted in Figure 5.14.

The cumulative hazard function suggests that 100% of the bolt population fails at 800 h.

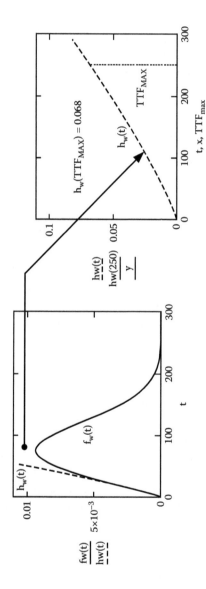

FIGURE 5.13
Single Weibull pdf and hazard function plots: hex bolts.

TABLE 5.2

Single Weibull Reliability Model Functions: Hex Bolt

Reliability Functions	General Expression	Reliability Model Functions
Probability density function of TTF	$f_w(t) = \dfrac{\beta}{\eta}\left(\dfrac{t}{\eta}\right)^{\beta-1} e^{-\left(\frac{t}{\eta}\right)^{\beta}}$	$f_w(t) = 9.0 \times 10^{-5} t^{1.2} e^{-\left(\frac{t}{98.7}\right)^{2.2}}$
Survival function	$S_w(t) = e^{-\left(\frac{t}{\eta}\right)^{\beta}}$	$S_w(t) = e^{-\left(\frac{t}{98.7}\right)^{2.2}}$
Reliability function	$R_w(\tau/t) = \dfrac{S_w(t/\tau)}{S_w(t)} = \dfrac{e^{-\left(\frac{t+\tau}{\eta}\right)^{\beta}}}{e^{-\left(\frac{t}{\eta}\right)^{\beta}}}$	$R_w(t) = \dfrac{e^{-\left(\frac{t+3}{98.7}\right)^{2.2}}}{e^{-\left(\frac{t}{98.7}\right)^{2.2}}}$
Hazard function	$h(t) = \dfrac{\beta}{\eta}\left(\dfrac{t}{\eta}\right)^{\beta-1}$	$h_w(t) = 9.0 \times 10^{-5} t^{1.2}$

But:

- The Weibull reliability functions, though statistically correct, do not pass the "common sense test" for engineering analysis or field data for the following reasons:
  - The assumption that a single failure mechanism causes bolt failure is probably wrong. As previously noted, statistical parameters and analytical methods can only be used for a single population of data—a lone failure mechanism. The statistical analysis is worthless if two or more failure mechanisms are at work on the bolt.

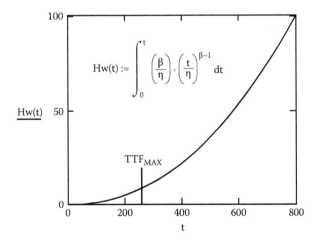

FIGURE 5.14
Cumulative Weibull hazard function.

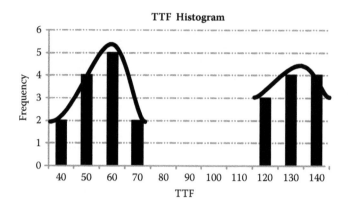

FIGURE 5.15
TTF data and histogram: hex bolt.

- The assumption that time to failure is the best independent variable to explain the frequency of failures is probably wrong. As previously noted, time does not cause failure; failure mechanisms cause failure. The statistical analysis is worthless if the independent variable is not the cause.

The reliability functions will be characterized using the existing time-to-failure data for multiple failure mechanisms, on the assumption that they are the only information available.

Multiple Failure Mechanism Weibull Model Approach

Plotting the frequency distribution of the TTF data at the beginning of the failure data analysis is the best-practice approach to reliability modeling. A frequency distribution of the TTF, as shown in Figure 5.15, provides visual evidence that the TTF data are bimodal. Two failure mechanisms are acting on the hex bolt.

A fault tree analysis (FTA) determines the identities of the failure mechanisms. The FTA for the bolt finds that shear and tensile stresses act on the bolt and cause it to fail, as shown in Figure 5.16. An operations and maintenance timeline developed from the historical data is another tool that provides understanding of the behavior of the bolt by showing operating time and maintenance downtime, as shown in Figure 5.17. Time to failure is the interval between maintenance downtime.

Estimations for mean time to failure, mean time to repair, mean downtime, and availability are based on the events portrayed in the time line. A table of time to failure by failure mode is provided in MS Excel, as shown in Figure 5.18. The historical TTF data for the two failure mechanisms are fit to Weibull distributions. The MS Excel TTF data for the two failure mechanisms, tension and shear, are imported to data tables in MathCAD, as shown in Figure 5.19. The data are rank ordered in MS Excel.

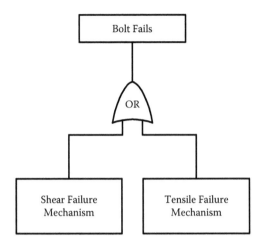

FIGURE 5.16
Fault tree analysis: bolt fails.

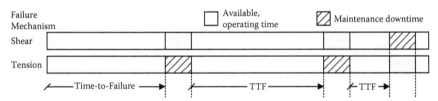

FIGURE 5.17
Failure mode time line: hex bolt.

TTF tension	TTF shear
39	122
47	131
55	127
55	119
56	115
63	129
56	135
61	139
37	126
53	140
49	118
41	
50	

FIGURE 5.18
Time-to-failure data by failure mode: hex bolt.

Rank Ordered TTF in Excel

Tension = _____   Shear = _____

	0
0	37
1	...

	0
0	122
1	...

FIGURE 5.19
Tension and shear rank ordered TTF data: hex bolt.

The independent variables for each failure mechanism, "ln(Tension)" and "ln(Shear)," are calculated in MathCAD in column matrices designated as $X_{tension}$ and X_{shear}, as shown in Figure 5.20. We do not need to show the two independent variable matrices as in the single failure mechanism Weibull model. The column matrices ranging from $1 - n$ and $1 - m$ are designated as "indextension" and "indexshear," respectively, in MathCAD, as shown in Figure 5.21.

The cumulative probability percent estimators are calculated in MathCAD in column matrices designated as "$F_{tension}$" and "F_{shear}," as shown in Figure 5.22. These are the y-axes for a probability paper used to evaluate sample data against the expected cdf, F, for a selected probability distribution (exponential, Weibull, normal, etc.). The values for n and m are calculated using the "length(Tension)" and "length(Shear)" commands. The dependent variables, $Y_{tension}$ and Y_{shear}, are calculated in column matrices designated as "$Y_{tension}$" and "Y_{shear}."

The parameters of the median ranks regressions for tension and shear, y-intercept, defined as $tensionb_0$ and $shearb_0$, and slope, defined as $tensionb_1$ and $shearb_1$, are calculated in MathCAD using the "intercept($X_{tension}, Y_{tension}$)" and "intercept(X_{shear}, Y_{shear})" and "slope($X_{tension}, Y_{tension}$)" and "(slope(X_{shear}, Y_{shear})" commands, as shown in Figure 5.23. The median rank regression is expressed as the linear equation, $Y = b_0 + b_1X$, where Y is expressed as "$MRR_{tension}(X_{tension})$" and "$MRR_{shear}(X_{shear})$." The median ranks regression is evaluated by the coefficients of correlation, "corr($X_{tension}, Y_{tension}$)" and "corr(X_{shear}, Y_{shear})," and determination, calculated by $Rsq_{tension} = $ "corr($X_{tension}, Y_{tension}$)^2" and $Rsq_{shear} = $ "corr(X_{shear}, Y_{shear})^2."

The coefficients of determination ($Rsq_{tension}$ and Rsq_{shear}) suggest that both failure mechanisms are good estimators of the hex bolt survival function, but that tension is a more statistically significant cause of failure than shear. Hex bolt reliability will be increased by selection of a hex bolt with a stronger tensile strength. The median ranks regression is plotted in Figure 5.24.

The median ranks regression plot shows the relative magnitudes of the survival of hex bolts due to tension and shear where tension is the statistically significantly lower of the two. Bolt reliability will be increased by shifting the line to the right. We continue with the characterization of the Weibull reliability models for tension and shear by calculating the Weibull parameters

$$Xtension := \ln(Tension) \qquad\qquad Xshear := \ln(Shear)$$

FIGURE 5.20
Tension and shear independent variable, X: hex bolt.

$$
\text{index tension} := \begin{pmatrix} 1 \\ 2 \\ 3 \\ 4 \\ 5 \\ 6 \\ 7 \\ 8 \\ 9 \\ 10 \\ 11 \\ 12 \\ 13 \end{pmatrix} \qquad \text{index shear} := \begin{pmatrix} 1 \\ 2 \\ 3 \\ 4 \\ 5 \\ 6 \\ 7 \\ 8 \\ 9 \\ 10 \\ 11 \end{pmatrix}
$$

FIGURE 5.21
Tension and shear index: hex bolts.

$$
\text{Ftension} := \frac{\text{indextension} - 0.3}{\text{length(Tension)} + 0.4} \qquad \text{Fshear} := \frac{\text{indexshear} - 0.3}{\text{length(shear)} + 0.4}
$$

$$
\text{Ytension} := \ln\left(\ln\left(\frac{1}{1 - \text{Ftension}}\right)\right) \qquad \text{Yshear} := \ln\left(\ln\left(\frac{1}{1 - \text{Fshear}}\right)\right)
$$

FIGURE 5.22
MathCAD tension and shear cdf and dependent variable, Y: hex bolt.

$\text{tensionb}_0 := \text{intercept(Xtension, Ytension)}$ $\text{shearb}_0 := \text{intercept(Xshear, Yshear)}$

$\text{tensionb}_1 := \text{slope (Xtension, Ytension)}$ $\text{shearb}_1 := \text{slope(Xshear, Yshear)}$

$\text{MRRtension(Xtension)} := \text{tensionb}_0 + \text{tensionb}_1 \cdot \text{Xtension}$

$\text{MRRshear(Xshear)} := \text{shearb}_0 + \text{shearb}_1 \, \text{Xshear}$

$\text{corr(Xtension, Ytension)} = 0.982$

$\text{Rsqtension} := \text{corr(Xtension, Ytension)}^2$ $\text{Rsqtension} = 0.964$

$\text{Corr(Xshear, Yshear)} = 0.862$

$\text{Rsqshear} := \text{corr(Xshear, Yshear)}^2$ $\text{Rsqshear} = 0.744$

FIGURE 5.23
MathCAD tension and shear median ranks regression parameters: hex bolt.

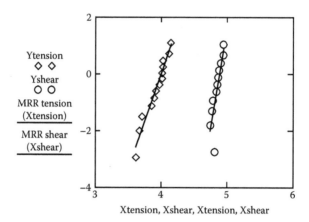

FIGURE 5.24
MathCAD tension and shear median ranks regression plots: hex bolt.

in MathCAD, as shown in Figure 5.25. The shape parameters, "β_{tension}" and "β_{shear}," are equal to the slopes of the respective median ranks regression lines. The characteristic life, "η_{tension}" and "η_{shear}," are calculated as the antilog$_e$ of the ratios of the respective y-intercept (b_0) and line slope (b_1).

The Weibull expressions for the reliability functions are summarized in Table 5.3 for each failure mechanism.

The characteristic lives, "η_{tension}" and "η_{shear}," suggest that the measures of central tendency for the hex bolt's useful life are 54 h under tension and 132 h under shear. The larger characteristic life for shear suggests that selection of a bolt with higher tensile strength will increase the survival of the bolt and decrease the instantaneous hazard rate of the bolt, thereby improving the bolt reliability. Conversely, selection of a bolt with higher shear strength will not significantly increase the survival, decrease the instantaneous hazard rate, or improve the bolt reliability.

The Weibull probability density functions for bolt failure in tension and shear are plotted in MathCAD in Figure 5.26. The Weibull pdf for bolt failure for tension or shear is calculated as the probability that the bolt fails, P(bolt), in tension, P(tension), OR fails in shear, P(shear):

$$P(\text{Bolt}) = P(\text{Tension OR Shear}) = P(\text{Tension}) + P(\text{Shear}) - P(\text{Tension})P(\text{Shear})$$

(5.29)

$$\beta\text{tension} := \text{tension}b_1. \qquad \beta\text{tension} = 6.685 \qquad \beta\text{shear} := \text{shear}b_1 \qquad \beta\text{shear} = 14.654$$

$$\eta\text{tension} := e^{-\left(\frac{\text{tension}b_0}{\beta\text{tension}}\right)} \qquad\qquad \eta\text{shear} := e^{-\left(\frac{\text{shear}b_0}{\beta\text{shear}}\right)}$$

$$\eta\text{tension} = 54.459 \qquad\qquad \eta\text{shear} = 131.762$$

FIGURE 5.25
MathCAD tension and shear Weibull parameters: hex bolt.

TABLE 5.3

Failure Mechanism Reliability Functions

Reliability Functions	Shear Failure Mode	Tension Failure Model
Probability density function	$f_{shear}(t) = 1.27 \times 10^{-3} t^{13.65} e^{-\left(\frac{t}{131.8}\right)^{14.65}}$	$f_{tension}(t) = 1.67 \times 10^{-11} t^{5.69} e^{-\left(\frac{t}{54.46}\right)^{6.69}}$
Survival function	$S_{shear}(t) = e^{-\left(\frac{t}{131.8}\right)^{14.65}}$	$S_{tension}(t) = e^{-\left(\frac{t}{54.46}\right)^{6.69}}$
Reliability function	$R_{shear}(t) = \dfrac{e^{-\left(\frac{t+3}{131.8}\right)^{14.65}}}{e^{-\left(\frac{t}{131.8}\right)^{14.65}}}$	$R_{tension}(t) = \dfrac{e^{-\left(\frac{t+3}{54.46}\right)^{6.69}}}{e^{-\left(\frac{t}{54.46}\right)^{6.69}}}$
Hazard function	$h_{shear}(t) = 1.27 \times 10^{-30} t^{13.65}$	$h_{tenion}(t) = 1.67 \times 10^{-11} t^{5.69}$

The pdf, $f_{tension}(t)$, is the continuous probability that the bolt fails in tension, and $f_{shear}(t)$ is the continuous probability that the bolt fails in shear. Therefore, the continuous probability that the bolt fails, $f_{bolt}(t)$, is expressed as

$$f_{bolt}(t) = (f_{tension}(t)) + (f_{shear}(t)) - (f_{tension}(t))(f_{shear}(t)) \qquad (5.30)$$

The bolt pdf, $f_{bolt}(t)$, exactly overlays the two failure mechanisms' probability density functions, $f_{tension}(t)$ and $f_{shear}(t)$. This failure behavior is multimodal and suggests that using the Weibull to model reliability functions by failure mechanism will still be an unacceptable approach.

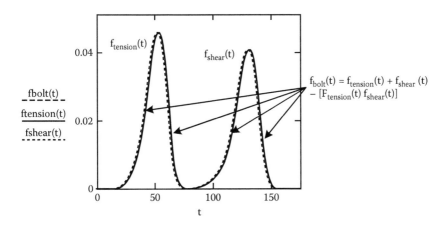

FIGURE 5.26
Tension and shear probability density functions and bolt pdf plots.

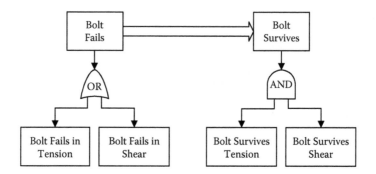

FIGURE 5.27
FTA for bolt failure and bolt survival.

The calculation of the bolt survival function, $S_{bolt}(t)$, is based on the fault tree analysis logic, as shown in Figure 5.27. The bolt can either fail or survive. The bolt survives when it survives in tension AND survives in shear.

The probability that the bolt survives is expressed as

$$P(Survive) = P(Survive\ Tension)P(Survive\ Shear). \qquad (5.31)$$

The continuous probability of bolt survival, $S_{bolt}(t)$, is calculated as the product of the survival function in tension, $S_{tension}(t)$, and the survival function in shear, $S_{shear}(t)$, as shown:

$$S_{bolt}(t) = (S_{tension}(t))(S_{shear}(t)) \qquad (5.32)$$

The bolt reliability function, $R_{bolt}(t)$, is the conditional probability that the bolt survives to the time to completion of the next mission, $t + \tau$, given that the bolt has survived to the beginning of the next mission, t (Figure 5.28). The

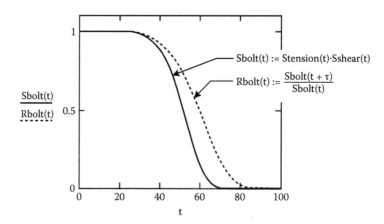

FIGURE 5.28
Bolt survival and mission reliability plots.

$$\theta\text{bolt} := \int_0^\infty (S\text{bolt}(t))\, dt$$

$$\theta\text{bolt} = 50.82$$

FIGURE 5.29
Bolt mean time between failure.

bolt continuous survival and reliability functions appear to be dominated by the design "weak link": tension.

The part mean time between failure (MTBF), θ, is defined as the indefinite integral of the survival function,

$$\theta = \int S(t)\,dt = \int_0^\infty S(t)\,dt \tag{5.33}$$

The lower boundary, $t = 0$, is based on the fact that a part cannot experience negative time. The MTBF for the bolt is calculated in MathCAD as shown in Figure 5.29. The value of the bolt MTBF confirms that the tension failure mechanism dominates the behavior of bolt failure.

The single failure mechanism hazard function, $h_{bolt}(t)$, is plotted with calculation in MathCAD and plots of the tension and shear hazard functions, $h_{tension}(t)$ and $h_{shear}(t)$, as shown in Figure 5.30. The Weibull bolt hazard function shown in the figure is dominated by the tension hazard function.

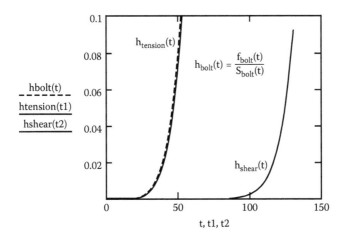

FIGURE 5.30
Single failure mechanism, tension, and shear hazard functions: hex bolt.

Comparative Evaluation of Exponential, Single Weibull, and Multiple Failure Mechanism Weibull Model Approaches using TTF Data

The survival functions for the exponential single failure mechanism, Weibull single failure mechanism, and Weibull multiple failure mechanisms provided significantly different results. Shortcomings of the exponential approach to characterize the behavior of failure for a part that experiences wear out as it ages have already been discussed. The exponential survival function is not shaped the way that we know a structural or dynamic material degrades.

Of the two Weibull approaches, the single failure mechanism survival function appears to overstate the useful life we would expect from the TTF data. The tension and shear failure mechanism survival function appears to be dominated by the tension failure mechanism. The respective mission reliability functions mirror the survival functions. There appears to be very little difference between the mission duration reliability functions for the exponential and the single failure mechanism Weibull approaches. An engineer should be concerned that there is such a large difference between the mission duration reliability functions for single failure mechanism Weibull approach and the tension and shear failure mechanisms' approach Weibull approaches.

Part of the answer can be found by comparing the characteristic lives for the two approaches. The single failure mechanism characteristic life, η_{bolt}, is between the tension and shear characteristics lives, $\eta_{tension}$ and η_{shear}. The location of η_{bolt} is directly related to the distribution and frequency of the tension and shear TTF data and can be viewed as a complex weighted average of the two. The value of η_{bolt} would be closer to $\eta_{tension}$ if the frequency of tension TTF data were doubled, holding the frequency of shear TTF data constant. The effects of both tension and shear TTF data are factored in the shape of the tension and shear failure mechanisms mission reliability function, $R_{bolt}(t)$.

The exponential and single failure mechanism Weibull hazard functions, λ and $h_w(t)$, do not rationally explain the behavior of hex bolt failure. The tension and shear failure mechanisms' Weibull hazard function is dominated by tension. Hazard functions are unique to failure mechanisms. A part with two or more failure mechanisms does not have a single hazard function.

The preceding three TTF-based reliability models show statistically significant differences between the findings. One might conclude that the Weibull TTF reliability model is more accurate than the exponential TTF reliability model, and that the Weibull TTF reliability model for the combined failure mechanisms (shear and tension) is more accurate than the Weibull TTF model for only the TTF data.

Or, all three may be wrong.

The failure pdf plots for the two failure mechanisms suggest that the shear failure mechanisms would never have happened; 100% of the bolts would have failed in tension. Logically, from the TTF statistical analysis, the only way shear failure can occur is that tensile failure does not occur. But almost as many hex bolts failed in shear (11) as in tension (13). The explanation of this paradox is simple; failure was not caused by time in use but rather by stresses exceeding strength. The parameters of the part reliability models should be stress and strength, rather than TTF.

Part Stress and Strength: Interference Theory

Interference theory posits that stress and strength are statistically distribute and therefore have an overlap on the stress-strength axis, as shown in the notional plot in Figure 5.31. The overlap characterizes the probability that stress exceeds strength, causing part failure.

The basis for interference theory is the spread between the points esti-mates for stress and strength. Engineers are familiar with the use of safety factors (k) in design analysis (Figure 5.32). Stress is given in the specification or allocated by analysis from a higher design level stress specification. A safety factor is determined from the specifications by system engineering. Design analysis uses the safety factor to determine the minimum strength property for material or part selection.

> *The safety factor implies that a part exceeds the design specification and will not fail. This implication is false!*

All stress estimates are point estimations of the mean or maximum loads applied by failure mechanisms. The strength of a material is a point

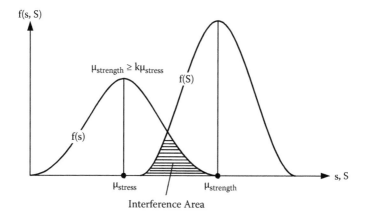

FIGURE 5.31
Interference theory.

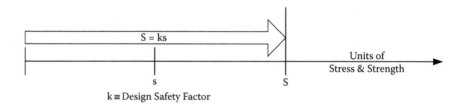

FIGURE 5.32
Factor of safety.

estimate of the material property that opposes the stress load. Each point estimate has a corresponding measure of dispersion. Therefore, stress and strength metrics are statistically distributed. Theoretically, strength (S) has a probability distribution of $f_S(s)$ and stress (s) has a probability distribution of $f_s(s)$. The reliability of a material is the probability that as stress loads are applied, the strength of the material will exceed the stress loads, as shown:

$$R = \int f_S(s) \left[\int f_s(s) ds \right] ds \tag{5.34}$$

The integral of the stress distribution is the cumulative probability distribution of stress, $F_s(s)$:

$$\int f_s(s) ds = F_s(s) \tag{5.35}$$

The material reliability expression can be simplified as follows:

$$R = \int f_S(s) F_s(s) ds \tag{5.36}$$

The justifying bases for interference theory are the theories of tensile, shear, bending, and torsional failure that stipulate that failure occurs when the stress load exceeds the strength of the material to oppose the load.

The most likely distributions for stress and strength are skewed distributions. Stress is a lower-is-best criterion and is empirically found to be positively skewed; strength is a higher-is-best criterion and is empirically found to be negatively skewed. Both are well represented by the Weibull probability distribution. But the literature often shows both to be treated as normally distributed. This can be explained by the ease of use of the normal distribution, as shown in Figure 5.33.

Normal Stress–Normal Strength

The ease of use is illustrated by the general expression for the material reliability that permits use of the standard normal probability tables. No

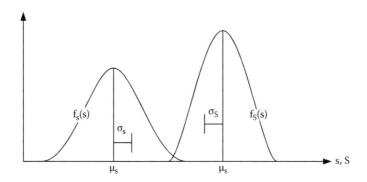

FIGURE 5.33
Normal stress–normal strength.

complicated math operations are required. Consider a material that has a mean strength (μ_S) of 20 kpsi, a standard deviation (σ_S) of 2.5 kpsi, and a mean stress ($\mu_{s\_}$) of 10 kpsi with a standard deviation (σ_s) of 2 kpsi. The z-statistic is calculated to be

$$z = \frac{\mu_S - \mu_s}{\sqrt{\sigma_S^2 + \sigma_s^2}} = \frac{20 - 10}{\sqrt{2.5^2 + 2^2}} = 3.123475$$

We find the fixed, deterministic unreliability, $U = \Phi(-z) = \Phi(-3.123475) = 0.000894$, using MS Excel; therefore, the fixed, deterministic $R = 1 - U = 0.9^31$.

The difference between the mean stress and strength implies a safety factor of 2. We can infer that from the unreliability of the material design specification (0.09%) that 1 part out of each 1,000 will fail on the first mission. How many of the survivors will fail on the second, third, tenth, one-hundredth mission? We do not know. The material properties of the material will have changed from wear-out modes due to successive exposures to failure mechanisms.

Another question that must be posed is how we knew the parameters of the normal probability distributions for stress and strength. Are the parameter estimates related directly to the conditions of use? Empirically characterized from historical or experimental data? Global estimators from a vendor catalogue? We do not know.

A common error is to use the central limit theory that states that the mean calculated from a sample size greater than 30 can be assumed to be normally distributed. The central limit theory applies only to the sampling distribution of the mean—not to the actual distribution of the population from which the samples were drawn. Interference theory is applied to the pdf of the sample—not the sampling distribution of the mean of the sample.

$\mu s := 10 \quad \sigma s := 2$ $\eta s := 10$

$\eta S := 20 \quad \beta S := 9.75$ $\beta s := 19$

$$fns(x) := \left(\frac{1}{\sigma s \times \sqrt{2 \times \pi}}\right) \cdot e^{\left(\frac{1}{2}\right)\left(\frac{x-\mu s}{\sigma s}\right)^2} \qquad fwS(x) := \left(\frac{\beta S}{\eta S}\right) \cdot \left(\frac{x}{\eta S}\right)^{\beta S-1} \cdot e^{-\left(\frac{x}{\eta S}\right)^{\beta S}}$$

FIGURE 5.34
Normal stress–Weibull strength.

We may also question whether the reliability is accurate because both distributions exist in the range [−∞, ∞]. But the interference area is assumed to exist only in the positive axis.

Normal Stress–Weibull Strength

An improvement on the assumption of the normal distribution is to fit the strength probability distribution to a Weibull probability distribution. Stress is still assumed to be normally distributed with $\mu_s = 10$ kpsi and $\sigma_s = 2$ kpsi. Empirical data fit strength to a Weibull probability distribution with parameters $\eta_S = 20$ kpsi and $\beta_S = 9.75$. The probability density functions for normal stress, $f_{ns}(t)$, and Weibull strength, $f_{wS}(t)$, are written in MathCAD, as shown in Figure 5.34.

The normal pdf for stress, $f_{ns}(x)$, and Weibull pdf for strength, $f_{wS}(t)$, are plotted in MathCAD, as shown in Figure 5.35. The independent variable for

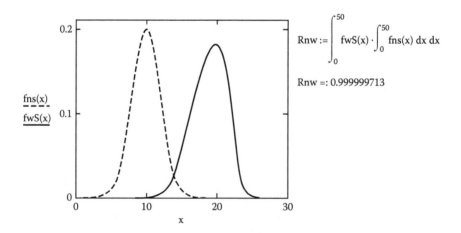

$$Rnw := \int_0^{50} fwS(x) \cdot \int_0^{50} fns(x) \, dx \, dx$$

$$Rnw =: 0.999999713$$

FIGURE 5.35
Normal stress–Weibull strength interference area plot.

$$\eta S := 20 \qquad\qquad\qquad \eta s := 10$$

$$\beta S := 9.75 \qquad\qquad\qquad \beta s := 1.9$$

$$fws(x) := \left(\frac{\beta s}{\eta s}\right) \cdot \left(\frac{x}{\eta s}\right)^{\beta s - 1} \cdot e^{-\left(\frac{x}{\eta s}\right)^{\beta s}} \qquad fwS(x) := \left(\frac{\beta S}{\eta S}\right) \cdot \left(\frac{x}{\eta S}\right)^{\beta S - 1} \cdot e^{-\left(\frac{x}{\eta S}\right)^{\beta S}}$$

FIGURE 5.36
Weibull stress–Weibull strength.

stress and strength is stated as $x \equiv$ units of load. The integral of $f_{wS}(t)$ over the integral of $f_{ns}(t)$ is shown with a fixed, deterministic solution for reliability, $R_{nw} = 0.9^67$.

Weibull Stress–Weibull Strength

The ideal approach is to fit the stress and strength probability distributions to a Weibull probability distribution. Empirical data fit stress to a Weibull probability distribution, $f_{ws}(t)$, with parameters $\eta_s = 10$ kpsi and $\beta_s = 1.9$, and strength to a Weibull probability distribution, $f_{wS}(t)$, with parameters $\eta_S = 20$ kpsi and $\beta_S = 9.75$, as shown in Figure 5.36.

The Weibull pdf for stress, $f_{ws}(x)$, and Weibull pdf for strength, $f_{wS}(t)$, are plotted in MathCAD, as shown in Figure 5.37. The integral of $f_{wS}(t)$ over the integral of $f_{ws}(t)$ is shown with a fixed, deterministic solution for reliability, $R_{ww} = 0.9^37$.

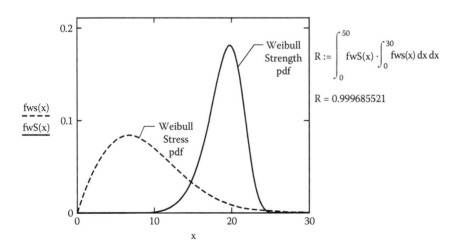

$$R := \int_0^{50} fwS(x) \cdot \int_0^{30} fws(x) \, dx \, dx$$

$$R = 0.999685521$$

FIGURE 5.37
Weibull stress–Weibull strength interference area plot.

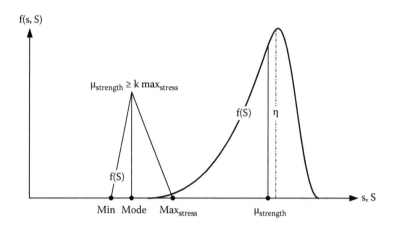

FIGURE 5.38
Triangular stress—Weibull strength interference plot.

Triangular Stress–Weibull Strength

Knowledge of stress loads is rarely ideal and stress cannot be sufficiently understood to characterize a normal distribution. Our understanding of stress loads can be characterized to fit a triangular distribution through understanding of the operating and empirical conditions of use or through Adelphi survey of subject matter experts. Material strength can always be empirically characterized and fit to a Weibull probability distribution. We can illustrate the relationship between stress and strength as shown in Figure 5.38.

The stress triangular distribution is closed form with pdf = 0 for stress less than min stress and for stress greater than max stress. The triangular pdf, $f_{ts}(t)$, takes on values over the range [Stress$_{Min}$, Stress$_{Max}$]. The parameters of the stress triangular distribution are Stress$_{Min}$, Stress$_{Mode}$, and Stress$_{Max}$. The strength Weibull distribution is bounded at stress = 0 and open at max strength. The strength Weibull pdf, $f_{wS}(t)$, takes on values over the range [0, ∞]. The interference area is bounded by the strength pdf from 0 to max stress and takes on values ranging from [0, Stress$_{Max}$].

Consider a design analysis where the tensile stress is expected to be x_{mode} = 20 kpsi, with peak loading at x_{max} = 30 kpsi and a system idle load of x_{min} = 15 kpsi. The safety factor is calculated at two times the peak load for a minimum strength of 60 kpsi. The selected material is empirically tested to characterize and fit the strength Weibull pdf with parameters of characteristic life, η = 60 kpsi, and shape parameter, β = 12.375 (Figure 5.39).

The plots for the stress and strength probability density functions are illustrated in Figure 5.40. The unreliability of the material can be closely approximated in two ways. First, the area of the inference area, U, is found in MathCAD by integrating the strength Weibull pdf, $f_{wS}(t)$, over the range of

$$\text{xmode} := 20$$
$$\text{xmax} := 30$$
$$\text{xmin} := 15$$

$$ft(x) := \begin{vmatrix} \dfrac{2 \cdot (x - \text{xmin})}{(\text{xmax} - \text{xmin}) \cdot (\text{xmode} - \text{xmin})} & \text{if } \text{xmin} \le x \le \text{xmode} \\[2ex] \dfrac{2 \cdot (\text{xmax} - x)}{(\text{xmax} - \text{xmin}) \cdot (\text{xmax} - \text{xmode})} & \text{if } \text{xmode} < x \le \text{xmax} \end{vmatrix}$$

$$\eta := 60$$
$$\beta := 12.375$$

$$fw(x) := \left(\frac{\beta}{\eta}\right) \cdot \left(\frac{x}{\eta}\right)^{\beta-1} \cdot e^{-\left(\frac{x}{\eta}\right)^{\beta}}$$

FIGURE 5.39
Triangular stress—Weibull strength example in MathCAD.

the interference area, $[0, x_{\text{max}}]$. The reliability of the material design is calculated as $R = 1 - U$.

Second, the unreliability is found by solving for the cumulative strength distribution for $\text{Stress}_{\text{Max}}$. The reliability is calculated as $R = 1 - U$, or $R = 1 - F_w(x_{\text{max}})$. $F_w(x_{\text{max}}) = 1 - e^{-\left(\frac{x}{\eta}\right)^{\beta}}$

We can characterize the reliability for a critical part where empirical strength data are not available; yet, an approximation for the material design reliability is required. The triangular distribution provides the placeholder that will allow an estimate that is within a reasonable order of magnitude.

Consider an as yet uncharacterized material. It is estimated to have a modal strength of 60 kpsi, a minimum strength of 35 kpsi, and a maximum strength of 66 kpsi, as shown in Figure 5.41. We illustrate the relationship between stress and strength and find that there is no interference area. The unreliability is zero and the reliability is one. But no such material exists, and this finding is not helpful. We can better estimate the strength distribution using MathCAD to fit an overlay Weibull distribution by setting the characteristic

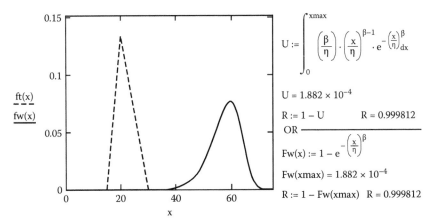

$$U := \int_0^{\text{xmax}} \left(\frac{\beta}{\eta}\right) \cdot \left(\frac{x}{\eta}\right)^{\beta-1} \cdot e^{-\left(\frac{x}{\eta}\right)^{\beta}} dx$$

$$U = 1.882 \times 10^{-4}$$

$$R := 1 - U \qquad R = 0.999812$$

OR

$$Fw(x) := 1 - e^{-\left(\frac{x}{\eta}\right)^{\beta}}$$

$$Fw(\text{xmax}) = 1.882 \times 10^{-4}$$

$$R := 1 - Fw(\text{xmax}) \quad R = 0.999812$$

FIGURE 5.40
Triangular stress–Weibull strength interference area plot.

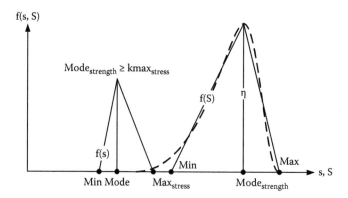

FIGURE 5.41
Triangular stress–triangular strength.

life (η) at the modal strength and varying the shape parameter (β) until we see a fit like the one illustrated in Figure 5.41.

The approximation of the unreliability (U) can be found by solving the strength cdf for max stress, and solving for the reliability from the complement of U, as performed in the preceding example.

Stress-Strength Reliability of the Bolt in Tension and Shear

The characterization of reliability for the bolt in tension and shear, as shown in the free body diagram in Figure 5.42, is performed for the combined stresses. A fault tree analysis defines the logic of the bolt reliability characterization, as shown in Figure 5.43. The bolt will fail in tension or in shear. The probability that the bolt will fail—the unreliability (U_{bolt})—is the sum of the probability for failure in tension ($U_{tension}$) plus the probability for failure in shear (U_{shear}) minus the probability for failure in tension and shear.

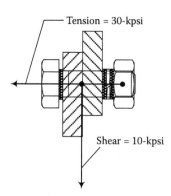

FIGURE 5.42
Hex bolt free body diagram.

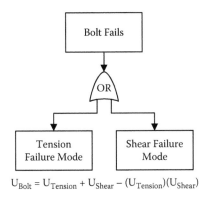

FIGURE 5.43
Bolt fault tree analysis.

The fit for stress and strength distributions for each failure mode is performed by finite element modeling, simulation, and analysis with source inputs from design analysis performed to specify part selection and acquisition. The characteristic life for bolt strength in tension is fit to the Weibull distribution with characteristic life of $\eta_{\text{tension}} = 60$ kpsi and shape parameter of $\beta_{\text{tension}} = 13.45$. The 95% maximum tensile stress ($L_{\text{max\_tension}}$) is determined to be 30 kpsi. The pdf for tensile strength, $f_{\text{tension}}(x)$, and the cumulative probability function for tensile strength, $F_{\text{tension}}(x)$, are written in MathCAD. The unreliability of the bolt in tension (U_{tension}) is found by solving the cumulative probability function for $L_{\text{max\_tension}}$. The reliability of the bolt in tension is calculated to be $R_{\text{tension}} = 1 - U_{\text{tension}} = 0.9^41$ (see Figure 5.44).

The characteristic life for bolt strength in shear is fit to the Weibull distribution with characteristic life of $\eta_{\text{shear}} = 45$ kpsi and shape parameter of $\beta_{\text{tension}} = 6.82$. The 95% maximum shear stress ($L_{\text{max\_shear}}$) is determined to be 10 kpsi. The pdf for tensile stress, $f_{\text{shear}}(x)$, and the cumulative probability

$$\eta\text{tension} := 60$$

$$\beta\text{tension} := 13.45 \quad f\text{tension}(x) := \left(\frac{\beta\text{tension}}{\eta\text{tension}}\right)^{\beta\text{tension}-1} .e^{-\left(\frac{x}{\eta\text{tension}}\right)^{\beta\text{tension}}}$$

$$L\text{max\_tension} := 30$$

$$F\text{tension}(x) := 1 - e^{-\left(\frac{x}{\eta\text{tension}}\right)^{\beta\text{tension}}}$$

$$F\text{tension}(L\text{max\_tension}) = 8.936 \times 10^{-5}$$
$$R\text{tension} := 1 - F\text{tension}(L\text{max\_tension})$$
$$R\text{tension} = 0.999911$$

FIGURE 5.44
Bolt reliability in tension.

$\eta Shear := 45$

$$\beta shear := 6.82 \qquad fshear(x) := \left(\frac{\beta shear}{\eta shear}\right) \cdot \left(\frac{x}{\eta shear}\right)^{\beta shear-1} \cdot e^{-\left(\frac{x}{\eta shear}\right)^{\beta shear}}$$

$Lmax\_shear := 10$

$$Fshear(x) := 1 - e^{-\left(\frac{x}{\eta shear}\right)^{\beta shear}}$$

$Fshear(Lmax\_shear) = 3.508 \times 10^{-5}$

$Rshear := 1 - Fshear(Lmax\_shear)$

$Rshear = 0.999965$

FIGURE 5.45
Bolt reliability in shear.

function for shear strength, $F_{shear}(x)$, are written in MathCAD. The unreliability of the bolt in shear (U_{shear}) is found by solving the cumulative probability function for L_{max_shear}. The reliability of the bolt in tension is calculated to be $R_{shear} = 1 - U_{shear} = 0.9^47$ (see Figure 5.45).

The plots for the tension and shear probability density functions and the location of the 95% tensile and shear loads are illustrated in Figure 5.46. The plots for the tension and shear probability distributions for bolt failure illustrate why failures in both modes occurred in the historical data. We observe that the conditions for bolt failure by either failure mechanism are exclusively functions of operational loading rather than to time in service.

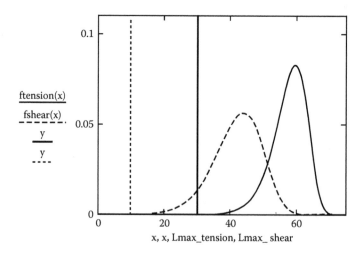

FIGURE 5.46
Tension and shear probability density functions with locations for respective 95% loads.

Ubolt := Ftension(Lm ax_tension) + Fshear(Lm ax_shear) – (Ftension(Lm ax_
tension)) (Fshear(Lm ax_shear))

Ubolt = 1 244 × 10⁻⁴

Rbolt := 1 – Ubolt

Rbolt = 0 999276

FIGURE 5.47
Bolt reliability calculation.

The deterministic, fixed reliability of the bolt is solved in MathCAD to be 0.999876, as shown in Figure 5.47. The reliability of the bolt is less than reliability of the bolt in tension and the reliability of the bolt in shear.

Nondeterministic, Variable Approach

The nondeterministic, variable characterization of reliability functions describes the most realistic analytical modeling of part behavior. The measure of central tendency for strength decreases over time, just as parts are observed to weaken through use. The measure of dispersion for strength increases over time as part variability increases due to part wear out. Concurrently, the stress remains fixed. A nondeterministic, variable interference area is shown graphically in Figure 5.48.

The nondeterministic, variable survival function requires a nonlinear degradation factor for central tendency and a linearly increasing factor for dispersion, unlike the straightforward deterministic, fixed survival function. Research is ongoing to develop and test a nondeterministic, variable stress-strength survival function algorithm.

Advantages and Disadvantages
for Stress-Strength Analysis Approach

The principal advantage for stress-strength-based reliability functions is the understanding of failure mechanisms and the guidance that understanding provides to mitigate failure during the performance of part design analysis. Stress-strength-based reliability functions are achieved through analysis, unlike time-to-failure experiments. It is the least expensive approach that stays on schedule and best achieves technical requirements.

Stress-strength-based reliability provides invaluable information for system sustainment by providing insights into the conditions of failure for critical parts. Condition-based maintenance is defined by the ability to measure and understand a metric for wear out; stress-strength-based reliability provides that insight directly. TTF-based reliability treats all parts with population statistics that provide no insight into part condition.

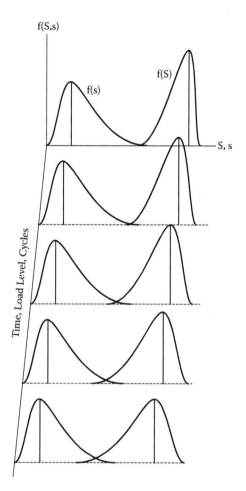

FIGURE 5.48
Nondeterministic, variable interference theory.

Stress-strength-based reliability provides quantified metrics of risk that can be used to define inspection intervals for time-directed maintenance. TTF-based reliability can only offer lower confidence limits of the population, which infuses high costs.

The deterministic, fixed approach relies heavily on the current state of the art for strength of materials applied by design engineers. Professional development courses that enable engineers to fit statistical stress-strength models are needed.

The nondeterministic, variable approach has one fatal disadvantage: It has not become practical yet. The algorithms have yet to be developed that pass scrutiny. The most salient conclusion is that reliability functions characterized from stress-strength analysis should become an essential aspect of the part to system design process. This will assure achievement of the best design solution.

Notes

1. Zaid, M. & R. P. Kolb. 1964. Section 15: Mechanics of materials. In *Mechanical design and systems handbook*. Rothbart, H. A., Ed. New York: McGraw–Hill.
2. Shigley, E. S. 1977. *Mechanical engineering design,* 3rd ed. New York: McGraw–Hill.
3. Shanley, F. R. 1967. *Mechanics of materials.* New York: McGraw–Hill.
4. Hudson, R. G. 1939. *The engineers' manual.* New York: John Wiley & Sons.
5. Brooks, G. E., Eonomides, L., & Winckowski, B.F. 1976. Section 4: Building systems engineering. In *Engineering manual.* New York: Perry, R. H., Ed. McGraw–Hill.
6. Collins, J. A. 1993. *Failure of materials in mechanical design,* 2nd ed. New York: John Wiley & Sons.
7. Ibid.
8. Collins posits that the distortion energy theory is an improvement over the total strain energy theory; current best practices appear to confirm this.
9. The data are copied in MS Excel. MathCAD imports data using the "insert," "data," and "table" commands. The data table is named TTF, and the copied data are pasted in the first cell.
10. Previous versions of MathCAD required placing the numerical solution below the expression; the current version allows the numerical solution tabbed to the right.
11. Statistical software will fit a Weibull with differing parameter estimators depending on the fit method employed, maximum likelihood estimator (MLE) rather than median ranks regression or graphical. Another source of differing parameters is the choice of median rank estimators; Bartlett's median rank estimator is used here.
12. Note that TTF and $S(t)$ are transformed to achieve a linear model but remained the independent and dependent variables.
13. The graphical solution using a Weibull paper constructs the median rank regression line. The shape parameter is calculated as the tangent of the angle of the line, $\tan(q)$, and the characteristic life is constructed by the intersection of the TTF that corresponds to the point of the line for $Y = 63.2\%$.

6

Reliability Engineering Functions from Stress-Strength Analysis

Simplicity is prerequisite for reliability.

Edsger Dijkstra

Introduction

Failure modes that are not characterized by time to failure are defined by failure mechanisms that are characterized by the four states of stress acting on a part: tension, compression, shear, and torsion. Failure mechanisms are physical loads, thermal loads, and chemical reactivity. Failure mechanisms cause changes in material geometry and properties that act to reduce the strength of the material. Successive applications of failure mechanisms age the material. Aging is characterized by reduction in the measure of central tendency of strength of the material and increases in the measure of dispersion of strength of the material, as illustrated in Figure 6.1.

As a part characterized by normally distributed strength ages, the measure of central tendency—the mean strength (μ_S)—decreases, and the measure of dispersion—the standard deviation (σ_S)—increases. As a part characterized by Weibull distributed strength ages, the measure of central tendency—the characteristic life (η_S)—decreases, and the measure of dispersion—the standard deviation (σ_S)—increases. The Weibull standard deviation is negatively correlated to the shape parameter (β) such that σ increases as β decreases. This relationship is illustrated for a characteristic life of $\eta = 1$ in Figure 6.2.

Frequency Distributions of the Mechanisms of Failure

Reliability analyses of failure mechanisms include the frequency distributions for stress and strength. There are two approaches to characterize the frequency distributions of stress and strength: design for reliability and field sustainment.

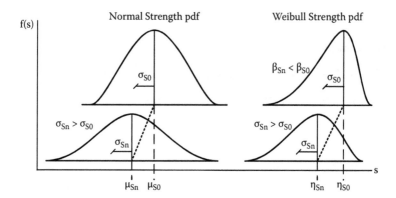

FIGURE 6.1
Measures of central tendency and dispersion as part ages: normal and Weibull pdf.

Design for Reliability

The frequency distribution of stress in design is a straightforward determination of the maximum stress: tensile load, compressive load, shear load, and torsional load (Figure 6.3). The best-practice design method recommends that the material strength must be at least the maximum stress times the safety factor, k: $S = ks_{MAX}$. This relationship is illustrated in Figure 6.4.

It is intuitively obvious that the design information provided by a strength-of-materials specification of at least the product of the maximum stress and a safety factor is insufficient to characterize reliability of the design. The calculated strength is no more than a guideline that is used to select a material.

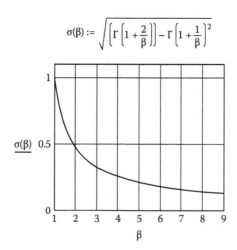

$$\sigma(\beta) := \sqrt{\left[\Gamma\left(1 + \frac{2}{\beta}\right)\right] - \Gamma\left(1 + \frac{1}{\beta}\right)^2}$$

FIGURE 6.2
Negative correlation of Weibull standard deviation and shape parameter.

FIGURE 6.3
Maximum stress load.

The strength distribution of the candidate materials is characterized by physical tests and evaluation.

Assume that a structural body is loaded in tension with $s_{MAX} = 33$ kpsi.[1] Two candidate materials are identified that meet the design requirements to include a safety factor of $k = 1.75$: 1 in. round steel bar and 1 in. square steel bar; both have a rated tensile strength of 58 kpsi. The rated tensile strength is not known to reflect the operating and ambient conditions of use for this design. The failure mechanisms acting on the design function served by the bar include steady-state vibration, $\omega_{MAX} = 8\ g_{RMS}$; random shock, $\Psi_{MAX} = 12\ g_{RMS}$, expected to not exceed six shocks per hour; maximum operating temperature determined by analysis to be $T_{MAX} = 190°F$; and initial maximum operating temperature shock, determined by analysis to be $\Delta T_{MAX} = 190°F - (-18°F) = 208°F$. [Vibration can be expressed by the square-root of the mean square of the acceleration acting on the body g_{RMS}. The mean square acceleration is the average of the square of the acceleration over time.]

An experiment is designed that simulates the wear-out behavior of the candidate materials by exposing them to the design operating and ambient conditions of use with a highly accelerated life test (HALT) chamber. A test cycle is specified, as shown in Table 6.1.

The cycle begins at the lab ambient conditions for temperature and humidity. Humidity is held constant throughout the experiment:

Step 1 exposes the candidate materials to the worst-case starting scenario—in this case, the temperature that will result in the largest change in temperature during system start-up.

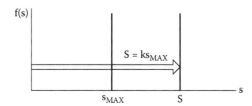

FIGURE 6.4
Deterministic stress strength design method.

TABLE 6.1

Combined Operating and Ambient Conditions of Use Experiment Design

Step	Duration (min)	Operating Temperature	Temperature Change	Vibration	Shock	Note
1	2	Lab ambient	Lab ambient to −18°F	$0\,g_{RMS}$	—	Establish worst-case starting ambient conditions
2	1	−18°F	—	$8\,g_{RMS}$	—	Initiate operating vibration
3	3	—	+208°F	$8\,g_{RMS}$	—	Ramp to maximum operating temperature
4	2	190°F	—	$8\,g_{RMS}$	—	Steady-state maximum operating conditions
5	0.25	190°F	—	$8\,g_{RMS}$	$12\,g_{RMS}$	Random shock
4	2	190°F	—	$8\,g_{RMS}$	—	Steady-state maximum operating conditions
5	0.25	190°F	—	$8\,g_{RMS}$	$12\,g_{RMS}$	Random shock
4	2	190°F	—	$8\,g_{RMS}$	—	Steady-state maximum operating conditions
5	0.25	190°F	—	$8\,g_{RMS}$	$12\,g_{RMS}$	Random shock
4	2	190°F	—	$8\,g_{RMS}$	—	Steady-state maximum operating conditions
5	0.25	190°F	—	$8\,g_{RMS}$	$12\,g_{RMS}$	Random shock
4	2	190°F	—	$8\,g_{RMS}$	—	Steady-state maximum operating conditions
5	0.25	190°F	—	$8\,g_{RMS}$	$12\,g_{RMS}$	Random shock
4	2	190°F	—	$8\,g_{RMS}$	—	Steady-state maximum operating conditions
5	0.25	190°F	—	$8\,g_{RMS}$	$12\,g_{RMS}$	Random shock
4	2	190°F	—	$8\,g_{RMS}$	—	Steady-state maximum operating conditions
6	0.25	190°F	—		—	Shutdown—vibration
7	3		190°F to (−18°F)t		—	Shutdown—worst case
8	2	−18°F				Ambient temperature
9	1		−18°F to lab ambient			Shutdown to lab ambient temperature

Step 2 exposes the candidate materials to the maximum steady-state vibration expected during system operations. The duration of step 2 is calculated to avoid physical shock.

Step 3 exposes the candidate materials to the maximum steady-state temperature expected during system operations. The duration of step 3 is calculated to avoid thermal shock.

Step 4 exposes the candidate materials to the expected maximum combined operating conditions of use.

Step 5 exposes the candidate materials to the expected random shock and is repeated six times.

Steps 6, 7, and 8 shut down the vibration and cool down the temperature exposure, returning the candidate materials to the starting ambient temperature.

Step 9 concludes the experiment by returning the candidate materials to the lab ambient conditions.

The candidate materials are installed in a test fixture that loads the materials in tension at the maximum stress during the experimental exposure cycles. The test cycle presented previously condenses a 1-h mission to less than 27 min of test time. Only 17.75 min represents continuous system operation; 9 min represents system start-up and shutdown. Steps 4 and 5 can be repeated to condense an 8-h shift in 150 min or condense 1 year in 26 test days. The former provides information to characterize the reliability of the candidate material in design and the latter sustainment.

The temperature cycle illustrated in Figure 6.5 shows the timeline from lab ambient temperature to the worst-case start-up ambient temperature where the environment provides the thermal stress. The temperature ramp from –18 to 190°F describes the thermal stress from system start-up where operational conditions replace ambient conditions. Theoretically, thermal strain will occur during this phase of system operations. The temperature soak at 190°F describes the maximum steady-state thermal stress from continued operations. Changes in material properties occur during this phase of system operations. The temperature ramp from 190 to –18°F describes the thermal stress from system shutdown where ambient conditions replace operational conditions. Theoretically, thermal strain will relax during this phase of system operations.

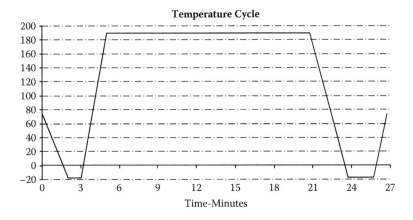

FIGURE 6.5
Experimental cycle: temperature.

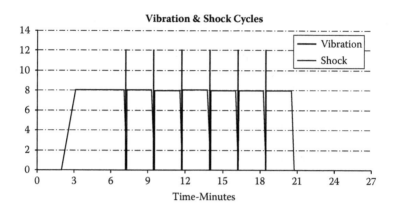

FIGURE 6.6
Experimental cycle: vibration and shock.

The vibration and shock cycles illustrated in Figure 6.6 show the time line from system start-up to shutdown for steady-state vibration and random cyclical shock. Theoretically, steady-state vibration causes deflections in the material that induce physical strain, and shock causes cracking and fracture of materials.

Stresses are defined experimentally as factors. The range of factors is defined as levels. Experimental levels can be set as fixed, discrete magnitudes or random, continuous magnitudes. Prime factor experiments that test one factor at fixed or random levels provide limited information about the true failure behavior of a material. The tensile strength quoted for the bars in this example is the result of a prime factor experiment. An unstressed sample of the material is placed in a universal materials test machine (as shown in Figure 6.7) and loaded to failure.

An experiment is designed to investigate the tensile strength of the square- and round-bar materials in three phases:

- phase I: test simulation period is 1 operating day
- phase II: test simulation period is 1 operating year
- phase III: test simulation period is 2 operating years

The experiment concurrently loads the candidate materials to the maximum operational tensile load with the expected ambient and operational thermal and physical loads. The candidate material is tested to failure following the test to characterize the tensile strength relevant to the design conditions of use.

Phase I: 1-Operational-Day Test Simulation Period

The data from the tensile tests are tabulated in Table 6.2. The parameters of the normal distribution for failure are characterized and evaluated.

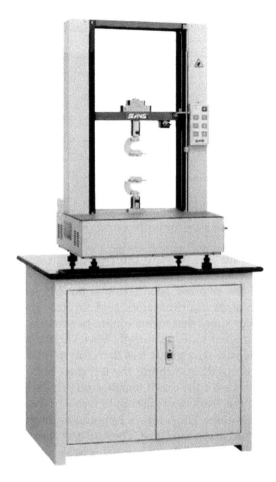

FIGURE 6.7
Universal strength of materials test machine.

TABLE 6.2

Tensile Test Data

	1 in. Round Bar (kpsi)	1 in. Square Bar (kpsi)
Sample 1	51	61
Sample 2	55.5	60.5
Sample 3	56.5	63
Sample 4	59	62.5
Sample 5	54	61.5

Tensile Test Data (kpsi)			Summary Statistics	Round	Square
	Round	Square		Round	Square
Sample 1	56	58	Mean	55	57.8
Sample 2	53	59	Standard Deviation	1.581	0.837
Sample 3	55	57	Sample Variance	2.50	0.70
Sample 4	57	57	Kurtosis	−1.20	−0.61
Sample 5	54	58	Skewness	0.00	0.51

FIGURE 6.8
One-day accelerated life test summary statistics.

Tensile strength test data are tabulated in MS Excel, as shown in the table in Figure 6.8. Summary statistics for both round and square bars are calculated in MS Excel.

The first opportunity to compare the two design alternatives seeks to determine whether there is a statistically significant difference between the tensile strength of the two bar geometries. If no difference exists, we can select one based on other discriminators, cost, fit, etc.

Summary statistics show that mean square-bar tensile strength is higher than the mean round-bar tensile strength; however, the standard deviation of the square-bar tensile strength is less than the standard deviation of the round-bar tensile strength. Assuming the test data to be normally distributed, test of hypothesis for the mean tensile strength for the two bar shapes is performed using the Student's t-distribution due to the small sample size, n_{round} and n_{square} (less than 30). A test of hypothesis for the variances of the bar stock is performed using the F-test. The null and alternate hypotheses for mean tensile strength are stated as

$$H_0: \mu_{square} = \mu_{round}$$
$$H_1: \mu_{square} \neq \mu_{round} \tag{6.1}$$

The null and alternate hypotheses for variances of tensile strength are stated as

$$H_0: \sigma^2_{square} = \sigma^2_{round}$$
$$H_1: \sigma^2_{square} \neq \sigma^2_{round} \tag{6.2}$$

The tests of hypotheses are performed in MS Excel at 95% confidence. The relationship between the test and critical statistics is evaluated: t_{test} and t_{crit} for means and F_{test} and F_{crit} for variances, as shown in the table in Figure 6.9. The tests of hypotheses suggest a statistically significant difference between the means of tensile strength for square and round bars at 95% confidence because $|t_{test}| > |t_{crit}|$, but not the variances because $F_{test} < F_{crit}$.[2]

t-Test Two-Sample Assume Unequal Variances			F-Test Two-Sample for Variances		
	Round	*Square*		*Round*	*Square*
Mean	55	57.8	df	4	4
Variance	2.5	0.7	Variance	2.5	0.7
Observations	5	5	F	3.571	
Hypothesized Mean Difference	0		P(F<=f) one-tail	0.123	
df	6		Critical one-tail	6.388	
tStat	−3.5				
P(T<=t) two-tail	0.013				
t Critical two-tail	2.447				

FIGURE 6.9
Test of hypothesis for means and variances: bar tensile strength.

The p-values for the two tests of hypotheses confirm the findings of the test and critical statistics and add that both null hypotheses would be accepted at 99% confidence level, such that $p = 0.013 > \alpha = 0.01$ for means and $p = 0.123$ for variances.

The normal distribution characterizes the tensile strength probability density function (pdf) for round bar, $f_{normal}(xrnd)$, and square bar, $g_{normal}(xsq)$, with parameters of mean tensile strength for round bar, 1-day simulation test (μrnd_d), and mean tensile strength for square bar, 1-day simulation test (μsq_d). The plots for the probability density functions for round and square bars are provided in Figure 6.10.

Tensile strength data for the two bar geometries are fitted to the Weibull distribution using median ranks regression, as shown in Figure 6.10. The

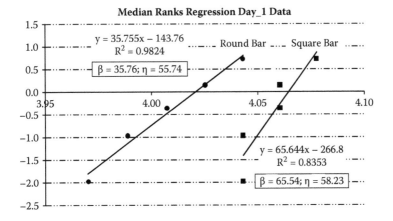

FIGURE 6.10
One-day median ranks regression.

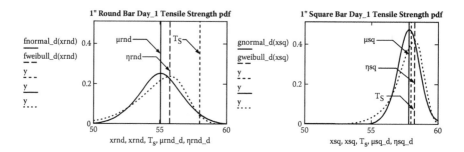

FIGURE 6.11
Normal and Weibull tensile strength pdf.

coefficients of determination find both median rank regressions to be suitably predictive. The parameters of the Weibull distribution are found for each bar geometry, a shape parameter is the slope of the equation, and the characteristic life is calculated as the antilogarithm for the quantity, $\ln^{-1}(-y_0/\beta)$.

The parameters of the Weibull distribution are used to express and plot the tensile strength probability density functions for the round- and square-bar stock and are plotted along with the normal probability density functions in Figure 6.11. The normal and Weibull probability plot shows the relationship between the measures of central tendency for square- and round-bar tensile strength and the rated tensile strength (T_s). It is not surprising to find empirical tensile strength to be less than the rated tensile strength; the empirical tensile strength is a response to the actual conditions of use to which the material will be subjected; the rated tensile strength is a laboratory norm. The empirical results for the round bar find its tensile strength to be much lower than the rated tensile strength and the square bar is much closer.

Phase II: 1-Operational-Year Test Simulation Period

The second phase of the experiment subjects the round- and square-bar test articles to a simulated 1-year operating test condition. The test data are documented in MS Excel, along with calculation of the summary statistics as shown in Figure 6.12.

The mean tensile strength and standard deviation from the summary statistics for the round and square bars are used to characterize the normal distribution probability density functions for tensile strength of the test articles. Parameter nomenclature uses "_d" for the 1-day simulation test, "_y1" for the 1-year simulation test, and "_y2" for the 2-year simulation test.

The tensile strength data are used to fit the Weibull parameters for the probability density functions for tensile strength using median ranks regression, as shown in Figure 6.13.

Tensile Test Data (kpsi)			Summary Statistics		Round	Square
	Round	Square			Round	Square
Sample 1	51	49	Mean		48	52 4
Sample 2	50	54	Standard Error		1.225	1.077
Sample 3	44	51	Standard Deviation		2.74	2.41
Sample 4	47	53	Sample Variance		7.5	5.8
Sample 5	48	55	Kurtosis		-0.133	-0.945
			Skewness		-0.609	-0.601

FIGURE 6.12
One-year accelerated life test summary statistics.

Phase III: 2-Operational-Year Test Simulation Period

The experiment concludes with phase III with a 2-year test simulation. The tensile strength data are tabulated in Figure 6.14 along with the summary statistics. The mean and standard deviation from the summary statistics are used to characterize the normal pdf for the round and square bars' tensile strength. The tensile test data are used to fit the Weibull distribution using the median ranks regression method, as shown in Figure 6.15.

The behavior of the normally distributed tensile strength material properties under load for 1-day, 1-year, and 2-year test simulations is illustrated in Figure 6.16. As hypothesized, the point estimate of the mean tensile strength for both geometries decreased with each successive test; the measure of dispersion, the standard deviation, increased over the same test intervals. The empirical evidence shows that the material weakens over time as manifested by a reduction in mean strength and material

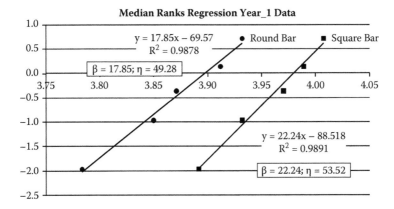

FIGURE 6.13
One-year median ranks regression.

Tensile Test Data (kpsi)			Summary Statistics	Round	Square
	Round	Square		Round	Square
Sample 1	39	44	Mean	37.2	41.4
Sample 2	41	38	Standard Error	1428	1.364
Sample 3	35	41	Standard Deviation	3.19	3.05
Sample 4	33	39	Sample Variance	10.2	9.3
Sample 5	38	45	Kurtosis	−1.344	−2.501
			Skewness	−0.301	0.162

FIGURE 6.14
Two-year accelerated life test summary statistics.

properties that control variability of the mean strength. The variability of mean strength is influenced by surface wear-out, oxidation, embrittlement, and internal cracks.

The behavior of the Weibull distributed tensile strength material properties under load for 1-day, 1-year, and 2-year test simulations is illustrated in Figure 6.17. The Weibull measure of central tendency, the characteristic life, decreases over the successive tests, as did the normally distributed mean tensile strength. The Weibull shape parameter also decreased over the successive tests, as we would expect because the shape parameter is negatively correlated to the measure of dispersion of the Weibull distribution.

The degradation of tensile strength for round and square bars using the Weibull distribution is shown in Figure 6.18. The top two graphs show the probability density functions for the round bar and square bar, respectively.

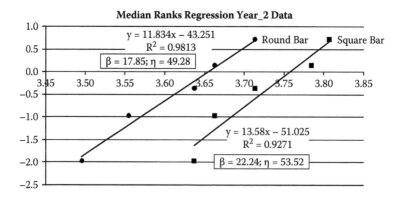

FIGURE 6.15
Two-year median ranks regression.

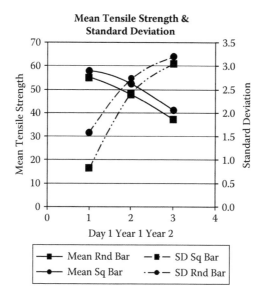

FIGURE 6.16
Tensile strength summary: normal distribution.

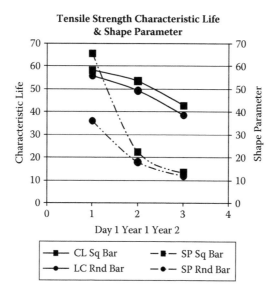

FIGURE 6.17
Tensile strength summary: Weibull distribution.

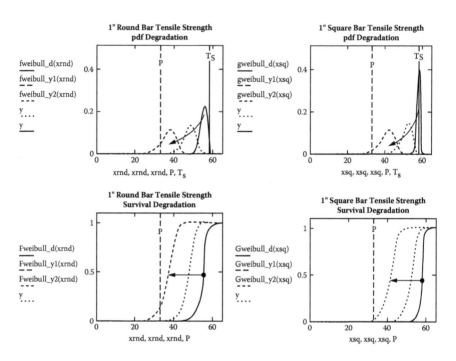

FIGURE 6.18
Tensile strength degradation—round and square bars: Weibull distribution.

The rated tensile strength (T_s) and the maximum expected load (P) are shown on the pdf graphs. The tensile strength pdf for the round bar degrades to include a much larger overlap below the maximum load than does the square bar. We can infer from the graphic evidence that the round-bar design degrades to a greater extent than the square-bar design under the specified conditions of use.

The bottom two graphs show the tensile strength survival function for the round bar and square bar, respectively. The maximum expected load (P) is shown on the survival function graphs. The round-bar tensile strength survival function also degrades farther below the maximum load than the square-bar tensile strength survival function. We can make the same inference from this graphic evidence that the round-bar design degrades to a greater extent than the square-bar design under the specified conditions of use.

From this experiment, we estimate the unreliability of the round bar to be equal to the cumulative distribution of tensile strength evaluated at the maximum load (P). Reliability of the round bar is equal to one minus the unreliability: $R = 1 - U$. The expressions for reliability for the round and square bars are presented in Figure 6.19. Note the use of a special MathCAD function for the cumulative probability distribution for the Weibull. Rather

Reliability Calculations		
1" Round Bar		
1 Day	$Rmd_1 := 1 - Fweibull\_d(P)$	$Rmd_1 = 0.999999993$
1 Year	$Rmd_2 := 1 - Fweibull\_yl(P)$	$Rmd_2 = 0.999221726$
2 Year	$Rmd_3 := 1 - Fweibull\_y2(P)$	$Rmd_3 = 0.857517599$
1" Square Bar		
1 Day	$Rsq_1 := 1 - Gweibull\_d(P)$	$Rsq_1 = 1.000000000$
1 Year	$Rsq_2 := 1 - Gweibull\_yl(P)$	$Rsq_2 = 0.999978643$
2 Year	$Rsq_3 := 1 - Gweibull\_y2(P)$	$Rsq_3 = 0.971513238$

FIGURE 6.19
Reliability calculations—round and square bars: Weibull distribution.

than write out the entire expression,

$$F(x) = \frac{\beta}{\eta}\left(\frac{x}{\eta}\right)^{\beta-1} e^{\left(\frac{x}{\eta}\right)^{\beta}}$$ (6.3)

MathCAD uses a paint function named Fweibull(x) where x is either a discrete value or an array of values.

The reliability values are plotted in Figure 6.20. The reliability plots show that the reliability of round and square bars is statistically the same at day 1 and year 1. However, reliability of the round bar drops off at year 2 and the square bar provides the highest reliability design.

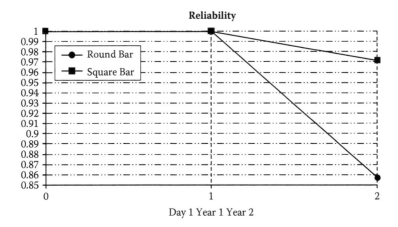

FIGURE 6.20
Reliability degradation—round and square bars: Weibull distribution.

Design for Reliability by Analysis

Interference theory can be applied by analysis to solve for critical design parameters, including dimensions and material strength properties. Current applications use the normal distribution; research is in progress to apply the Weibull distribution.

In procedure 2, reliability is known; solve for the design parameter that achieves the reliability:

1. Determine z for the reliability requirement.
2. Express strength in terms of the known design parameters.
3. Express stress in terms of known and unknown design parameters.
4. Solve for the unknown design parameters.

Material in Tension

An example of design for reliability by analysis is the round bar from the previous empirical design-for-reliability analysis. Design analysis of the round bar requires specification of the radius.

Step 1: Determine z for the reliability requirement. The reliability allocation for the round bar is 0.950, or 99.999%. The interference area represents the cumulative probability of failure, the unreliability, equal to $1 - R$, or 0.00001. The value of z that corresponds to a cumulative probability of 0.00001 is -4.26489. The expression for z is given by

$$z = \frac{\mu_S - \mu_s}{\sqrt{\sigma_S^2 + \sigma_s^2}} = -4.26489 \tag{6.4}$$

Step 2: Express strength in terms of known design parameters. The point estimate of the mean strength and the standard deviation are known values from empirical analysis; mean strength (μ_S) was found empirically to be 55 kpsi, with a standard deviation (σ_S) of 1.581 kpsi. Two of the four variables for z are known numerically.

Step 3: Express stress in terms of known and unknown design parameters. The other two variables for z are symbolically expressed in terms of the geometry of the round bar. A free body diagram of the round bar is prepared as shown in Figure 6.21.

The following are parameters of tensile strength and stress:

$P \equiv$ load (lb)

$\sigma_P \equiv$ standard deviation of the load (lb)

μ_S used for $S \equiv$ ultimate yield strength (lb)

$\sigma_S \equiv$ standard deviation of strength (lb)

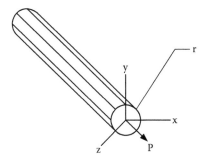

FIGURE 6.21
Free body diagram, round bar in tension.

μ_s used for $s \equiv$ tensile stress (lb)

$\sigma_s \equiv$ standard deviation of the tensile stress (lb)

$A \equiv$ area of circular cross-section of shaft (in.^{2})

$d \equiv$ diameter of solid shaft (in.)

$r \equiv$ radius of solid shaft (in.)

$\sigma_r \equiv$ standard deviation of the radius (in.)

$\alpha \equiv$ tolerance fraction of the radius

The mean tensile stress is expressed as the mean load $(\bar{P})$ divided by the mean area $(\bar{A})$:

$$s = \frac{\bar{P}}{\bar{A}} = \frac{\bar{P}}{\pi \bar{r}^2} \tag{6.5}$$

The standard deviation[3] for stress is just a tad less convenient. The variance of tensile stress (σ_s^2) is the sum of the variances of the load (P) and the area (A) and is given by

$$\sigma_s^2 = \sigma_P^2 \left(\frac{1}{\bar{A}} \right)^2 + \sigma_A^2 \left(\frac{\bar{P}}{\bar{A}^2} \right)^2 \tag{6.6}$$

The mean load $(\bar{P})$ is known from empirical data to be 31 kpsi, with a standard deviation (σ_P) of 5.5 kpsi. The mean area $(\bar{A})$ is expressed as $\pi \bar{r}^2$. The standard deviation of the area of a circle is given by

$$\sigma_A = 2\pi \bar{r} \sigma_r = \left(\frac{\alpha}{3} \right) 2\pi \bar{r}^2 \tag{6.7}$$

$$r^4 - 23.88r^2 + 96.4 \text{ solve} \longrightarrow \begin{pmatrix} -4.328323004419159603 \\ 4.328323004419159603 \\ -2.2683959022657177424 \\ 2.2683959022657177424 \end{pmatrix}$$

FIGURE 6.22
Roots of radius (*r*).

The tolerance of the radius (α) is given to be 10% and is determined to be ±3 standard deviations of the radius ($\pm 3\sigma_r$). Therefore,

$$3\sigma_r = \alpha\bar{r}$$

$$\sigma_r = \left(\frac{\alpha}{3}\right)\bar{r} \tag{6.8}$$

The variance of tensile stress is further developed by substitution for *A* and the variance of the area:

$$\sigma_s^2 = \frac{\sigma_P^2}{(\pi\bar{r}^2)^2} + \frac{\sigma_A^2\bar{P}^2}{(\pi\bar{r}^r)^4} = \frac{\sigma_P^2 + \left(\frac{4}{9}\right)\alpha^2\bar{P}^2}{\pi^2\bar{r}^4} \tag{6.9}$$

The expression for *z* can be stated as

$$z = -\frac{\mu_S - \mu_s}{\sqrt{\sigma_S^2 + \left(\frac{\sigma_P^2 + \left(\frac{4}{9}\right)\alpha^2\bar{P}^2}{\pi^2\bar{r}^4}\right)}} \tag{6.10}$$

Substitution for known numerical values gives us

$$-4.26489 = -\frac{55000 - \frac{31000}{3.1416\bar{r}^2}}{\sqrt{1581^2 + \frac{5500^2 + (0.444)(0.01)\times 31000^2}{9.87\bar{r}^4}}} \tag{6.11}$$

which reduces to

$$\bar{r}^4 - 0.36\bar{r}^2 + 0.011 = 0 \tag{6.12}$$

MathCAD solved the fourth-order equation for the four roots for radius (*r*) shown in Figure 6.22. The analysis recommends a bar radius greater than 0.57125 in. or machined to 0.57122 in.

Notes

1. kpsi = 1,000 pounds per square inch.
2. The absolute value for t_{test} is 3.5; t_{crit} is 2.44, and 3.5 > 2.44.
3. The topic for many papers found at symposia and technical journals is expressions for variability of design parameters.

7

Failure Modeling Based on Failure Mechanisms

Always do what you are afraid to do.

Ralph Waldo Emerson

Introduction

The design engineer can estimate a part's reliability in its new condition. Part reliability based on time-to-failure data using the exponential probability distribution assumes that part reliability is constant over its useful life—a best-practices approach for electronics and digital parts. Weibull failure distributions based on time-to-failure data suppose that the degradation of structural and dynamic part reliability can be modeled over the useful life; however, one must ask what practical application that information serves. More important to the design and sustainment engineer is the blunt fact that time does not cause part failure: Failure mechanisms cause part failure!

Part reliability should be evaluated in design as a starting point in understanding the behavior of the failure mechanism on the part; that understanding is conveyed to sustainment engineers to enable the user organization to develop a maintainability program focused on the failure mechanisms experienced in their specific ambient and operational conditions of use. For example, a design engineer for an air brake used on construction and mining off-road vehicles should evaluate the part reliability based on vibration stress, physical shock, thermal stress, thermal shock, and corrosion. Organizations that operate the system with the air brake will realize different useful lives based on the unique conditions of use for their operating scenario.

Contrast the differences in vibration stress and physical shock between a construction site with ever-changing road conditions and a mine with well-graded haul roads; thermal stress and shock between a coal mine in Arizona and a taconite mine in Minnesota; corrosion from oxidation between a copper mine in Papua, New Guinea, and a uranium mine in Wyoming; corrosion from salt spray between a sand pit in Florida and a rock quarry in Kansas; and corrosion from biological agents on a road construction project in the Philippines and a road construction project in Brazil. The same model of truck and same air brake experience widely varying conditions of use.

Design and sustainment engineers also understand that not all failure mechanisms act the same way. Consider corrosion from oxidation: Parts wear out at different rates between exposure to humidity in storage and exposure in operations. Parts exposed to physical loads within the elastic limits at different frequencies of occurrence wear out at different rates.

At best, part reliability is an estimate of the ability of a part to withstand a failure mechanism fresh off the drawing board. The burden on sustainment personnel is to slow the wear-out effects from the failure mechanisms and modes that attack the part once it is put into use. Design engineers owe sustainment engineers the qualitative and quantitative factors of the reliability estimate to give them an understanding of the starting point from which they will maintain the part to the longest technically and economically feasible useful life.

Failure mechanisms and material strength properties can be described in one of three ways: positively skewed, negatively skewed, and symmetrical. Skewed behavior of failure and strength is well modeled by the Weibull distribution from historical or empirical data, where failure is typically positively skewed (tail of the distribution extends to the right on the x-axis, away from the measure of central tendency) and strength is typically negatively skewed (tail of the distribution extends to the left on the x-axis, away from the measure of central tendency). The triangular distribution is a good estimator of the skewed shape of failure and strength in the absence of data, and it serves as a good placeholder in the early stages of design. Often the triangular distribution placeholder remains as the only estimator for part reliability when the economic priorities of the design project prevent further failure analysis.

Using the normal probability distribution, symmetrical failure and strength are the most common quantitative approach to characterize part reliability, but are too often wrong! The normal distribution is well understood and convenient; the Weibull and triangular distributions are less understood and viewed as inconvenient. But the advent of computers with statistical and engineering software should change that perception. It is a case of switching our analytical approach from what was convenient—and wrong—to what is correct.

Consider what the normal distribution describes when applied to a parameter of failure mechanisms or strength material properties. Normally distributed metrics have an equal likelihood to take on values on either side of the mean value. Empirical measures of a stress will cluster about the measure of central tendency but will be more likely to have extreme values above the central metric. The cluster will appear very nearly normally distributed, but the tail of higher values (positive skew) is not insignificant and must be included in the failure analysis. Likewise, strength behaves in a mirror image of stress. Treating the two as normally distributed will overstate the reliability of the part.

Examples of reliability analysis are presented for the following relationships between stress loads and material strength properties:

- normal distribution stress–normal distribution strength
- normal distribution stress–Weibull distribution strength
- Weibull distribution stress–Weibull distribution strength
- triangular distribution stress–triangular distribution strength

Normal Distribution Stress–Normal Distribution Strength[1]

Consider the design of a hydraulic actuator. A test fixture is designed to simulate tensile load acting on a double-acting cylinder rod. Measurements of the peak tensile load are tabulated and summary statistics are calculated in MS Excel. Concurrently, 15 samples of the proposed design of a cylinder rod are tested to failure on a tensile test machine. The sample data and summary statistics are calculated in MS Excel (see Figure 7.1).

The difference between the mean load and median (27.85 – 26.95) is positive and suggests a positively skewed distribution of stress; the skewness factor (0.212) quantifies a slight positive skew. The same assessment for strength suggests a nearly symmetrical distribution; the skewness factor is very close to zero. Kurtosis for both load and strength differs sufficiently from zero to suggest that both distributions are not really normally distributed and the use of standard normal tables will introduce error in the estimations. The

Tensile Load					Tensile Strength				
26 9	24.6	34.9	Mean	27.85	43.8	44.0	Mean		45.22
23.5	25.3	31.0	Standard Error	0.691	46.3	44.2	Standard Error		0.519
28.9	20.6	36.3	Median	26.95	46.3	43.9	Median		44.77
32.5	26.0	25.7	Standard Deviation	4.20	44.0	44.8	Standard Deviation		2.01
32.2	26.6	33.8	Sample Variance	17.66	41.7		Sample Variance		4.03
34.1	28.5	22.4	Kurtosis	-0.662	48.7		Kurtosis		-0.800
20.4	26.7	29.9	Skewness	0.212	47.7		Skewness		0.017
27.2	26.9	31.2	Range	15.96	42.6		Range		7.03
31.8	26.7	34.7	Minimum	20.4	45.9		Minimum		41.7
24.2	32.7	27.7	Maximum	36.3	47.3		Maximum		48.7
25.6	27.7	26.2	Sum	1,030.4	47.1		Sum		678.3
22.1	27.3	26.2	Count	37			Count		15
21.5									

FIGURE 7.1
Normal stress–normal strength data and summary statistics.

μload:=27.85   μstrength:=45.22

σload:=4.20   σstrength:=2.01

$$f(\text{load}) := \left(\frac{1}{\sigma\text{load} \cdot \sqrt{2\pi}} \right) \cdot e^{-\left(\frac{1}{2}\right)\left(\frac{\text{load}-\mu\text{load}}{\sigma\text{load}}\right)}$$

$$g(\text{strength}) := \left(\frac{1}{\sigma\text{strength} \cdot \sqrt{2\pi}} \right) \cdot e^{-\left(\frac{1}{2}\right)\left(\frac{\text{strength}-\mu\text{strength}}{\sigma\text{strength}}\right)^2}$$

FIGURE 7.2
Normal probability density functions for load, *f*(load), and strength, *g*(load).

use of the normal distribution for both load and strength is used to characterize the reliability of the design.

The mean for stress and strength (μload and μstrength) and the standard deviation for stress and strength (σload and σstrength) are entered in MathCAD. The normal probability density function (pdf) expressions for failure, *f*(load), and strength, *g*(strength), are written in MathCAD, as shown in Figure 7.2.

The normal probability density functions, *f*(load) and *g*(strength), are plotted in MathCad. The shapes of the probability density functions illustrate the locations of the measures of central tendency (μload and μstrength), the proximity between the measures of central tendency, and the spreads of the measures of dispersion (σload and σstrength). The proximity between the measures of central tendency provides an empirical safety factor (*SF*) calculated as the ratio of μstrength to μload (*SF* = 1.624) above the design specification of 1.5. The stress load has a wider dispersion than the strength.

Because both distributions are normally distributed, we can calculate the difference between the distributions that is also normally distributed. The parameter of the difference between the two distributions, *z*, is defined as the difference between the means divided by the square root of the sum of the two variances:

$$z = \frac{\mu\text{strength} - \mu\text{load}}{\sqrt{\sigma\text{strength}^2 + \sigma\text{load}^2}} \tag{7.1}$$

The pdf for the difference between the two distributions is

$$f(z) = \left(\frac{1}{\sqrt{2\pi}} \right) e^{\frac{z^2}{2}} \tag{7.2}$$

The reliability of the design based on the relationship between the two probability density functions is the integral (evaluate from −*z* to infinity) of *f*(z), as shown in Figure 7.3.

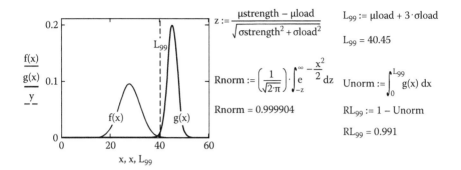

$$z := \frac{\mu \text{strength} - \mu \text{load}}{\sqrt{\sigma \text{strength}^2 + \sigma \text{load}^2}}$$

$$L_{99} := \mu \text{load} + 3 \cdot \sigma \text{load}$$

$$L_{99} = 40.45$$

$$\text{Rnorm} := \left(\frac{1}{\sqrt{2 \cdot \pi}}\right) \cdot \int_{-z}^{\infty} e^{-\frac{x^2}{2}} dz$$

$$\text{Unorm} := \int_{0}^{L_{99}} g(x) \, dx$$

$$\text{Rnorm} = 0.999904$$

$$RL_{99} := 1 - \text{Unorm}$$

$$RL_{99} = 0.991$$

FIGURE 7.3
Normal stress–normal strength reliability using interference theory.

The reliability of the design is calculated to be 0.999904, or 9^40, referred to "4 9's reliability." An alternative approach to characterize reliability is to evaluate the area of the strength distribution above the nth percentile for load (L_n). The 99th percentile (L_{99}) is approximated by the mean plus three standard deviations.[2] We can calculate the unreliability (U) of the design by calculating the integral of the strength pdf, g(strength), from zero to L_{99}, as shown in Figure 7.3. The L_{99} reliability (RL_{99}) is equal to $1 - U$. The L_{99} reliability can be calculated directly as the integral of $g(x)$ from L_{99} to infinity—six of one, half a dozen of the other, as the expression goes. The L_n approximation of reliability is a lower confidence limit of the design reliability. These approximations are also useful to compare two or more part selection alternatives by normalizing the reliability-based assessment of risk.

The reliability calculated from interference theory is the starting point for the part before it begins operations and is subjected to failure mechanisms. We can say that the reliability of this design is at least 0.9^21 the first time it is put into use. Reliability will degrade as the part experiences wear out.

Normal Distribution Stress–Weibull Distribution Strength

Consider the design of a materials processing machine that uses a plunger to press fit a shaft into a bushing. A test fixture is designed to simulate the compression loads on the plunger. Measurements of the peak compression loads and summary statistics are tabulated in MS Excel. Concurrently, 15 samples of the proposed design of a plunger are tested to failure on a compression test machine. The sample data are tabulated in MS Excel (see Figure 7.4).

The compression load is assumed to be normally distributed. The magnitude of the difference between the mean load and the median is near zero,

Compression Load						Strength
14.4	9.6	22.3	17.9	Mean	15.32	21
15.2	15.0	12.4	12.6	Standard Error	0.411	22
12.5	17.9	12.7	9.6	Median	15.04	23
16.8	13.8	16.0	18.1	Standard Deviation	2.69	25
14.4	12.4	14.0	14.6	Sample Variance	7.26	26
16.9	16.8	12.3	15.3	Kurtosis	0.336	27
15.3	13.5	18.0	15.9	Skewness	0.291	23
20.0	14.4	12.7		Range	12.75	30
14.6	13.2	16.5		Minimum	9.60	30
19.7	20.1	16.6		Maximum	22.34	31
16.1	14.2	14.7		Sum	658.7	31
19.1	15.7	14.9		Count	43	32
						32
						33
						34

FIGURE 7.4
Normal stress–Weibull strength data and summary statistics.

suggesting a symmetrical distribution. The low values for the skewness and kurtosis factors are also acceptably small, suggesting that the normal distribution is a good fit. However, the frequency distribution for strength has a negative skew that suggests that the normal distribution will introduce too much error, as shown in Figure 7.5.

The strength data are rank ordered and the data index is tabulated in MS Excel and imported to MathCAD. The median ranks regression is calculated to estimate the parameters of the Weibull for strength. The parameters of the median ranks regression are the independent variables X, defined as the natural log of strength, and Y, the natural log of the natural log of one

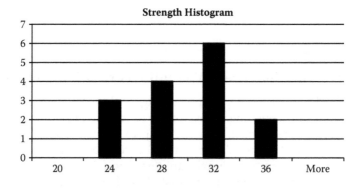

FIGURE 7.5
Strength frequency distribution.

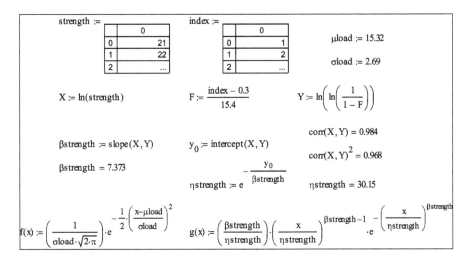

FIGURE 7.6
Normal pdf for load, $f(x)$, and Weibull pdf for strength, $g(x)$.

divided by one—the estimator for the cumulative probability distribution, F. Bartlett's index is used to approximate F.

The parameters of the Weibull distribution are calculated in MathCAD, the shape parameter (βstrength) is estimated as the slope of the median ranks regression line, and the characteristic life (ηstrength) is estimated by the antilog of the quantity, $-y_0/\beta$, where y_0 is the y-intercept of the median ranks regression line. The coefficients of correlation and determination are sufficiently high to suggest that the median ranks regression is both highly correlative and predictive. The load mean and standard deviation (μload and σload) are entered in MathCAD.

The expressions for the normal pdf for load, $f(x)$, and the Weibull pdf for strength, $g(x)$, are written in MathCAD,[3] as shown in Figure 7.6. The normal load pdf and the Weibull strength pdf are plotted in Figure 7.7. The proximity between the measures of central tendency provides an empirical safety factor calculated as the ratio of ηstrength to μload ($SF = 1.968$) above the design specification of 1.75. The stress load has a narrower dispersion than the strength. The strength distribution shows the negative skew expected from the frequency distribution. The normal cumulative probability distribution for load, $F(x)$, is calculated as the integral of the normal pdf, $f(x)$, from zero to x.[4] The reliability of the design is calculated as the indefinite integral of the pdf of strength, $g(x)$, times the cumulative density function (cdf) of load, $F(x)$.

The reliability of the example design is computed in MathCAD to be 0.987587. The L_{99} reliability is calculated using the strength cdf:

$$G(x) = 1 - e^{\left(\frac{x}{\eta\text{strength}}\right)^{\beta\text{strength}}} \qquad (7.3)$$

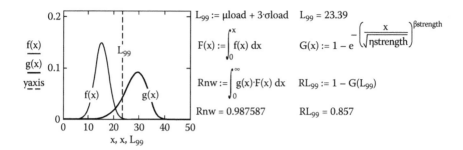

FIGURE 7.7
Normal stress–Weibull strength reliability using interference theory.

Solving for the strength cdf, $G(x)$, calculates the area under $G(x)$ from zero to L_{99}, the unreliability, U. The L_{99} reliability, RL_{99}, is found by

$$1 - G(L_{99}) = 0.857$$

Weibull Distributed Stress–Weibull Distribution Strength

Consider the design of a shaft that is located in proximity to a diesel engine. The shaft is exposed to thermal shock as the engine warms up from start-up ambient temperature to peak operating temperature. The thermal shock varies from $\Delta T = 100°F$ in summer to $\Delta T = 200°F$ in winter, resulting in thermal strain that exerts torsion loads on the rotating shaft. A test fixture is designed to simulate torsion loads on the shaft. Measurements of the peak torsion loads are tabulated in MS Excel. Concurrently, 18 samples of the proposed design of the shaft are tested to failure on a torsion test machine. The sample data are tabulated in MS Excel; see Figure 7.8.

No assumption is made about the distribution of the two data sets; both are fit to a Weibull distribution. The data are rank ordered and index numbers are tabulated in MS Excel and imported to MathCAD. Median ranks regression is calculated in MathCAD for load and strength data, as shown in Figure 7.9. The coefficients of correlation and determination for load and strength are calculated in MathCAD. The high values suggest that the median ranks regression for load and strength is highly correlated and predictive. The Weibull expressions for pdf for load, $f(x)$, and strength, $g(x)$, are written in MathCAD. The Weibull expression for cumulative probability

Load	Strength
8	21
9	25
11	25
12	26
13	27
13	29
13	30
14	31
IS	31
15	32
17	32
18	32
18	33
19	33
20	34
21	35
23	35
	37

FIGURE 7.8
Weibull stress–Weibull strength data.

distribution for load, $F(x)$, is also written. The value for L_{99} is found by solving $F(x)$ for values for x until $F(x) = 0.99$.

The Weibull load pdf, $f(x)$, and the Weibull strength pdf, $g(x)$, are plotted in Figure 7.10. The proximity between the measures of central tendency provides an empirical safety factor calculated as the ratio of ηstrength to μload ($SF = 1.92$) just below the design specification of 2.00. The stress load is very nearly normally distributed, as suggested by the value of the shape parameter.[5] It is worth noting that the Weibull will always be a better model of the load and strength no matter how close the shape parameter is to fitting the exponential and normal distributions. The precision of the reliability functions will always be more precise using the Weibull than if the other distribution is used. One might revert to the exponential if the shape parameter is very close to $\beta = 1$—say, $\beta = 1.06$. The difference in reliability might differ on the third or fourth decimal place, yet the error will be compounded when the system reliability is comprised of many part reliability models that are factored to the system level. Also note that there is no clear definition of how much difference is small enough.

load :=

	0
0	8
1	9
2	...

Strength :=

	0
0	21
1	25
2	...

load_index :=

	0
0	1
1	...

Strength_index :=

	0
0	1
1	...

$\text{Xload} := \ln(\text{load})$

$\text{Fload} := \dfrac{\text{load\_index} - 0.3}{17.4}$

$\text{Yload} := \ln\left(\ln\left(\dfrac{1}{1 - \text{Fload}}\right)\right)$

$\text{corr}(\text{Xload}, \text{Yload}) = 0.991$

$\text{corr}(\text{Xload}, \text{Yload})^2 = 0.982$

$\beta\text{load} := \text{slope}(\text{Xload}, \text{Yload})$

$\beta\text{load} = 3.921$

$\text{yload}_0 := \text{intercept}(\text{Xload}, \text{Yload})$

$\text{yload}_0 = -11.067$

$\eta\text{load} := e^{-\left(\frac{\text{yload}_0}{\beta\text{load}}\right)}$

$\eta\text{load} = 16.823$

$f(x) := \left(\dfrac{\beta\text{load}}{\eta\text{load}}\right) \cdot \left(\dfrac{x}{\eta\text{load}}\right)^{\beta\text{load}-1} \cdot e^{-\left(\frac{x}{\eta\text{load}}\right)^{\beta\text{load}}}$

$F(x) := 1 - e^{-\left(\frac{x}{\eta\text{load}}\right)^{\beta\text{load}}}$

$\text{Xstrength} := \ln(\text{strength})$

$\text{Fstrength} := \dfrac{\text{strength\_index} - 0.3}{18.4}$

$\text{Ystrength} := \ln\left(\ln\left(\dfrac{1}{1 - \text{strength}}\right)\right)$

$\text{corr}(\text{Xstrength}, \text{Ystrength}) = 0.988$

$\text{corr}(\text{Xstrength}, \text{Ystrength})^2 = 0.976$

$\beta\text{strength} := \text{slope}(\text{Xstrength}, \text{Ystrength})$

$\beta\text{strength} = 7.841$

$\text{ystrength}_0 := \text{intercept}(\text{Xstrength}, \text{Ystrength})$

$\text{ystrength}_0 = -27.45$

$\eta\text{strength} := e^{-\left(\frac{\text{ystrength}_0}{\beta\text{strength}}\right)}$

$\eta\text{strength} = 16.823$

$g(x) := \left(\dfrac{\beta\text{strength}}{\eta\text{strength}}\right) \cdot \left(\dfrac{x}{\eta\text{strength}}\right)^{\beta\text{strength}-1}$

$\times e^{-\left(\frac{x}{\eta\text{strength}}\right)^{\beta\text{strength}}}$

$L_{99} := 25$

FIGURE 7.9
Weibull pdf for load, $f(x)$, Weibull cdf, $F(x)$, Weibull pdf for strength, $g(x)$, and L_{99}.

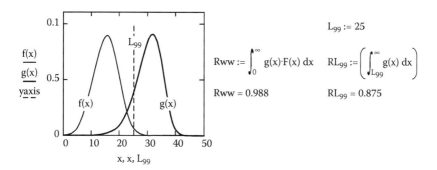

FIGURE 7.10
Weibull stress–Weibull strength reliability using interference theory.

The stress load has a similar dispersion to the strength. The strength distribution shows the negative skew. The reliability of the design is calculated as the indefinite integral of the pdf of strength, $g(x)$, times the cdf of load, $F(x)$. The reliability of the design is solved to be 0.988. The L_{99} reliability (RL_{99}) is calculated directly by the integral of the strength pdf, $g(x)$, from L_{99} to infinity to be 0.875.

Triangular Distribution Stress–Triangular Distribution Strength

No failure and strength data are available early in the system design analysis. The triangular distribution enables the design engineer to characterize the reliability of a part without empirical data. An Adelphi survey of subject matter experts is queried to characterize the minimum, mode, and maximum values for load and strength. Minimum is also called worst-case or least likely value; mode is frequently called the most likely value, and maximum is also called best case. However, these terms need to be avoided because they do not apply to load or strength magnitudes. One might interpret the worst-case load as the maximum. "Least likely" conjures up extreme values without direction; they could be high or low. "Best case" is equally confusing.

The results of the Adelphi survey are entered in MathCAD and the expressions for the probability density functions for load, $f(x)$, and strength, $g(x)$, are written in MathCAD, as shown in Figure 7.11. The plots of the triangular probability density functions, $f(x)$ and $g(x)$, are shown in Figure 7.12. The unreliability is calculated directly as the area of the interference between the two distributions by the double integral of the strength pdf, $g(x)$, and the

loadmin := 12   strengthmin := 20
loadmode := 15   strengthmode := 30
loadmax := 22   strengthmax := 35

$$f(x) := \begin{vmatrix} \dfrac{2 \cdot (x - \text{loadmin})}{(\text{loadmax} - \text{loadmin}) \cdot (\text{loadmode} - \text{loadmin})} & \text{if loadmin} \leq x \leq \text{loadmode} \\[2ex] \dfrac{2 \cdot (\text{loadmax} - x)}{(\text{loadmax} - \text{loadmin}) \cdot (\text{loadmax} - \text{loadmode})} & \text{if loadmode} < x \leq \text{loadmax} \end{vmatrix}$$

$$g(x) := \begin{vmatrix} \dfrac{2 \cdot (x - \text{strengthmin})}{(\text{strengthmax} - \text{strengthmin}) \cdot (\text{strengthmode} - \text{strengthmin})} & \text{if strengthmin} \leq x \leq \text{strengthmode} \\[2ex] \dfrac{2 \cdot (\text{strengthmax} - x)}{(\text{strengthmax} - \text{strengthmin}) \cdot (\text{strengthmax} - \text{strengthmode})} & \text{if strengthmode} < x \leq \text{strengthmax} \end{vmatrix}$$

FIGURE 7.11
Triangular data and probability density functions for load, $f(x)$, and strength, $g(x)$.

load pdf, $f(x)$, over the range from the minimum strength (strength$_{\min}$) to the maximum load (load$_{\max}$). The triangular distribution makes no provision for the L_{99} reliability. The reliability is calculated to be 0.988.

   In summary, we can say that the characterization of part design reliability can be found using interference theory to solve for the probability that stress will exceed strength, causing part failure. The reliability characterization is valid for a single stress–strength relationship. A part design that includes two or more failure mechanisms will need a reliability characterization for each stress and its corresponding material strength property. Consider a part design that has the failure mechanism acting in tension and shear. Part reliability is a function of both failure mechanisms. The part will fail if one failure

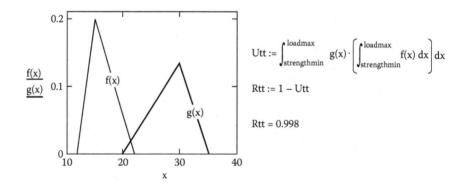

$$Utt := \int_{\text{strengthmin}}^{\text{loadmax}} g(x) \cdot \left[\int_{\text{strengthmin}}^{\text{loadmax}} f(x) \, dx \right] dx$$

$$Rtt := 1 - Utt$$

$$Rtt = 0.998$$

FIGURE 7.12
Triangular stress–triangular strength reliability using interference theory.

mechanism or the other occurs. We can solve for the probability of failure, the unreliability of the part design. The probability that the part will fail is equal to the sum of the unreliability from tension plus the unreliability from shear plus the probability that the part will fail in tension and shear together:

$$U\text{part} = U\text{tension} + U\text{shear} - U\text{tension} \times U\text{shear}. \tag{7.4}$$

We first inquire whether the failure mechanisms are independent and mutually exclusive. If the part can fail in tension without effects from shear, and vice versa, then the failure mechanisms are independent. If failure in tension and shear cannot occur simultaneously, then they are mutually exclusive. In that case, the term $U\text{tension} \times U\text{shear}$ is null, equal to zero. Assume $R\text{tension} = 0.99$ and $R\text{shear} = 0.98$. The respective unreliabilities are $U\text{tension} = 0.01$ and $U\text{shear} = 0.02$. If the two failure mechanisms are mutually exclusive, the unreliability of the design is found by

$$U\text{part} = U\text{tension} + U\text{shear} = 0.01 + 0.02 = 0.03$$

and the reliability of the design is 0.97.

If the two failure mechanisms are not mutually exclusive, the unreliability of the design is found by

$$U\text{part} = U\text{tension} + U\text{shear} - U\text{tension} \times U\text{shear} = 0.01 + 0.02 - (0.01)(0.02)$$
$$= 0.0298$$

and the reliability of the design is 0.9702.

What if there are three failure mechanisms for which three reliabilities are characterized? Let the unreliabilities be designated $U1 = 0.01$, $U2 = 0.02$, and $U3 = 0.03$. For mutually exclusive failure mechanisms, the unreliability of the design is found by

$$U\text{part} = U1 + U2 + U3 = 0.01 + 0.02 + 0.03 = 0.06$$

and the reliability of the design is 0.94.

If the three failure mechanisms are not mutually exclusive, the unreliability of the design is found by

$$U\text{part} = U1 + U2 + U3 - U1 \times U2 - U1 \times U3 - U2 \times U3 - U1 \times U2 \times U3$$
$$= + 0.02 + 0.03 - (0.01)(0.02) - (0.01)(0.03) - (0.02)(0.03) - (0.01)(0.02)(0.03)$$
$$= 0.058894$$

and the reliability of the design is 0.941106.

Notes

1. Although this example is data based, the use of the normal–normal characterization of reliability is recommended when only the estimators for the mean and standard deviation are known.
2. The 95th percentile, L_{95}, is approximately the mean plus two standard deviations.
3. The variable x is used for both pdf expressions for the obvious reason that both expressions are in terms of the same metric: force. This is a different approach than used in the normal stress–normal strength example. Both approaches to express the independent variable symbolically are correct.
4. The limits of the integral zero to infinity are explained by the fact that there is no negative time in reliability characterization. This does not pose an error for the normal distribution because the interference area exists only in the positive range of the independent variable.
5. The Weibull distribution reduces to the exponential distribution when the shape parameter $\beta = 1$ and the normal when $\beta = 3.6$.

8

Reliability Modeling for Assembly Design Levels

> Keep inwardly calm and clear in the midst of chaos; do not forget the possibility of disorder in times of order. When your life is on the line make use of all your tools.
>
> **Miyamoto Musashi,** *The Book of Five Rings,* **1643**

Introduction

Reliability modeling is perceived as abstract mathematical computations involving software programming, IT specialists, and sophisticated analysis. Jan and Paul Pukite claim that a reliability model is an abstract mathematical and graphical representation of system reliability characteristics used for predicting mission reliability and maintainability.[1]

This book suggests that there are two levels of reliability modeling: part-level failure modeling and systems integration modeling. Part-level failure modeling constructs probabilistic bricks that make up the structure of systems integration modeling. Each probabilistic brick characterizes the behavior of part failure using a statistical probability distribution (i.e., normal, Weibull, triangular). Systems integration modeling assembles the individual probabilistic bricks into assemblies, assemblies into subsystems, and subsystems into a system.

A part-level failure model is not an abstraction; it is a mathematical description of the behavior of part failure caused by failure mechanisms and manifested by failure modes. This model describes the measure of central tendency of a metric that describes failure and a measure of dispersion of that metric.

A system integration model combines two or more part failure models to describe the effects of one or more part failures on the assembly created. Systems integration models combine assembly reliability models into subsystem reliability models that describe the effects of one or more part failures on the subsystem created. System design hierarchies can have many levels of combinations up to the single system design configuration. In every case, the system integration model describes the effects of one or more part failures on the design configuration modeled.

Indeed, the power of a system integration model is its capability to estimate top-level design configuration effects of part failure events.

Reliability Allocation

Design for reliability and sustainment engineering require design specifications to define the functional and material limits of a part. A statement of work to design a pump and piping system to move a liquid is meaningless unless design specifications include the quantity rate to be moved (gallons per minute), the distance and change in elevation that the liquid will be moved (1,000 ft laterally and 100 ft vertically), the description of the liquid (water [specific gravity = 1], slurry [water and solid material in suspension, specific gravity > 1], process chemical solution [specific gravity < 1]), etc. Part reliability is just such a design specification, but it is typically expressed at the system level: 90% probability that the system will function without failure for a mission duration equal to 10 h, under stated conditions of use, or the system will have a mean time between downing event (MTBDE) greater than or equal to 80 h.

Reliability allocation is facilitated with conversion of the system work breakdown structure (WBS) to a reliability block diagram (RBD), as illustrated in Figure 8.1. The RBD translates the WBS from a part list to a process

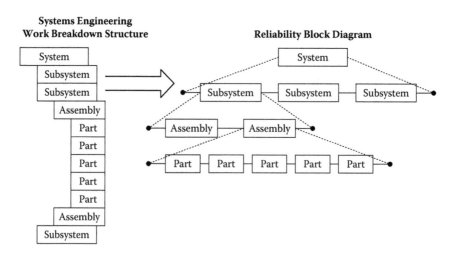

FIGURE 8.1
Work breakdown structure to reliability block diagram.

flow chart that shows the relationship between the subsystems to perform the function of the system. Each subsystem is flow charted to show the relationship between the assemblies to perform the function of the subsystem. Ultimately, the RBD shows the design configuration of the parts within each assembly.

Systems engineering allocates the system reliability down through the design configuration to the part. The reliability allocation process demands close coordination between design and systems engineering because not all parts require a reliability allocation and reliability allocations can be achieved by design configuration and design for maintainability. The design-for-reliability guideline presented earlier recognizes that part failure analysis performed by failure mechanisms, modes, and effects analysis identifies parts that do not cause system downing events. The iterative approach to part failure analysis yields updated failure modes and effects analysis (FMEA) that evaluate how part design analysis instate mitigation of failure effects on assemblies and high-level design configurations. Part maintainability analysis during design also instates mitigation of failure effects. But design or maintainability cannot mitigate what is not known: the reliability allocation.

Reliability allocation for a system given reliability expressed as a probability (90%) has a straightforward allocation procedure, equal allocation, or the nth root approach. The design hierarchy is expressed in a WBS, such that the system is described by its subsystems, subsystems are described by its assemblies, and assemblies are described by its parts. Initially, the design configuration is presented as a serial design unless the design specification requires a redundant parallel configuration. The reliability allocation from the system to subsystems is performed by calculating the nth root of the system reliability, where n is the number of subsystems. A system comprising three subsystems allocates the cube root,

$$\sqrt[3]{\text{Rsys}}$$

of the system reliability specification to each subsystem. A subsystem comprising five assemblies allocates the fifth root,

$$\sqrt[5]{\text{Rsubsystem}}$$

of the subsystem allocation to each assembly. An assembly composed of 20 parts allocates the 20th root,

$$\sqrt[20]{\text{Rassembly}}$$

of the assembly allocation to each part. An example of reliability allocation logic is illustrated in Figure 8.2.

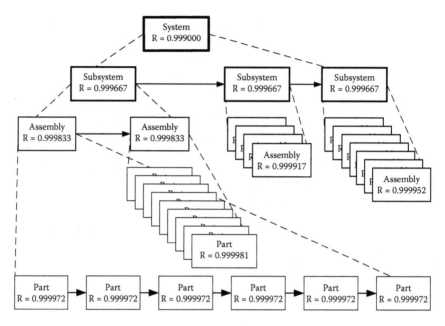

FIGURE 8.2
Reliability allocation.

Reliability allocation for a system given reliability expressed as a static (MTBDE = 80 h) also uses an equal allocation method. The reliability allocation from the system to subsystems is performed by calculating the product of the system MTBDE and n, where n is the number of subsystems.[2] A system composed of three subsystems allocates 3 (MTBDEsystem) = 240 h to each subsystem. A subsystem composed of five assemblies allocates 5 (MTBDEsubsystem) = 1,200 h to each assembly. N assembly composed of 20 parts allocates 20 (MTBDEassembly) = 24,000 h to each part.

Reliability allocations are not intractable specifications or requirements. They should be treated as guidelines or budgets that can be managed to allocate reliability or MTBDE units from parts that have an excess to parts that require additional allocation.

Reliability Math Model

Part reliability math models have been covered in previous chapters. The recommended model for part failure is the Weibull probability distribution. Common practice also uses the exponential probability model. The survival, mission reliability, and hazard functions are the reliability functions

TABLE 8.1

Part Reliability Model Expressions

Reliability Function	Exponential	Weibull
Survival Function	$S(t) = e^{-t/\theta} = e^{-\lambda t}$	$S(t) = e^{-(t/\eta)^\beta}$
Mission Reliability Function	$R(\tau/t) = e^{-\tau/\theta} = e^{-\lambda \tau}$	$R(\tau/t) = \dfrac{e^{-((t+\tau)/\eta)^\beta}}{e^{-(t/\eta)^\beta}}$
Hazard Function	$h(t) = \dfrac{1}{\theta} = \lambda$	$h(t) = \left(\dfrac{\beta}{\eta}\right)\left(\dfrac{t}{\eta}\right)^{\beta-1}$

commonly used as the building block to assembly, subsystem, and system integration reliability models, as shown in Table 8.1.

System reliability integration includes characterization of the MTBDE (θ); mean time to repair (MTTR), μ; mean downtime; and availability, A_i and A_o. Mean time between maintenance (MTBM) applies to a part and is sometimes erroneously applied to assemblies and above. It is true that MTBDE equals MTBM for a serial design; maintenance actions performed for a failed part place the system in a down state. Failed redundant parts require maintenance actions following a mission but do not place the system in a down state. Mean time between downing event is calculated as the indefinite integral of the system survival function:

$$\theta = \int S(t)\,dt \tag{8.1}$$

Consider a simple serial assembly comprising two parts: Part_1 failure is exponentially modeled with failure rate (λ) equal to 0.00055. Part_2 failure is Weibull modeled with characteristic life (η) equal to 320 h and shape parameter (β) equal to 2.67. The survival function is expressed as

$$\text{Sassy}(t) = \text{Spart1}(t)\ \text{Spart2}(t) = (e^{-\lambda t})e^{-\left(\frac{t}{\eta}\right)^\beta} = (e^{-0.00055t})e^{-\left(\frac{t}{320}\right)^{2.67}} \tag{8.2}$$

and the MTBDE is expressed as

$$\theta = \int \left[(e^{-\lambda t})e^{-\left(\frac{t}{\eta}\right)^\beta}\right]dt \tag{8.3}$$

MathCAD is used to solve for MTBDE (θ) as shown in Figure 8.3.

Recall that exponentially distributed part mean time between failure (MTBF) is the inverse of the failure rate (λ). The Weibull hazard function is defined to be the part instantaneous failure rate; therefore, the inverse of the hazard function is the continuous part MTBF, as shown in Figure 8.4.

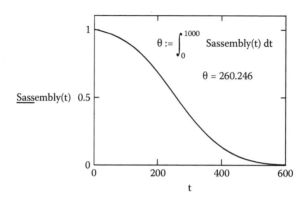

FIGURE 8.3
MTBF serial design.

Continuous part MTBF and system MTBDE are logically sound as they age. With the presumption of continued survival, the mean time to part failure or system downing event becomes smaller. Consider the choice between two cars; they are the same model, but one is new and the other is a used 5-year-old vehicle. The used car MTBDE is intuitively less than the new car MTBDE.

Mean time to repair (μ) is characterized by analysis, empirical methods, and judgment. Sufficient anthropomorphic studies are published to enable analytical characterization of maintenance actions. The analysis approach is required where part failure and maintenance actions pose system and personal risk, and it is beneficial to assess access to system maintenance points, fatigue, and personal lifting and carrying loads. An intangible benefit of the analytical approach is the thought process it requires. The maintenance task must be identified in specific terms.

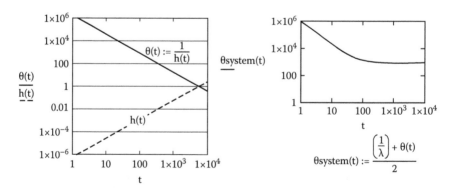

FIGURE 8.4
Part and system continuous MTBF.

Empirical methods are better suited to characterize MTTR. Maintenance actions for the actual part and system are developed as a hypothesis that is tested in the shop or the field. The results are evaluated and the hypothesis is modified. No analytical method can do that. But empirical methods are relevant only if qualified individuals perform the maintenance tasks. Consider the basic task of using an oxyacetylene torch to remove a bolt as a step in a maintenance procedure. The likelihood is high that the experiment will reflect an accurate result. Experienced, skilled trades folk make such tasks look easy.

Judgment is a close second to empirical methods for maintenance actions that are not complex or innovative. But judgment must be qualified; notice that the term "engineering judgment" was not used. Instead, the judgment is sought from skilled personnel from the trades, who have experience performing similar maintenance tasks.

The MTTR for a system is calculated as the weighted average of the MTTR for each part using the parts' hazard functions as the weight factors. The exponential failure rate is the commonly used example of the equation for system MTTR, as shown:

$$\mu_{\text{system}} = \frac{\sum \lambda_i \mu_i}{\sum \lambda_i} \tag{8.4}$$

where the subscript identifies the *i*th part failure rate and MTTR.

The exponential approach yields a constant value for system MTTR. The Weibull distribution for system MTTR is presented as

$$\mu_{\text{system}}(t) = \frac{\sum h_i(t)\mu_i}{\sum h_i(t)} \tag{8.5}$$

The Weibull form for system MTTR yields a continuous function that suggests system MTTR varies with age. System MTTR is calculated and plotted in Figure 8.5 by including part MTTR values of 2 and 10 h for parts 1 and 2,

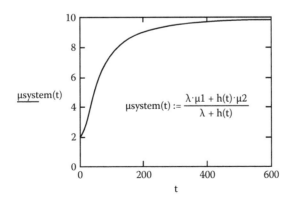

FIGURE 8.5
System MTTR.

respectively, with the known exponential failure rate and Weibull hazard functions.

The system MTTR curve is equal to the exponential part_1 MTTR at $t = 0$ because the Weibull hazard function is zero. The system MTTR climbs from 2 (the MTTR for part_1) to 10 (the MTTR for part_2) as the system ages. The curve suggests that system repair time increases as the system ages.

System availability is defined as the probability that a system will be able to begin a mission. The mission can be scheduled or carried out on demand. Inherent, operational, and achieved availability[3] (A_i, A_o, and A_a, respectively) are three common system availability measures. Inherent availability is a design metric developed by design engineers. Operational availability is an operations forecast developed by the system owner. Achieved availability is a benchmark measured of actual system performance over a period of time (i.e., latest week, month, quarter, or year).

Availability is a prediction of system uptime divided by predicted total time. Total time is the sum of uptime and downtime. Downtime is defined as maintenance time. Variations on availability are due to the components that define downtime. System uptime is consistently defined as the average time between maintenance, MTBDE. Design for reliability provides enough information to calculate system inherent availability—MTTR. The expression for inherent availability is provided in the following equation:

$$Ai = \frac{\theta}{\theta + \mu} \tag{8.6}$$

where $\theta^4 \equiv$ system MTBDE[4] and $\mu \equiv$ system MTTR.

Only systems that use the exponential failure distribution have constant availability as the system ages. Inherent availability based on the Weibull failure distribution is expressed in the following equation:

$$Ai = \frac{1/h_{system}(t)}{\left(1/h_{system}(t)\right) + \frac{\Sigma h_i(t)\mu_i}{\Sigma h_i(t)}} \tag{8.7}$$

System inherent availability is calculated in MathCAD using the system MTBDE and MTTR with part_1 and part_2 data. Inherent availability decreases over time, as shown in Figure 8.6. It is not a constant. A system becomes less available because frequency and duration of repair time increase as a system ages.

Sustainment engineering possesses sufficient information to calculate operational and achieved availability. Operational availability is a prediction of system uptime using MTBDE and predicted total time. As the system owner and maintainer, predictions for uptime and downtime are much more relevant than the predictions by design engineers for inherent availability. However, the downtime includes more than system MTTR; it can include mean calculations for pre- and postmaintenance logistical time, pre- and

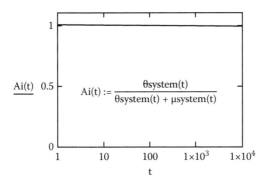

FIGURE 8.6
System inherent availability.

postmaintenance service time, administrative downtime, waiting for a part, waiting for labor, waiting for a facility, etc. All of the additional sources of downtime are added to MTTR, which decreases the value for availability from inherent to operational. Inherent availability done right is a system's ideal availability; it cannot get any better than that. Operational availability is a forecasting tool to plan operations capacity for a stated period of time (i.e., next week, month, quarter, year):

$$Ao = \frac{\theta}{\theta + \mu + \text{ALD} + \sum \text{Wait} + X} \tag{8.8}$$

where ALD is mean administrative downtime; $\sum Wait$ is the sum of all sources of time waiting for a part, labor, and facilities; and X is any other expected source of downtime associated with conducting business (e.g., regulatory inspections, absentees on the first day of hunting season, etc.).

Achieved availability is the actual system availability for the preceding period of time. Uptime is reported from operational records; downtime is reported from maintenance records. Sound maintenance management trends operational and achieved availability. The denominator of operational availability contains the elements of downtime that must be controlled and reduced to improve system achieved availability.

$$Aa = \frac{\theta_{actual}}{\theta_{actual} + \text{MDT}_{actual}} \tag{8.9}$$

where MDT $\equiv$ mean downtime.

Mention has been made about design configurations in preceding parts of this book, particularly the serial design configuration. Focus for design for reliability and sustainment has been on reliability models for individual parts. Reliability system integration combines individual parts into assemblies, assemblies into subsystems, and subsystems into the system. Survival, reliability, maintainability, and availability math models are developed to predict the behavior of assemblies, subsystems, and systems based on the behavior of part failure.

Design configurations used to integrate parts into assemblies include:

- serial design
- active parallel redundancy design
  - serial-in-parallel design
  - parallel-in-serial design
- *n*-provided; *r*-required design
- shared load design
- standby design
- equal hazard function, perfect switch
- unequal hazard function, perfect switch
- equal hazard function, imperfect switch
- unequal hazard function, imperfect switch

Design configurations used to integrate assemblies and subsystems to the next higher design level include:

- serial design
- active parallel redundancy design
  - serial-in-parallel design
  - parallel-in-serial design

The following design configurations are used to a lesser degree:

- standby design
  - equal assembly/subsystem, perfect switch
  - unequal assembly/subsystem, perfect switch
  - equal assembly/subsystem, imperfect switch
  - unequal assembly/subsystem, imperfect switch

Math Modeling for Design Configurations of Assemblies

Serial design is the most basic approach to configuring parts for an assembly. It consists of the minimum number of parts that will perform the specified function of the assembly. Each part has a linear interface with parts before and after it in the assembly, as shown in Figure 8.7.

Serial Design Configuration

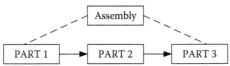

FIGURE 8.7
Serial design configuration.

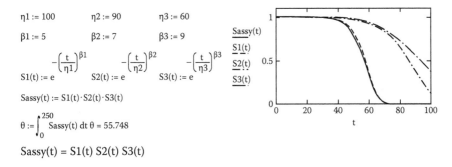

$\eta1 := 100 \qquad \eta2 := 90 \qquad \eta3 := 60$

$\beta1 := 5 \qquad \beta2 := 7 \qquad \beta3 := 9$

$S1(t) := e^{-\left(\frac{t}{\eta1}\right)^{\beta1}} \qquad S2(t) := e^{-\left(\frac{t}{\eta2}\right)^{\beta2}} \qquad S3(t) := e^{-\left(\frac{t}{\eta3}\right)^{\beta3}}$

$Sassy(t) := S1(t) \cdot S2(t) \cdot S3(t)$

$\theta := \int_0^{250} Sassy(t)\, dt \quad \theta = 55.748$

$Sassy(t) = S1(t)\, S2(t)\, S3(t)$

FIGURE 8.8
Serial survival function and MTBF.

$$\tau := 8$$

$$Rassy(t) := \frac{Sassy(t + \tau)}{Sassy(t)}$$

$$Rassy(\tau) = 0.999886$$

FIGURE 8.9
Assembly reliability function.

$\mu1 := 4 \qquad \mu2 := 3 \qquad \mu3 := 1$

$\theta1 := \int_0^{250} S1(t)dt \qquad \theta2 := \int_0^{250} S2(t)dt \qquad \theta3 := \int_0^{250} S3(t)dt$

$\theta1 = 91.817 \qquad\qquad \theta2 = 84.189 \qquad\qquad \theta3 = 56.818$

$$\mu assy := \frac{\left(\dfrac{\mu1}{\theta1}\right) + \left(\dfrac{\mu2}{\theta2}\right) + \left(\dfrac{\mu3}{\theta3}\right)}{\left(\dfrac{1}{\theta1}\right) + \left(\dfrac{1}{\theta2}\right) + \left(\dfrac{1}{\theta3}\right)}$$

$\mu assy = 2.398$

FIGURE 8.10
Assembly MTTR.

$$\text{Ai:} = \frac{\theta}{\theta + \mu\text{assy}} \qquad \text{Ai} = 0.959$$

FIGURE 8.11
Assembly inherent availability.

The assembly serial survival function is expressed as the product of the part survival functions:

$$\text{Sassy}(t) = \prod \text{Spart}_i(t) = \text{Spart}_1(t) \times \text{Spart}_2(t) \times \cdots \times \text{Spart}_n(t) \qquad (8.10)$$

The survival function of the assembly is less than the lowest part survival function in the serial design. Serial design has the advantage of low cost and simplicity. It has the disadvantage that every part is a single point of failure for the assembly. Every part in a serial design is critical to the next higher design configuration.

The simple active redundant design combines two or more serial designs in an assembly that perform the same function as the initial serial design. Active redundancy describes a combination of serial designs that are fully operational; none is idle. The combination of two or more serial designs adds complexity to the next higher design configuration but introduces benefits that offset the disadvantages. Simple active redundant design has two forms:

- Parallel redundant design combines two or more of the same serial designs comprising the same parts.
- Redundant design combines two or more serial designs that differ in parts and composition.

Individual parts in one of the serial designs are not critical to the next higher design configuration.

Parallel design configurations are illustrated in Figure 8.12. The top two reliability block diagrams show a simple two-path redundancy of two different parts, 1 and 2, and a parallel redundancy of two equal parts, 1A and 1B. The bottom two reliability block diagrams show simple *n*-paths' redundancy of unequal parts, 1, 2, ..., *n*, and a parallel redundancy of *n* equal parts, 1A, 1B, ..., 1*n*.

The redundant survival function is expressed as shown in the following equation:

$$\text{Sassy}(t) = 1 - (1 - \text{Spart}_1(t))(1 - \text{Spart}_2(t)) \cdots (1 - \text{Spart}_n(t)) \qquad (8.11)$$

The two-path expression for the assembly survival function for equal parts is

$$\text{Sassy}(t) = 1 - (1 - \text{Spart}(t))^2 \qquad (8.12)$$

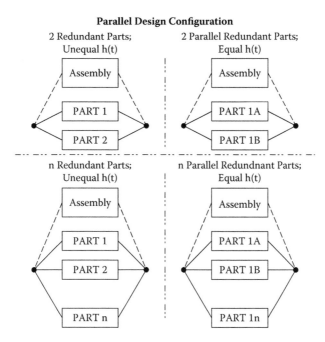

Parallel Design Configuration

FIGURE 8.12
Parallel design configuration.

The *n*-path expression for the assembly survival function for equal parts is

$$\text{Sassy}(t) = 1 - (1 - \text{Spart}(t))^n \tag{8.13}$$

The redundant design survival function is greater than the maximum survival function of the individual serial designs. This is an important distinction. Consider a serial design with a power supply (PS1) that is the weak link in an assembly. Reliability improvement for the assembly dictates that the power supply must be improved. A design modification that adds a redundant power supply (PS2) will improve the reliability of the assembly. But the redundant power supply need not be the same as the first.

Failure analysis for PS1 yields a characteristic life of $\eta1 = 90$ h and a shape parameter of $\beta1 = 6.6$. A second power supply is selected with characteristic life of $\eta2 = 75$ h and shape parameter of $\beta2 = 5.33$. The survival functions are written in MathCAD for PS1, PS1 and PS2 in redundancy, and PS1 in parallel redundancy. The plots for the redesigned configurations illustrate that the survival function for PS1 is the lowest of the three; PS1 and PS2 in redundancy are higher than PS1 alone, and PS1 in parallel redundancy is the highest of the three (Figure 8.13). The power supply example also illustrates

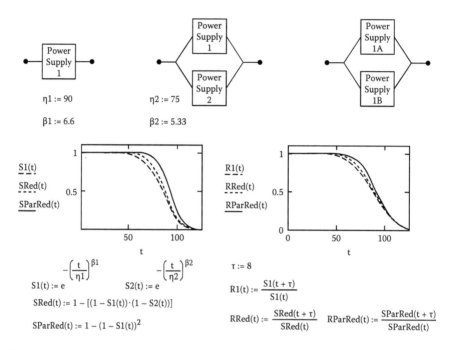

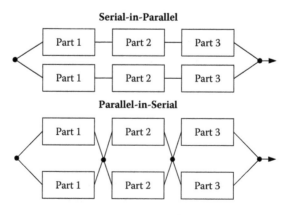

FIGURE 8.13
Redundant and parallel redundant survival and reliability functions.

that the reliability function is higher for the redundant unequal power supply and highest for the parallel redundant equal power supply.

Redundant designs are configured as either serial in parallel or parallel in series, as shown in Figure 8.14. The reliability for an assembly that is configured as a series-in-parallel design configuration is calculated in two steps, as

FIGURE 8.14
Serial in parallel: parallel in serial.

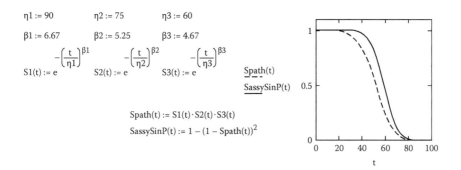

$$\eta1 := 90 \qquad\qquad \eta2 := 75 \qquad\qquad \eta3 := 60$$

$$\beta1 := 6.67 \qquad\qquad \beta2 := 5.25 \qquad\qquad \beta3 := 4.67$$

$$S1(t) := e^{-\left(\frac{t}{\eta1}\right)^{\beta1}} \qquad S2(t) := e^{-\left(\frac{t}{\eta2}\right)^{\beta2}} \qquad S3(t) := e^{-\left(\frac{t}{\eta3}\right)^{\beta3}}$$

$$\text{Spath}(t) := S1(t)\cdot S2(t)\cdot S3(t)$$

$$\text{SassySinP}(t) := 1 - (1 - \text{Spath}(t))^2$$

FIGURE 8.15
Serial-in-parallel survival function.

shown in Figure 8.15. The procedure for evaluating the survival function for a serial-in-parallel design configuration is presented in the following two steps:

1. Solve for the reliability of each serial path.
2. Solve for the reliability of the assembly treating each path as two redundant parts.

The procedure is illustrated in Figure 8.15 using an example for three parts: part one with a characteristic life, $\eta1 = 90$, and shape parameter, $\beta1 = 6.67$; part two with a characteristic life, $\eta2 = 75$, and shape parameter, $\beta2 = 5.25$; and part three with a characteristic life, $\eta3 = 60$, and shape parameter, $\beta3 = 4.67$. The survival function for the path, Spath(t), is calculated for the serial design configuration as the product of the survival functions of the three parts. Then the survival function of the assembly, SassySinP(t), is calculated for the parallel design configuration of the two paths.

The reliability for an assembly that is configured as a parallel-in-series design configuration is calculated in two steps:

1. Solve for the reliability of each redundant combination of parts.
2. Solve for the reliability of the assembly treating each redundant combination of parts as a serial path.

The procedure is illustrated in Figure 8.16, using the same example for three parts. The survival function for each part in parallel redundancy is calculated for the three parts. The survival function of the assembly is calculated for the three parallel parts in series.

A variation of redundant design configuration includes a combination of two or more parts with a stipulated minimum number of failed parts. An active redundant design is functional until all parts fail. An assembly can

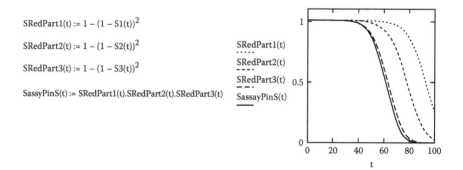

$\text{SRedPart1}(t) := 1 - (1 - S1(t))^2$

$\text{SRedPart2}(t) := 1 - (1 - S2(t))^2$

$\text{SRedPart3}(t) := 1 - (1 - S3(t))^2$

$\text{SassyPinS}(t) := \text{SRedPart1}(t).\text{SRedPart2}(t).\text{SRedPart3}(t)$

SRedPart1(t)
SRedPart2(t)
SRedPart3(t)
SassayPinS(t)

FIGURE 8.16
Parallel in serial.

experience a down state even if one or more parts remain functional. The redundant design configurations include the following.

n-Provided, r-required is a design configuration where at least *r* parts out of *n* parts remain functional before an assembly downing event occurs, as shown in Figure 8.17. Each part functions at its design level independently of the failure of other parts. This design configuration at the assembly or higher level allows one or more assemblies to be taken off-line for maintenance for systems that are scheduled for continuous service. The assembly reliability is calculated using the binomial probability distribution:

$$\text{Sassy}(t) = \sum_{i=r}^{n} \left(\frac{n!}{i!(n-i)!} \right) \text{Spart}(t)^{i} (1 - \text{Spart}(t))^{n-i} \qquad (8.14)$$

There are two special cases of the *n*-provided, *r*-required design configuration: (1) the *n*-provided, (*n* − 1)-required; and (2) the *n*-provided, (*n* − 2)-required. The expressions for these two design configurations are included in Figure 8.17.

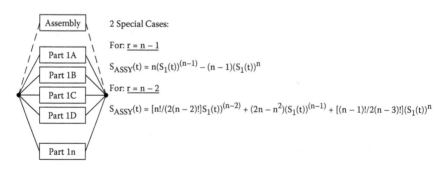

2 Special Cases:

For: r = n − 1

$S_{\text{ASSY}}(t) = n(S_1(t))^{(n-1)} - (n-1)(S_1(t))^n$

For: r = n − 2

$S_{\text{ASSY}}(t) = [n!/(2(n-2)!]S_1(t))^{(n-2)} + (2n - n^2)(S_1(t))^{(n-1)} + [(n-1)!/2(n-3)!](S_1(t))^n$

FIGURE 8.17
n-Provided, *r*-required design configuration.

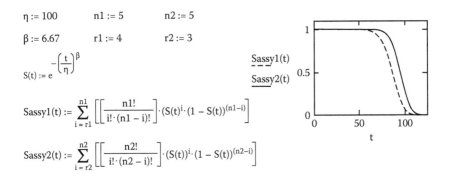

$\eta := 100$ $n1 := 5$ $n2 := 5$

$\beta := 6.67$ $r1 := 4$ $r2 := 3$

$$S(t) := e^{-\left(\frac{t}{\eta}\right)^{\beta}}$$

$$\text{Sassy1}(t) := \sum_{i=r1}^{n1} \left[\left[\frac{n1!}{i! \cdot (n1-i)!} \right] \cdot (S(t))^{i} \cdot (1 - S(t))^{(n1-i)} \right]$$

$$\text{Sassy2}(t) := \sum_{i=r2}^{n2} \left[\left[\frac{n2!}{i! \cdot (n2-i)!} \right] \cdot (S(t))^{i} \cdot (1 - S(t))^{(n2-i)} \right]$$

FIGURE 8.18
n-Provided, *r*-required 5:3 and 5:4 design configuration.

An example for the *n*-provided, *r*-required design configuration survival function is presented in the illustration in Figure 8.18. A design makes use of five parts in parallel redundancy. One design alternative requires that four parts be functional during a mission; the second alternative requires this of only three parts. The part has a characteristic life, $\eta = 100$, and a shape parameter, $\beta = 6.67$. The expressions for the survival function for the two design configurations are written and plotted in MathCAD. The survival function of three parts required is higher than that for four parts required. As the number of parts required reduces to one, the survival function increases. The design configuration becomes an active parallel redundancy when the number of parts required equals one. The design configuration reduces to a serial design when the number of parts required equals the number of parts provided.

Shared load is a design configuration where surviving parts must bear the load of failed parts until the load exceeds the strength of the surviving parts, as shown in Figure 8.19. Part function is not independent of the failure of other parts. This design configuration is often mistaken for *n*-provided, *r*-required.

$$\text{Sassy}(t) = e^{2\lambda 1 t} + \frac{2\lambda 1}{2\lambda 1 - \lambda 2}(e^{-\lambda 2 t} - e^{-2\lambda 1 t}) \tag{8.15}$$

where $\lambda 1 \equiv$ failure rate for each part when both are functioning[5]:

$$\text{Sassy}(t) = e^{-2\lambda 1 t} + 2\lambda 1 t e^{-\lambda 2 t} \tag{8.16}$$

where $2\lambda 1 = \lambda 2$ (the prior expression is undefined when $2\lambda 1 - \lambda 2 = 0$).

An example of assembly survival functions, Sassy12(*t*) and Sassy22(*t*), for the two shared load design configurations is shown in Figure 8.19. The solid line plot illustrates the assembly survival function where the failure rate of the surviving part is not twice the failure rate of the two parts, Sassy12(*t*).

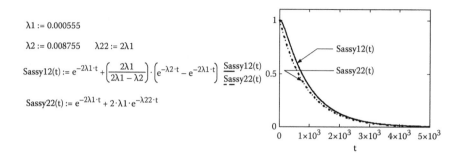

$\lambda 1 := 0.000555$

$\lambda 2 := 0.008755 \qquad \lambda 22 := 2\lambda 1$

$\text{Sassy12(t)} := e^{-2\lambda 1 \cdot t} + \left(\dfrac{2\lambda 1}{2\lambda 1 - \lambda 2} \right) \cdot \left(e^{-\lambda 2 \cdot t} - e^{-2\lambda 1 \cdot t} \right)$

$\text{Sassy22(t)} := e^{-2\lambda 1 \cdot t} + 2 \cdot \lambda 1 \cdot e^{-\lambda 22 \cdot t}$

FIGURE 8.19
Shared load assembly survival functions.

The dashed line plot illustrates the assembly survival function where the failure rate of the surviving part is equal to twice the failure rate of the two parts, Sassy22(*t*). The current published work for shared load uses the exponential probability distribution and is limited to two parts. A Weibull distribution option with more than two parts is presented later in this chapter.

Standby is a design configuration where a primary part is provided with an inactive standby part that operates when the primary part fails, as shown in Figure 8.20. Standby design adds a new part to the assembly: a switch mechanism that redirects the load from the primary part to the standby part. Standby design can take one of four forms.

Equal failure rates–perfect switch: the standby part is identical to the primary part; the reliability of the switch is assumed to be perfect, *Rsw* = 1. The logic for the reliability of a standby assembly with equal parts and perfect switch is shown in Figure 8.21. The assembly will function if the primary part functions **OR** if

$$S(t) = e^{-\lambda t}(1 + \lambda t) \tag{8.17}$$

where $\lambda \equiv$ failure rate of the primary and standby parts.

An example of a standby assembly survival function is shown in Figure 8.22. The primary and secondary parts are the same and have equal reliability with an exponential failure rate of $\lambda 1 = 0.000555$ failure per hour.

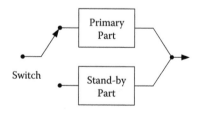

FIGURE 8.20
Basic standby design configuration.

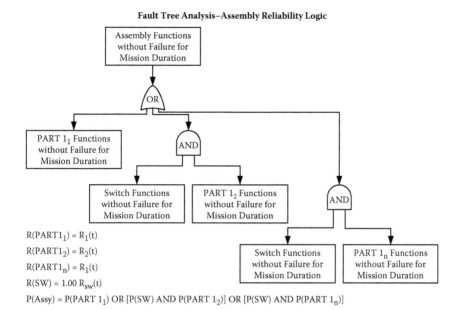

$R(PART1_1) = R_1(t)$
$R(PART1_2) = R_2(t)$
$R(PART1_n) = R_1(t)$
$R(SW) = 1.00 R_{sw}(t)$

$P(Assy) = P(PART\ 1_1)\ OR\ [P(SW)\ AND\ P(PART\ 1_2)]\ OR\ [P(SW)\ AND\ P(PART\ 1_n)]$

FIGURE 8.21
Fault tree logic for parts in standby with switch mechanism.

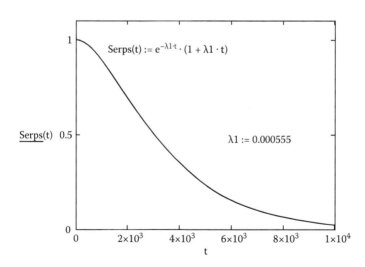

FIGURE 8.22
Standby assembly survival function: equal reliability, perfect switch.

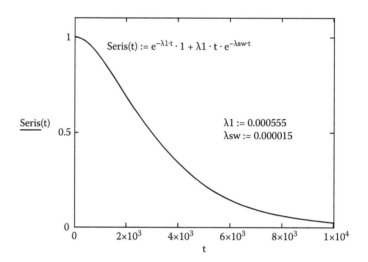

FIGURE 8.23
Standby assembly survival function: equal reliability, imperfect switch.

Equal failure rates–imperfect switch: the standby part is identical to the primary part; the switch is known to have failure mechanisms and modes:

$$S(t) = e^{-\lambda t}(1 + \lambda t e^{-\lambda swt}) \tag{8.18}$$

where $\lambda \equiv$ failure rate of the primary and standby parts and $\lambda sw \equiv$ failure rate of the switch.

An example of the standby survival function for the same assembly with an imperfect switch is shown in Figure 8.23. The switch has an exponential failure rate of $\lambda sw = 0.000015$ failure per hour.

Unequal failure rates–perfect switch: the standby part is different from the primary one and has a lower reliability; the reliability of the switch is assumed to be perfect, $Rsw = 1$:

$$S(t) = e^{-\lambda 1 t} + \frac{\lambda 1}{\lambda 2 - \lambda 1}(e^{-\lambda 1 t} - e^{-\lambda 2 t}) \tag{8.19}$$

where $\lambda 1 \equiv$ failure rate of the primary part and $\lambda 2 \equiv$ failure rate of the standby part.

The next example of a standby assembly survival function has a secondary part with an exponential failure rate of $\lambda 2 = 0.008755$ failure per hour (Figure 8.24).

Unequal failure rates–imperfect switch: the standby part is different from the primary and has a lower reliability; the switch is known to have failure mechanisms and modes:

$$Sassy(t) = e^{-\lambda 1 t} + e^{-\lambda swt}\left(\frac{\lambda 1}{\lambda 2 - \lambda 1}\right)(e^{-\lambda 1 t} - e^{-\lambda 2 t}) \tag{8.20}$$

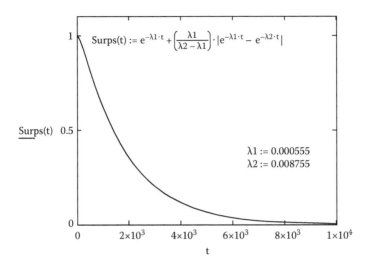

FIGURE 8.24
Standby assembly survival function: unequal reliability, perfect switch.

where $\lambda 1 \equiv$ failure rate of the primary part; $\lambda 2 \equiv$ failure rate of the standby part; $\lambda sw \equiv$ failure rate of the switch.

The last standby assembly survival has the same parts as in the previous example and an imperfect switch (Figure 8.25).

The utility of theoretical expressions for reliability, maintainability, and availability functions has drawbacks to practical systems integration. First

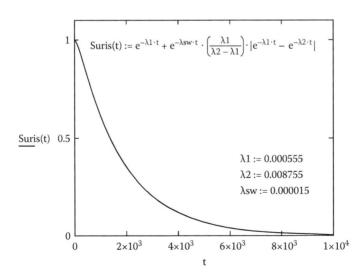

FIGURE 8.25
Standby assembly survival function: unequal reliability, imperfect switch.

and foremost is the complexity of the expressions because the number of parts, assemblies, and subsystems is large. The math becomes monstrous, unwieldy, and downright impossible. Second is the inability to factor in part replacement during the life cycle of a system. Consider a single assembly composed of three parts. The survival function of each part and the assembly follows the theoretical expressions as long as none of the parts are replaced following failure. When a part is replaced, its survival function starts over and is evaluated for operating time of zero. Meanwhile, the other parts continue to age as the theoretical survival functions show. But the assembly survival function adjusts to the combination of aging parts and the new part. Then, another part fails and then another. The survival functions for the parts and assembly are illustrated in Figure 8.26.

Theoretical reliability, maintainability, and availability expressions cannot include part dependency. Consider a serial design with five parts. When one part fails, the remaining four cease to age until the maintenance action restores the serial path. The reliability, maintainability, and availability expressions for each part predict downtime due only to part failure and assume that the remaining parts continue to function. This error extends to assemblies and subsystems that are treated as functioning when actual performance shows that they become idle. Reliability simulation software factors part downtime due to failure of dependent parts.

The same problem exists for the hazard function for a single part, as shown in Figure 8.27. The hazard function on the left describes the instantaneous failure rate until the part fails, at which time the function returns to the initial state of $t = 0$, as shown on the right. The effect is also experienced for the assembly availability, as shown in Figure 8.28.

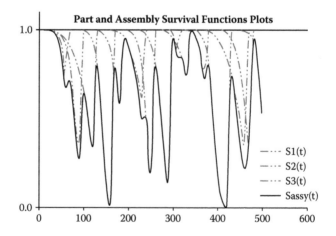

FIGURE 8.26
Part and assembly actual survival function.

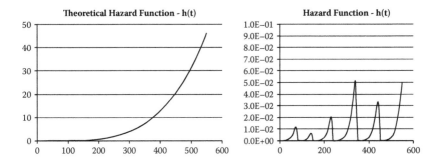

FIGURE 8.27
Theoretical versus actual hazard function.

System reliability integration is impossible to perform unless reliability software is used. Such software utilizes Monte Carlo methods to simulate the behavior of part failure on system reliability, maintainability, and availability functions. Two operating scenarios are simulated: mission duration and life cycle.

Mission duration simulation runs system operation for one mission and serves to evaluate one parameter: system reliability. The number of runs to achieve a statistically significant simulation result is calculated as a function of the minimum reliability specification and the confidence level, as follows:

$$n = \frac{\ln(\alpha)}{\ln(R)} \tag{8.21}$$

where α, the level of significance, is equal to $1 - C$, the confidence level, and R is the minimum system reliability specification.

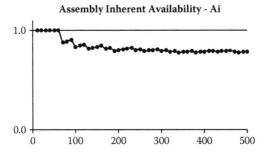

FIGURE 8.28
Actual inherent availability.

		Reliabilily				
Confidence	Alpha	90%	95%	99%	99.90%	99.99%
90%	0.10	22	45	229	2,301	23,025
95%	0.05	28	58	298	2,994	29,956
99%	0.01	44	90	458	4,603	46,049

FIGURE 8.29
Mission duration number of runs.

The number of mission duration simulation trials is tabulated in Figure 8.29 for 90, 95, and 99% confidence levels and reliability requirements from 90 to 99.99%. Reliability software can run thousands of trials in a few hours or less depending on the capability of the computer. The simulation output for mission duration will tabulate the minimum, mean, and maximum values for reliability and will include the standard deviation, as shown in Figure 8.30.

Reliability is a higher-is-best metric, so the lower confidence limit (LCL) is calculated from the mean, number of trials, standard deviation, and the standard normal z-score as follows:

$$LCL = \mu + z\left(\frac{\sigma}{\sqrt{n}}\right)$$

where $z = -1.645$, $n = 298$, $\sigma = 0.00245$, and $\mu = 0.999958$.

Life-cycle reliability simulation runs for the system expected life and serves to compute MTBDE, mean number of system downing events, MTTR, MDT, part consumption, and availability. The number of runs to achieve a statistically significant simulation result is determined from observing the standard deviations of the life-cycle parameters until they converge on a steady-state number, typically after 50–75 runs. A single life-cycle run can last 15–30 min. The number of mission duration runs can take 10s of hours, while 50 runs may take only several hours (life-cycle runs are often allowed to run overnight). The simulation output for mission duration will tabulate the minimum, mean, and maximum values for the maintainability and availability parameters and will include the standard deviation, as shown in Table 8.2.

Number of trials			298			
Min	Mean	Maximum	SD	LCL	$\alpha =$	0.05
0.999001	0.999958	0.999992	0.00245	0.999725	$z =$	−1.645

FIGURE 8.30
Mission duration simulation output table.

TABLE 8.2

Life-Cycle Simulation Output Table

Parameter	Min	Mean	Max	Number of Trials 298 SD	LCL	UCL	$\alpha = 0.05$
Availability	0.935	0.977	0.992	0.00245	0.976767		$z = -1.645$
MTBDE	165	226	235	35.1	222.66		−1.645
MDT	2.33	4.51	8.22	0.95		4.60	1.645
MTTR	1.25	3.74	5.67	0.85		3.821	1.645
System failures	35	42	46	1.33		42.13	1.645

Availability and MTBDE are higher-is-best metrics, so the LCL is calculated from the mean, number of trials, standard deviation, and the standard normal z-score, as for the reliability LCL. Mean downtime, MTTR, and the number of system failures per life-cycle run are lower-is-best metrics, so the upper confidence limit (UCL) is calculated from the mean, number of trials, standard deviation, and the standard normal z-score, as shown:

$$UCL = \mu + z\left(\frac{\sigma}{\sqrt{n}}\right)$$

where $z = +1.645$.

Next, practical applications for the use of reliability simulation software are presented for the basic design configurations.

Series Design Configuration

The first practical application views a hydraulic pump and actuator assembly as shown in Figure 8.31. The design configuration consists of a reservoir, pump, and actuator. There are two approaches to developing the reliability block diagram form of this design configuration: The first is to use each of the three primary parts as line replaceable units (LRUs), or parts, and the second is to break out the parts that consist of the interfaces between the reservoir and the pump and the actuator, as shown on the right-hand side of Figure 8.31. There is ample anecdotal evidence of the common mistake of failing to include interface parts in a serial design configuration.

Failure analysis provides failure and repair parameters, including the mean time between failures (μ) for parts that are exponentially distributed, characteristic life (η), shape parameter (β) for parts that are Weibull distributed, and mean time to repair (μ) (Figure 8.32).

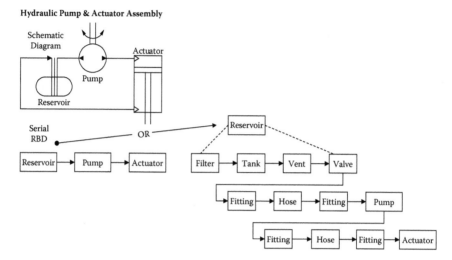

FIGURE 8.31
Serial reliability design configuration.

The data are entered into a reliability block diagram drawn in a reliability software program called RAPTOR. A mission duration simulation is run 1,000 times for a mission duration of 16 h, and the summary output table is printed as shown in Table 8.3.

Analysis of the output data tells us that at least one mission failed for a reliability of zero, at least one mission did not fail for reliability of one, and

Serial Failure and Repair Parameters				
	θ	η	β	μ
Filter	60			1
Tank		4000	9.6	3
Vent	1000			2
Valve		900	8 33	2
Fitting		750	7.25	1
Hose		600	6.67	1.5
Fitting		750	7.25	1
Pump		160	4.8	4
Fitting		750	7.25	1
Hose		600	6.67	1.5
Fitting		750	7.25	1
Actuator		2000	9.33	6

FIGURE 8.32
Serial reliability design and repair data.

TABLE 8.3

Serial Mission Duration Simulation Report

Results from 1,000 Runs of Sim Time: 16; z = 2.053749					
Parameter	Min	Mean	Max	SD	SEM	95% LCL
Reliability	0	0.761	1	0.426686139	0.013493	0.733289

TABLE 8.4

Serial Life-Cycle Simulation Report

Results from 50 Runs of Sim Time: 1,000; z = 2.053749						
Parameter	Min	Mean	SD	SEM	95% LCL	95% UCL
Availability	0.935201169	0.948235882	0.CO5592973	0.000790966	0.946611436	
MTBDE	24.610557	32.468289	4.580918	0.64784	31.13778831	
MDT	1.55929	1.748193	0.1001	0.014156		1.777266
System failures	22	29.76	4.068596	0.575386		30.9417

the mean and reliability for the assembly is 0.761. This software program provides us with a standard deviation and the standard error of the mean (SEM). The SEM is equal to the standard deviation divided by the square root of sample size. The sample size is equal to the number of runs (1,000). The standard normal z-score for 95% confidence is found to be 2.05. Refine the lower confidence limit of reliability to be equal to the mean minus see times a standard error of the measurement, 0.733289.

Next, a life-cycle simulation is run for 50 runs of 1,000 h. We found that the standard deviation of the parameters converged at 50 runs (Table 8.4). Analysis of the output data estimates the availability of the assembly to be greater than 0.9466 with 95% confidence. The assembly has a mean time between downing events of at least 31.14 h. The assembly experience has a mean downtime of no more than 1.777 h. System failures can be expected not to exceed 30.9, or 31, with 95% confidence.

Parallel Design Configuration

The next practical application is a crushing process plant that is a serial-in-parallel design configuration. The serial path consists of a primary crusher, a conveyor belt, ball mill, conveyor belt, hammer mill, elevator, and covered tank, as shown in Figure 8.33.

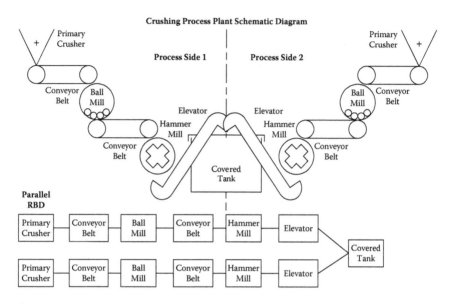

FIGURE 8.33
Parallel reliability design configuration.

The reliability block diagram is drawn in the reliability software program, and the heart failure and repair parameters are entered into the blocks. All of the parts are Weibull distributed and have a mean time to repair as shown in Table 8.5. The mission duration simulation is run for 1,000 runs of 8 h each. The output data for the mission duration simulation are shown in Table 8.6.

Analysis of the mission duration output shows that the reliability is at optimum for minimum mean and maximum time. It is unreasonable to expect that any system has 100% reliability all of the time.[6] However, the parallel design configuration assures us that at least one of the two paths is operating continuously during the mission duration of 8 h. Note that, at this point, reliability does not equal capacity. The reliability shown here means that both paths did not experience a downing event at the same time. A life-cycle simulation of 50 runs for 1,000 h is run for the process plant; its output is presented in Table 8.7.

Analysis of the life-cycle simulation shows that the availability is indeed much lower than the reliability because we have a mean time between downing events for a path less than the mission duration of 8 h. Mean downtime of a path is almost 2 h, and system failures are in the hundreds. We can readily evaluate a single path by running a serial simulation. The output of a serial simulation for mission duration is shown in Table 8.8. Analysis of the serial path mission duration simulation data shows that the reliability is at least 98%.

TABLE 8.5

Parallel Failure and Repair Data

	θ	η	β	μ
Primary crusher		40	9.45	6
Conveyor belt		16	8.15	2
Ball mill		24	7.9	4
Conveyor belt		16	8.15	2
Hammer mill		32	7.25	3
Elevator		48	5.67	5

TABLE 8.6

Parallel Mission Duration Simulation Report

	Results from 1000 Runs of Sim Time: 8				
Parameter	Min	Mean	Max	SD	SEM
Reliability	1	1	1	0	0

TABLE 8.7

Parallel Life-Cycle Simulation Report

	Results from 50 Runs of Sim Time: 1,000; $z = 2.053749$					
Parameter	Min	Mean	SD	SEM	95% LCL	95% UCL
Availability	0.739642373	0.766918599	0.011275867	0.001594648	0.763644	
MTBDE	4.774589	5.947719	0.415328	0.058736	5.82709	
MDT	1.568618	1.804427	0.106711	0.015091		1.83542
System failures	115	129.5	8.264134	1.168725		131.9003

TABLE 8.8

Serial Path Mission Duration Simulation Report

	Results from 1,000 Runs of Sim Time: 8; $z = 2.053749$				
Parameter	Min	Mean	SD	SEM	95% LCL
Reliability	0	0.988	0.108939744	0.003444977	0.980925

n-Provided, *r*-Required Redundancy

The next practical exercise is for the design configuration of a cyclone system used to capture particulate exhaust material from a dryer. The design configuration has five cyclones provided and three required, as shown in Figure 8.34. The consequences from cyclone failure have no impact on the process plant functionality. However, regulatory consequences of discharging particular matter into the atmosphere carry heavy penalties. This design allows one cyclone to be taken off-line for maintenance while the process plant functions. Design of the dryer and cyclones requires that three cyclones function in order for the plant to meet air exhaust standards.

The reliability block diagram is drawn in the reliability software and the data for failure to repair parameters are entered into the blocks as shown in Table 8.9. The output for a mission duration simulation of 1,000 runs for a 16-h mission is shown in Table 8.10. Again we find its design achieves a high reliability. Taking a cyclone off-line does not reduce plant capacity—unlike the previous example for the crushing process plant.

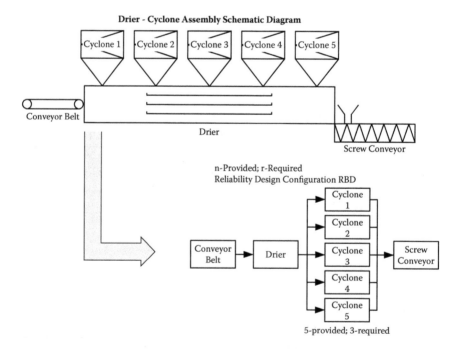

FIGURE 8.34
n-Provided, *r*-required reliability design configuration.

TABLE 8.9

n-Provided, *r*-Required Failure and Repair Data

	θ	η	β	μ
Conveyor belt		16	8.15	2
Drier	320			
Cyclone		60	6.67	6
Screw conveyor		80	9.33	5

TABLE 8.10

n-Provided, *r*-Required Mission Duration Simulation Report

Results from 1,000 Runs of Sim Time: 16					
Parameter	Min	Mean	Max	SD	SEM
Reliability	1	1	1	0	0

The applet for the life-cycle simulation of 25 runs of 1,000 h is shown in Table 8.11.

The high availability demonstrates the effectiveness of the design of the dryer and cyclones. The mean time between downing events is well in excess of the typical mission duration. The mean downtime is a parameter management that can track to implement process improvements, not unlike a pit crew during a NASCAR event. The number of system failures provides input to sustainment engineers for possible selection of upgraded parts or components.

TABLE 8.11

n-Provided, *r*-Required Life-Cycle Simulation Report

Results from 25 Runs of Sim Time: 1,000; $z = 2.053749$						
Parameter	Min	Mean	SD	SEM	95% LCL	95% UCL
Availability	0.971731068	0.989232132	0.006275401	0.00125508	0.986655	
MTBDE	88.339188	296.499978	242.903178	48.580636	196.7275	
MDT	0.434283	2.326862	1.063758	0.212752		2.763801
System failures	1	4.96	2.621704	0.524341		6.036865

Standby Redundancy

The use of reliability software to simulate standby redundancy design configurations is presented next.

Equal Reliability: Perfect Switch

This practical application shows a hammer mill that has a primary power supply and a secondary power supply. An electronic switch mechanism is able to sense the loss of power from the primary unit and instantaneously switch over to the secondary power supply. The electronic switch mechanism is assumed to be perfect. Consequences from loss of power to the hammer mill extend well beyond the failure and maintenance action to repair the power supply. A hammer mill that stops functioning with a full load has to be emptied by hand. The downtime to the system is lengthy and unacceptable. The secondary power supply is the same unit as the primary to assure uninterrupted power (Figure 8.35).

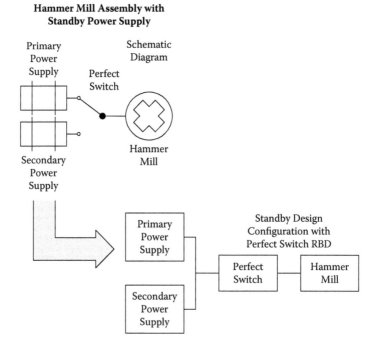

FIGURE 8.35
Standby equal reliability design configuration with perfect switch.

TABLE 8.12

Standby Equal Reliability, Perfect Switch Failure, and Repair Data

	θ	η	β	μ
Primary power supply	900			8
Secondary power supply	900			8
Perfect switch				0
Hammer mill		32	7–25	3

TABLE 8.13

Standby Equal Reliability, Perfect Switch Mission Duration Simulation Report

Results from 1,000 Runs of Sim Time: 8					
Parameter	Min	Mean	Max	SD	SEM
Reliability	1	1	1	0	0

The power supply failure repair data are presented in Table 8.12. A mission duration simulation for the standby assembly consisting of the primary and secondary power supplies and the perfect switch is performed with the reliability software. The mission duration simulation output is presented in Table 8.13. The hammer mill is not included in this simulation.

Analysis of the mission duration simulation data shows that the power supply has a nearly perfect reliability. The output data from the life-cycle simulation of 250 runs for 1,000 h are presented in Table 8.14. Analysis of the life-cycle simulation data shows that the occurrence of system failures is a rarity. The incomplete data in the table for mean time between downing events are the result of the lack of system failures to provide a statistically significant estimator for the mean.

TABLE 8.14

Standby Equal Reliability, Perfect Switch Life-Cycle Simulation Report

Results from 250 Runs of Sim Time: 1,000; $z = 2.053749$						
Parameter	Min	Mean	SD	SEM	95% LCL	95% UCL
Availability	0.992610302	0.99992325	0.000706219	0.000044665	0.999831519	
MTBDE	992.610302	>999.923250	N/A	N/A		
MDT (three runs)	5.044688	6.395816	1.212636	0.700116		7.833678
System failures	0	0.012	0.109104	0.0069		0.026171

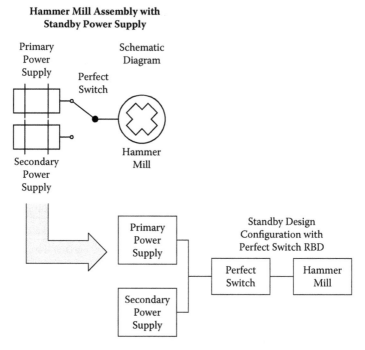

FIGURE 8.36
Standby unequal reliability design configuration with perfect switch.

Unequal Reliability: Perfect Switch

The next practical application is the same design of the power supply and hammer mill, except that a trade-off study is performed to evaluate the reliability functions for a secondary power supply that is not the same as the primary power supply (Figure 8.36).

The failure and repair parameters presented in Table 8.15 show that the secondary power supply has a significantly lower characteristic life than the primary power supply. The output data for a mission duration simulation of a thousand runs of 8 h are presented in Table 8.16. Analysis of the mission duration simulation data shows that the reliability of the design alternative is as good as the use of the same power supply for the secondary power supply. The output data from the life-cycle simulation for 250 runs of 1,000 h are presented in Table 8.17. Note that a null standard deviation causes a valve entry in the confidence limit cell.

Analysis of the life-cycle simulation data shows that the difference between the two design alternatives is statistically insignificant. The data suggest that the less expensive, unequal secondary power supply is a cost-effective approach that will not increase the risk in operations.

TABLE 8.15

Standby Unequal Reliability, Perfect Switch Failure and Repair Data

Standby Unequal Reliability Perfect Switch Failure and Repair Parameters				
	θ	η	β	μ
Primary power supply	900			8
Secondary power supply	300			5
Perfect switch				0
Hammer mill		32	7.25	3

TABLE 8.16

Standby Unequal Reliability, Perfect Switch Mission Duration Simulation Report

	Results from 1,000 Runs of Sim Time: 8				
Parameter	Min	Mean	Max	SD	SEM
Reliability	1	1	1	0	0

Equal Reliability: Imperfect Switch

The next practical application is a pump system composed of two equal pumps with an imperfect switch. The imperfect switch mechanism is known to have a statistically significant failure mechanism and mode that can be expected to interrupt pumping operations. The purpose of the switch mechanism is to switch from the primary pump to the secondary pump when the primary pump fails (Figure 8.37).

The failure and repair parameters for the primary and secondary pumps and the imperfect switch are presented in Table 8.18. The reliability block diagram entered into the reliability software includes a power supply, the switch, and the two pumps. The output data for a mission duration simulation of 1,000 runs of 8 h is presented in Table 8.19. Analysis of the mission duration

TABLE 8.17

Unequal Reliability, Perfect Switch Life-Cycle Simulation Report

	Results from 250 Runs of Sim Time: 1,000; z = 2.053749					
Parameter	Min	Mean	SD	SEM	95% LCL	95% UCL
Availability	0.995364333	0.999880616	0.000655728	0.000041472	0.999795	
MTBDE	995.364333	>999.880616	N/A	N/A		
MDT (nine runs)	1.029046	3.31623	1.206834	0.402278		4.142408
System failures	0	0.036	0.186664	0.011806		0.060247

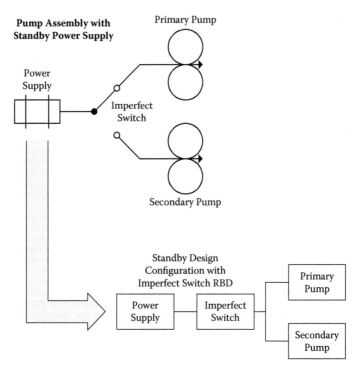

FIGURE 8.37
Standby equal reliability design configuration with imperfect switch.

TABLE 8.18

Standby Equal Reliability, Imperfect Switch Reliability, and Repair Data

Standby Equal Reliability, Imperfect Switch Failure, and Repair Parameters				
	θ	η	β	μ
Power supply	900			8
Imperfect switch	550			3
Primary pump		160	4.8	4
Secondary pump		160	4.8	4

TABLE 8.19

Standby Equal Reliability, Imperfect Switch Mission Duration Report

Results from 1,000 Runs of Sim Time: 8				
Parameter	Min	Mean	Max	SD	SEM
Reliability	1	1	1	0	0

TABLE 8.20

Standby Equal Reliability, Imperfect Switch Life-Cycle Simulation Report

		Results from 250 Runs of Sim Time: 1,000; $z = 2.053749$				
Parameter	Min	Mean	SD	SEM	95% LCL	95% UCL
Availability	0.995364333	0.999880616	0.000655728	0.000041472	0.999795	
MTBDE	995.364333	>999.880616	N/A	N/A		
MDT (nine runs)	1.029046	3.31623	1.206834	0.402278		4.142408
System failures	0	0.036	0.186664	0.011806		0.060247

simulation data shows a high reliability for the design. The output table for the life-cycle simulation of 250 runs for 1,000 h is provided in Table 8.20.

The life-cycle simulation data show a high availability and an MTBDE that exceeds the mission duration by a large margin. Mean downtime and system failures are opportunities for improvement.

Unequal Reliability: Imperfect Switch

The last reliability simulation is for unequal pumps with an imperfect switch mechanism, as shown in Figure 8.38. The failure and repair data

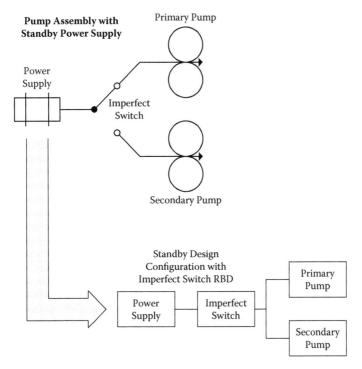

FIGURE 8.38
Standby unequal reliability design configuration with imperfect switch.

TABLE 8.21

Standby Unequal Reliability, Imperfect Switch Reliability, and Repair Data

Standby Unequal Reliability, Imperfect Switch Failure, and Repair Parameters				
	θ	η	β	μ
Power supply	900			8
Imperfect switch	550			3
Primary pump		160	4.8	4
Secondary pump		75	3.33	2

TABLE 8.22

Standby Unequal Reliability, Imperfect Switch Mission Duration Report

Results from 1,000 Runs of Sim Time: 8					
Parameter	Min	Mean	Max	SD	SEM
Reliability	1	1	1	0	0

for the pump assembly are provided in Table 8.21 and the output data for a mission duration simulation of 1,000 runs for 8 h are presented in Table 8.22. Analysis of the mission duration simulation shows a high reliability. The life-cycle simulation was performed for 250 runs of 1,000 h, as shown in Table 8.23.

The availability dropped significantly from 0.9998 to 0.975 by changing to a less reliable pump. The MTBDE was dropped from over 990 h to 159 h. System failures increased from near zero to over six. Clearly, this design poses significant maintainability and availability loses that suggest continued use of equal pumps.

TABLE 8.23

Standby Unequal Reliability, Imperfect Switch Life-Cycle Simulation Report

Results from 250 Runs of Sim Time: 1,000; $z = 2.053749$						
Parameter	Min	Mean	SD	SEM	95% LCL	95% UCL
Availability	0.96597644	0.97525311	0.002989382	0.000189065	0.974865	
MTBDE	120.747055	158.961964	17.930769	1.134041	156.6329	
MDT	3.49558	3.985655	0.222956	0.014101		4.014615
System failures	5	6.212	0.693527	0.043863		6.302084

Shared Load Redundancy[7]

Shared load redundancy describes an assembly where two or more parts bear a load that is distributed equally and that load is shared by surviving parts when one or more parts fail. Reliability software does not simulate shared load redundancy. Analysis of parts that fail piecemeal as each surviving part experiences larger loads is performed in software programs like MathCAD. An example is a pressure vessel end cap design with eight fasteners, as shown in Figure 8.39.

Empirical investigation of the load on the fasteners shows that the Weibull failure parameters for each bolt change as the shared load increases, as shown

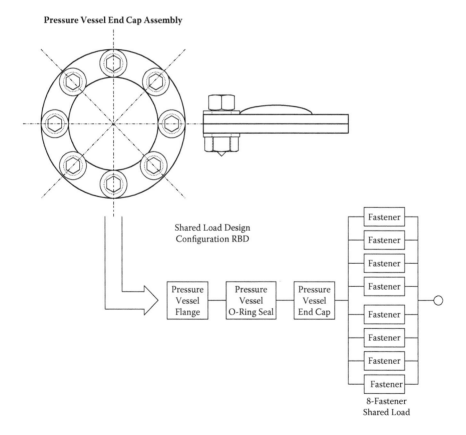

FIGURE 8.39
Shared load reliability design configuration.

TABLE 8.24

Shared Load Reliability Data

Bolts	η	β
8	56.50	6.85
7	55.09	6.68
6	50.41	6.11
5	40.58	4.92
4	24.55	2.98
3	7.49	1
2	1.35	1
1	0	1

in Table 8.24. The characteristic life for surviving fasteners decreases as the number of failed fasteners increases. From the bottom up, the tabulated data show that one to three surviving fasteners will fail rapidly.

The survival function for each number of surviving fasteners is written in MathCAD, as shown in Table 8.25. The expression S81(x) is the part survival function for one fastener with all eight fasteners intact. The expression Sassy8(x) is the assembly survival function for eight fasteners intact. The

TABLE 8.25

Shared Load Individual and Assembly Survival Functions and Assembly Hazard Functions

$$S81(x) := e^{-\left(\frac{x}{\eta 8}\right)^{\beta 8}}$$

$$S71(x) := e^{-\left(\frac{x}{\eta 7}\right)^{\beta 7}}$$

$$S61(x) := e^{-\left(\frac{x}{\eta 6}\right)^{\beta 6}}$$

$$S51(x) := e^{-\left(\frac{x}{\eta 5}\right)^{\beta 5}}$$

$$S41(x) := e^{-\left(\frac{x}{\eta 4}\right)^{\beta 4}}$$

$$S31(x) := e^{-\left(\frac{x}{\eta 3}\right)^{\beta 3}}$$

$$S21(x) := e^{-\left(\frac{x}{\eta 2}\right)^{\beta 2}}$$

$$Sassy8(x) := 1-(1-S81(x))^8$$

$$Sassy7(x) := 1-(1-S71(x))^7$$

$$Sassy6(x) := 1-(1-S61(x))^6$$

$$Sassy5(x) := 1-(1-S51(x))^5$$

$$Sassy4(x) := 1-(1-S41(x))^4$$

$$Sassy3(x) := 1-(1-S31(x))^3$$

$$Sassy2(x) := 1-(1-S21(x))^2$$

$$hassy8(x) := \frac{(f8(x))^8}{Sassy8(x)}$$

$$hassy7(x) := \frac{(f7(x))^7}{Sassy7(x)}$$

$$hassy6(x) := \frac{(f6(x))^6}{Sassy6(x)}$$

$$hassy5(x) := \frac{(f5(x))^5}{Sassy5(x)}$$

$$hassy4(x) := \frac{(f4(x))^4}{Sassy4(x)}$$

$$hassy3(x) := \frac{(f3(x))^3}{Sassy3(x)}$$

$$hassy2(x) := \frac{(f2(x))^2}{Sassy2(x)}$$

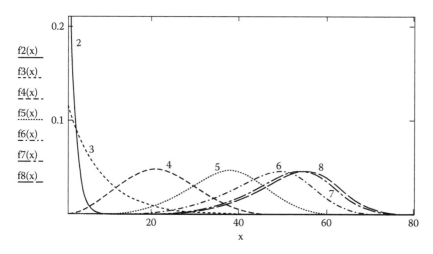

FIGURE 8.40
Shared load pdf by surviving bolts.

expression $f8(x)$ is the part probability density function (pdf) for one fastener with all eight fasteners intact. The expression hassy8(x) is the assembly hazard function with all eight fasteners intact.

The fastener probability density functions for surviving fasteners is plotted in Figure 8.40 and illustrates the probability density function of the respective material strength properties for eight through two survivors. The pdf labeled eight is the pdf for all eight fasteners intact. One can see the progression of the probability density functions as fasteners fail. The part survival functions for surviving fasteners are plotted in Figure 8.41 and illustrate the

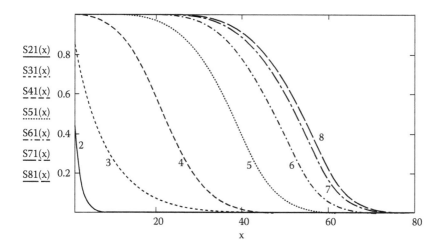

FIGURE 8.41
Survival function by surviving bolts.

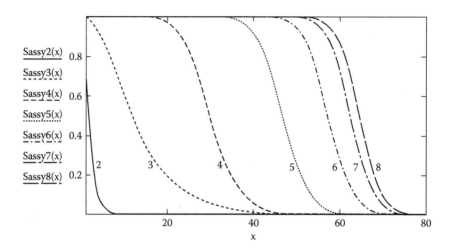

FIGURE 8.42
Assembly survival functions by surviving bolts.

decrease in capability of each survivor to withstand the shared load as fasteners fail.

The assembly survival function for surviving fasteners is plotted in Figure 8.42 and illustrates the decrease in capability of survivors to withstand the shared load as the number of fasteners fails.

The assembly hazard function is the goal to understanding shared load redundancy. This function is compared to the maximum stress load and the organization's risk threshold as shown in Figure 8.43. First, we apply the

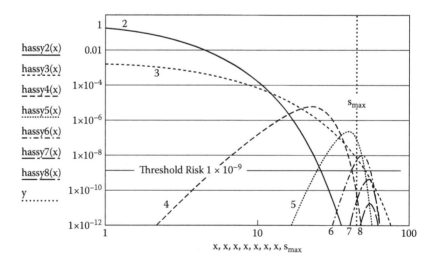

FIGURE 8.43
Assembly hazard function by surviving bolts with threshold risk.

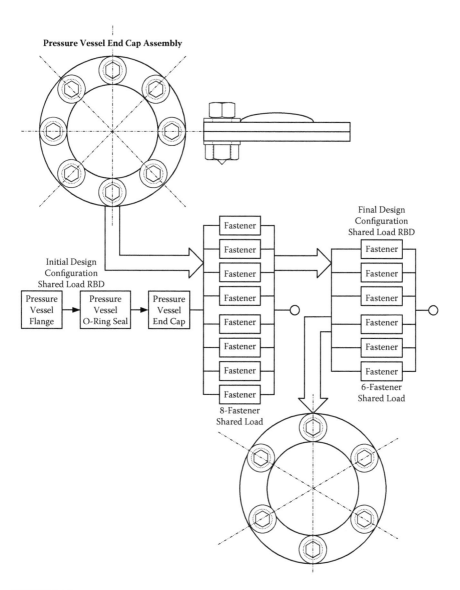

FIGURE 8.44
Design-for-reliability influence on final design.

failure theories that state that failure occurs when stress exceeds strength. The plot illustrates that two fasteners do not meet the requirement and that eight fasteners exceed the requirement. We see that three, four, and five fasteners are not acceptable and that seven and six fasteners are acceptable. Assume that the organization demands that the hazard function not exceed an instantaneous probability of failure above 1×10^{-9}. The risk threshold eliminates six fasteners and suggests that seven fasteners are sufficient to meet the design requirement.

Notes

1. Pukite, J., and P. Pukite. 1998. *Modeling for reliability analysis.* New York: IEEE Press.
2. System MTBDE is equal to the sum of subsystem MTBDE divided by the number of subsystems.
3. Instantaneous availability is a network electronics and digital equipment metric that is relevant for brief mission durations—less than a few hours, often minutes. It reduces to operational availability as mission time increases.
4. θ is also used to define part MTBF.
5. *The reliability engineer primer,* 3rd ed. 2002. Terre Haute, IN: Quality Council of Indiana.
6. RAPTOR limits data to nine decimal places and rounds to one for values greater than 0.9^91.
7. The shared load procedures presented here are the results of my research and are in progress as this book is being written. Many of the hypotheses presented as survival and hazard functions are in varying stages of investigation. I welcome comments and discussion on the content of this work in progress.

9

Reliability Analysis for System of Systems

> We can't solve problems by using the same kind of thinking we used when we created them.
>
> **Albert Einstein**

Introduction

A "system of systems" comprises two or more end systems. An end system is one that stands alone as it performs specific functions. It has a mission, mission duration, and conditions of use (both ambient and operational). It has unique reliability systems integration and sustainment requirements.

"System" is a vague term. It depends on an organization's perspective—the integrator versus the end user/O&M.[1] System complexity can confuse reliability integration and sustainment analysis. A conventional system is complex due to redundancy and is put in a degraded design state due to part failure (a parallel subsystem degrades to a serial design configuration) or is put in a down state due to part failure (a single point of failure). A system of systems integrates the functions of individual systems to achieve a capability or capacity. A system of systems is placed in a degraded mode of capability or capacity when a system is in a degraded or down state. Failure of a part cannot put a system of systems in a down state. That single fact makes the system of systems a special reliability case.

Reliability of a system is defined as the probability that the system will perform a specific mission without failure or down state. Mission analysis provides understanding and identification of functional failures or down states. It identifies the failure modes that cause the failure that prevents mission performance. A system mission is specific and singular, and it defines sustainment requirements that mitigate failure modes that directly prevent performing the mission. However, a system of systems exists when multiple missions with unique respective failure modes exist and different sustainment requirements are needed to prevent all mission failure events.

The confusion and vagaries can be cleared up by determination of the three mission scenarios performed by the system of systems.

System of Systems Logistical Support

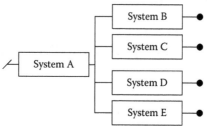

FIGURE 9.1
System of systems multiple missions.

Multiple missions is a system of systems composed of two or more systems that perform independent and distinctly different system missions. The system of systems may include one or more systems that support all or some of the individual systems; the system of systems provides a logistical support capability for the individual systems. The logic for a multiple-missions systems of systems is illustrated in Figure 9.1.

The illustrated system of systems comprises individual systems A through E with a common logistical support. Systems B through E require system A. The downing event for any system B through E has no impact on the functionality of the surviving systems. Operational scheduling is provided independently for systems B through E; operational scheduling for system A responds to demand from systems B through E. The system of systems provides four independent capabilities.

Simple single mission is a system of systems composed of systems that perform the same system mission. Systems may be identical or varied by manufacturer and age; the system of systems provides a logistical support capability for the individual systems. The logic for a simple single-mission system of systems is illustrated in Figure 9.2.

System of Systems Logistical Support

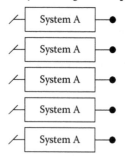

FIGURE 9.2
Simple single-mission system of systems.

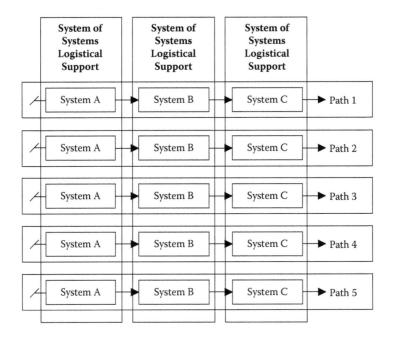

FIGURE 9.3
Complex single-mission system of systems.

The illustrated system of systems comprises a quantity (five) of individual systems A, with a common logistical support. The downing event for any system A has no impact on the functionality of the surviving systems. The system of systems achieves operational capability with a capacity that is a function of the number of surviving systems A.

Complex single mission is a system of systems comprising two or more functional paths where each path has a series of systems that performs a unique mission. The logistical support for the individual systems applies across the process paths. The logic for a complex single-mission system of systems is illustrated in Figure 9.3.

The illustrated system of systems is composed of five paths. Each path comprises three systems, A through C. System of systems logistical support is provided separately for systems A, B, and C. Operational scheduling is performed for paths 1 through 5. The operational capacity of the system of systems is defined by the paths.

Paths are either exclusive or shared. Exclusive paths restrict work flow from origin to output along the same path. All work in the path stops when a process system is in a down state. Shared paths allow transfer of work across paths. Work can change paths to bypass a process system that is in a down state. The exclusive and shared path work flows are illustrated in Figure 9.4.

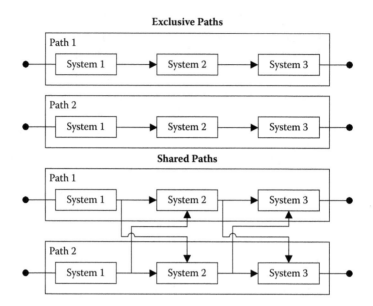

FIGURE 9.4
Exclusive and shared paths for complex single-mission system of systems.

Multiple-Missions System of Systems

Multiple-missions system of systems configuration is illustrated by the sea-based X-band radar (SBX). The U.S. Department of Defense (DoD) Missile Defense Agency (MDA) employs a system of systems to detect and destroy missile attacks on the United States. It acquired a land-based X-band radar system (XBR), installed it on an offshore oil well platform, and integrated it with other systems into the SBX. The SBX is a system of systems comprising[2] a number of systems.

The *X-band radar system* is composed of the following subsystems:

- tens of thousands of transmit/receivers in a thin-phased array subsystem that transmit and receive an X-band wave length
- the air-supported radome subsystem that protects phased array radar components from weather conditions
- the electronics equipment unit subsystem that controls the transmit signal, captures the received signal, and performs algorithms to analyze the data
- the liquid conditioning and cooling subsystem that captures and removes heat from the phased arrays
- operations control equipment

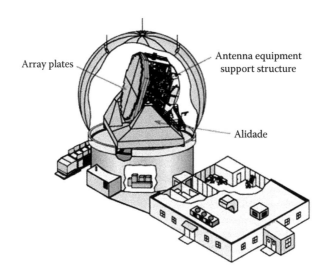

FIGURE 9.5
X-band radar system.

The land-based XBR system design boundaries assume that electrical power is provided by a local power utility or a dedicated power generation and control system and that water supply for the liquid cooling subsystem is acquired locally. The land-based XBR system is illustrated in Figure 9.5.

The X-band radar has a single mission: track targets. Each subsystem has a single mission that is essential to the XBR system mission no matter how complex the design configuration is. The XBR is not a system of systems because none of the subsystems can stand alone as an independent system and perform an end function. Conversely, the XBR system will experience a downing event if one subsystem is in a down state. The other SBX subsystems include:

- the electrical power plant system, which provides electrical power at rated demand and serves all systems on the SBX
- the weather radar system, which describes weather conditions surrounding the platform and serves platform leadership at sea and in dock
- the guidance radar system, which calculates position at sea and serves platform leadership at sea
- the propulsion and stabilization system, which moves the platform at a demand speed and direction to maintain a stable deck and serves the XBR system and helipad operations

FIGURE 9.6
Sea-based X-band radar.

- the liquid pump and distribution system, which pumps and processes sea water to the platform and the XBR liquid conditioning and cooling subsystem and serves habitation and XBR systems
- the habitation system, which serves operations, maintenance, and personnel facility needs

The SBX system of systems is illustrated in Figure 9.6. The XBR radome is the prominent structure on the SBX. The weather and guidance radar systems are also visually evident on the platform surface. The electrical power plant, liquid pump and distribution, and habitation systems are located within the several decks, along with the remaining subsystems of the XBR—weather and guidance radar systems. The propulsion and stabilization system is located within the SBX structure and externally below the waterline. All integrated operations and maintenance facilities are located within the SBX structure. Each system can stand alone and perform an end function independently of the other systems. They were originally designed to function on land, rather than to be integrated on the SBX.

The weather and guidance radar systems can be in a down state while the XBR performs missions, just as the XBR can be in a down state while the weather radar and guidance radar perform missions. Only a downing event of the electrical power plant system is able to put the XBR, weather, and guidance radar systems in a down state, while the habitation system will experience a degraded mode and the propulsion and stabilization system will be fully functional. Each system functions for different and independent mission durations and requires unique maintainability policies and practices. Any operational and maintainability overlapping is by chance and well beyond the scope of the individual systems' designers and integrators.

Ostensibly, the SBX mission is to transport the XBR. But other SBX missions exist. For example, the SBX has the mission to travel at sea. Its mission duration is 24/7—continuous operation for the entire deployment, lasting several months. The mission of the XBR is to scan the skies for a target. Its mission duration is finite, measured in minutes, and performed repeatedly, possibly daily, over part or all of the deployment. Can the SBX perform its mission if the XBR is in a down state? Yes! Can the XBR perform its mission if the SBX cannot move in the water? Yes!

Are the sustainment requirements for the SBX systems identical? Do maintenance technicians have the same skills mix? Do they use the same specialty tooling and facilities? Do they use the same spare parts and spare parts strategies? No, on all counts. Hence, the SBX systems are independent and only share a common platform.

The power process plant system mission is unique because it is inextricably tied to the missions of the XBR, as well as every system on the platform. It serves in a serial design configuration to each SBX system. Its mission duration is 24/7 for the entire deployment. Its sustainment requirements are unique to its mission and common to the interfaces to the SBX systems. The system of systems reliability block diagram (RBD) for the SBX is illustrated in Figure 9.7.

The logic of the SBX reliability block diagram shows that all resident systems will be in a down state if the platform infrastructure is put in a down state. Such an event would be a catastrophic consequence from a fire, listing, or sinking. The power generation system is essential to the SBX systems, although demand on it is variable: Different resident systems have different demand times and loads. The propulsion and stability system is an end system and essential to XBR and helipad aviation systems at different demand times and loads. The liquid cooling system is an end system and is essential to the XBR system at different demand times and loads. The remaining end systems are independent end systems.

The focus of SBX systems integration was primarily on fitting systems on the platform by volumn-metric analysis and balancing weight distribution, as well as the elimination of unnecessary duplication of logistical support requirements. The fault tree analysis for an SBX downing event is elusive because the SBX is a community of disparate systems with disparate functions. The AND and OR logic does not apply to the SBX system of systems. Mission reliability demands an unambiguous mission statement, and the SBX does not have one.

For example, one could state that the mission for the current day is to move from the dock to a location at sea and the mission reliability of the propulsion and stabilization system would be relevant. The next day the mission changes to scan the horizon for a test launch of a target and

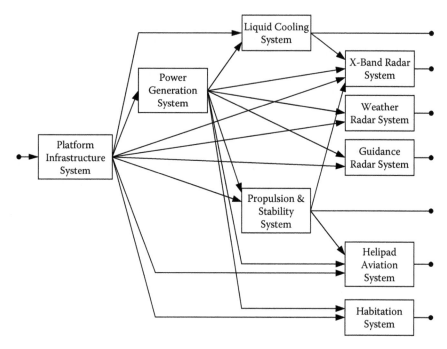

FIGURE 9.7
SBX reliability block diagram.

the mission reliability of the XBR would be relevant. Consider the weather radar for the two operating scenarios. Can the weather radar downing event cause the propulsion and stabilization or XBR systems to be in a downing state? No! However, the capability to predict the onslaught of violent weather fronts that can interrupt propulsion and stabilization or XBR operations is lost.

Maintainability fault tree analysis can be performed for the SBX. The goal is to characterize the demand for maintenance actions from the expected incidence of maintenance events, as shown in Figure 9.8. The parameters of maintainability are the demand for maintenance, mean time between downing event (MTBDE), and mean time to repair (MTTR). Reliability software performs analysis for each system to provide the 95% lower confidence limit (LCL) for MTBDE ($LCL_{95\%}$-$MTBDE_i$) and the 95% upper confidence limit (UCL) for MTTR ($UCL_{95\%}$-$MTTR_i$). Recall that the MTBDE is calculated as the indefinite integral of the system survival function. The system MTTR is calculated using Monte Carlo simulation of all part failures.

A maintenance policy that seeks to prevent an unscheduled system failure during a mission deployment will stipulate a system maintenance action prior to a system downing event. The mean time between maintenance actions (MTBMA) is initially based on the $LCL_{95\%}$-MTBDE. Notional data for

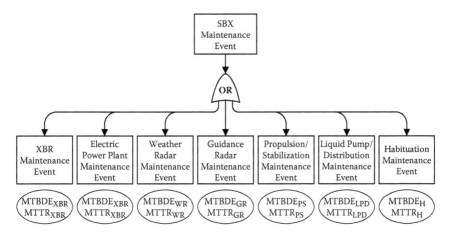

FIGURE 9.8
SBX fault tree analysis for maintenance events.

the SBX systems' MTBMA (MTBMA$_i$) and MTTR (MTTR$_i$) are tabulated by columns in Table 9.1.

Caution: The SBX mean time between maintenance actions (MTBMA$_{SBX}$) does not exist in practice and neither does an SBX MTTR.

System mission duration (τ) and deployment duration are two separate metrics used to evaluate system survival, reliability, and maintainability

TABLE 9.1

SBX Systems and SBX MTBDE, MTTR, and Mission Duration

System	MTBMA$_i$ Hours	MTTR$_i$ Hours	τ Hours	Operating Days	No. of Missions	Minimum Maintenance Events
XBR	56	6	2	16...164	150	5.36/6
EPP	1,008	36	4,328	−4...184	1	4.29/5
WR	198	20	4,320	1...180	1	21.82/22
GR	244	12	360	1...15; 165...180	2	2.95/3
PS	320	48	4,328	−4...184	1	13.50/14
LPD	4,320	18	4,328	−4...184	1	1.00/1[a]
H	4,320	192	4,328	−4...184	1	1.00/1[a]
SXB	N/A	N/A	4328	−4...184	1	2.89

[a] The number of maintenance actions for the LPD and H are 1 but are scheduled for after the completion of the deployment.

functions. The SBX performs 180-day deployments (4,320 h) and must be self-sufficient during that time. The SBX travels to a deployment location in the first 15 days of the deployment and returns to port on the last 15 days, during which the guidance radar system operates. The XBR performs a daily mission of 2 h duration during the time the SBX is on location, from day 16 to day 164. The electric power plan (EPP), power supply (PS), liquid pump and distribution (LPD), and habitation (H) systems operate continuously from 4 h before the deployment to 4 h after the deployment. The guidance radar (GR) system operates continuously from day 1 through day 15, and from day 165 through day 180. The weather radar (WR) system operates continuously from the first through the last day of the deployment.

The last column in the table is the minimum number of maintenance events for the deployment. The minimum number of maintenance actions expected to be performed during the deployment for each system is the time calculated to evaluate the survival function divided by the MTBMA and then rounded up to the next integer. The number of maintenance actions is the input to the logistical support analysis that specifies the maintenance skills, the number of maintenance personnel, the spare parts inventory, specialty tools, and facilities. (A calculated minimum number of maintenance actions of 4.01 is greater than 4, and there is no logical expectation of a fractional maintenance action.)

The survival function for each system is a measure of life-cycle or deployment risk that is evaluated for the total time that a system will function during a deployment equal to the product of the mission duration and the number of missions. The XBR survival function is evaluated for mission duration ($\tau = 2$ h) × 150 missions (300 h); the GR survival function is evaluated for $t = 360$ h × 2 missions (720 h); the WR survival function is evaluated for 4,320 h; and the remaining systems are evaluated for 4,328 h. The reliability function for each system is a measure of mission risk that is evaluated for the mission duration (τ).

From a design perspective, the engineers who design a system are not able to know every user organization's conditions of use. The original design conditions of use assumed ground benign ambient environments and only subsystem operational environments. Mil-HDBK-217 defines *ground benign* as

> Nonmobile, temperature and humidity controlled environments readily accessible to maintenance; includes laboratory instruments and test equipment, medical electronic equipment, business and scientific computer complexes, and missiles and support equipment in ground silos. Moderately controlled environments such as installation in permanent racks with adequate cooling air and possible installation in unheated buildings; includes permanent installation of air traffic control radar and communications facilities.

TABLE 9.2

Notional SBX Systems Integration Comparison

System	Conditions of Use	
	Ambient	Operational
Platform infrastructure	Naval unsheltered	Unique to subsystems operation
XBR	Naval unsheltered	
EPP	Naval sheltered	
WR	Naval unsheltered	
GR	Naval unsheltered	
PS	Naval unsheltered	
LPD	Naval sheltered	
H	Naval sheltered	

Mil-HDBK-217 defines naval sheltered and unsheltered as "includ[ing] sheltered or below deck conditions on surface ships and equipment installed in submarines ... unprotected surface shipborne equipment exposed to weather conditions and equipment immersed in salt water."

The ambient conditions as used on the SBX expose some systems to naval unsheltered environments and others to naval sheltered environments. The goal for reliability analysis is to provide the user with a known baseline of achieved reliability and the corresponding conditions of use constraints. The SBX system of systems integrator compares the design conditions of use to the actual conditions of use, as shown in Table 9.2. The SBX system of systems design must identify and mitigate stresses that the design did not consider (e.g., vibration from the propulsion system and salt spray corrosion).

The demand and practices for maintainability are dictated by the conditions of use. Comparison of the frequency and duration of maintenance actions describes the need for subsystem redundancies not originally provided that enable a system that functions continuously to be maintained without interruption of operation.

Simple Single-Mission System of Systems

A fleet of mine haul trucks and taxi cabs are examples of the simple single-mission system of systems. Caterpillar Tractor, Euclid, Komatsu, Liebherr, Mack, Atlas-Copco, and ESCO, to name a few, design and manufacture off-road haul trucks used in mining and construction. The haul truck is a system (see Figure 9.9).

FIGURE 9.9
Haul truck being loaded by excavator.

Mining organizations acquire off-road haul trucks and integrate them into a fleet of haul trucks used to haul overburden and ore (Figure 9.10). The fleet of haul trucks at a specific mine location is an example of the simple single-mission system of systems design configuration where a group of systems performs one or more common missions, mission durations, and conditions of use. Each system also has a combination of common and unique sustainment requirements. Different sustainment requirements result from varying missions, mission durations, and conditions of use. The system of systems RBD for the haul truck fleet is illustrated in Figure 9.11.

A fleet of mining haul trucks is rarely homogeneous (e.g., seldom are all trucks identical—that is, the same model from the same manufacturer distributed by the same vendor acquired the same year). Production capacity is measured in tons of ore moved per hour from the mine to the process plant. The life of a mine typically exceeds the useful life of a haul. Haul trucks are replaced when they age to the point of not being economically acceptable. A survey of any randomly selected mine operation will find a fleet that varies in the consumed useful life and capacity. In the latter case, mine production goals may change, requiring larger haul trucks than were initially acquired.

The ambient conditions of use for the systems comprising the system of systems are identical. The same cannot be said for operational conditions of use. Production scheduling will directly result in different operational conditions of use. Haulage missions can consist of overburden removal and

FIGURE 9.10
Fleet of haul trucks.

ore production. Overburden is excavated to expose ore and transported to other locations to reclaim mined out parts of the mine. Ore is excavated and transported to the process/preparation plant. Overburden removal mission duration is typically a single shift (8 h) or can be longer to take advantage of the sunlight.[3] Production ranges from straight time, 5 days/week, to three shifts/day, 5–7 days/week.

Overburden cycles are typically longer than production cycles; this means varying operational conditions of use on the haul truck subsystems. Overburden cycles load the power train for more time than production

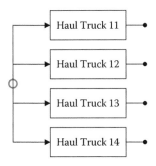

FIGURE 9.11
Haul truck fleet RBD.

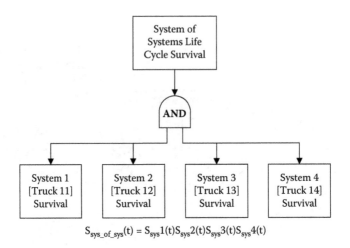

FIGURE 9.12
System of systems survival function logic.

cycles; production cycles load the hydraulic dump cycles more than over-burden cycles. Therefore, operational conditions of use for the systems comprising the system of systems can and will vary at the subsystem level.

The age of each system places different demand rates for spare parts. A mixed fleet of truck manufacturers also has an impact on spare part inventory requirements. The single-mission system of systems has a survival and reliability function, unlike the multiple-missions system of systems. The survival function is a measure of life cycle, capital recovery risk; the reliability function is a measure of system of systems mission capacity risk. The survival function logic is expressed in Figure 9.12.

Consider a mine haul truck fleet that has four trucks numbered as shown in Figure 9.11. Trucks 11 and 12 are scheduled for overburden removal; trucks 13 and 14 are scheduled for production. Overburden removal is scheduled for the 12 h of sunlight/day, 5 days/week. Ore production is scheduled for two 8-h shifts/day, 5 days/week. The overburden cycle is 3 miles and the production cycle is 1 mile. Excavation and truck loading time is equal for overburden and production at 10 min. Dumping time is equal for overburden and production at 5 min. Average truck speed is 10 mph. We can calculate truck and fleet capacity by scheduled activity as shown in Table 9.3.

The tabulated truck and fleet capacity assumes 100% survival and reliability for the week and day. Predicted actual capacity for the week and day is calculated as the product of the survival function evaluated for hours per week and the reliability function evaluated for hours per day and the respective capacity.

TABLE 9.3

Haul Truck Capacity Analysis

	Truck Capacity					
	Schedule		Cycles (Load)			
	t (h)	T/wk (h)	(min)[a]	(Per Day)	(Per Week)[b]	Fleet Capacity[c]
Production	16	80	21	45 7	225	450 Loads/wk
Overburden	12	60	33	21.8	105	210 Loads/wk

[a] Cycle time in minutes: [(1 mile/10 mph) * 60 min/h] + 10 min + 5 min = 21 min; [(3 mile/10 mph] * 60 min/h) + 10 min + 5 min = 33 min.
[b] Cycles/week based on cycles/day rounded down: min * (cycles/day) * 5 = cycles/wk
[c] Fleet capacity = truck cycles/wk * number of trucks.

Complex Single-Mission System of Systems

A production process line is an illustration of a complex single-mission system of systems. A production process is defined by three automated robotics systems that (1) pick up work at the origin and place it in a fixture; (2) prime, paint, and seal the work; and (3) pick up the work and place it on a conveyor that moves it to the next process line. Epson Robotics designs and manufactures pick-and-place robotics used in manufacturing. A pick-and-place robot is a system (see Figure 9.13).

The three robotics systems comprise a process path that has a single mission. Several process paths are employed and define a system of systems, as illustrated in Figure 9.14. Manufacturing organizations install multiple process lines to increase production capacity of one or more products. It is atypical that two or more process lines are homogeneous, although the production mission is the same. Process lines have systems of differing ages and manufacturers, as with the haul truck example.

Survival and reliability functions for complex system of systems are two-dimensional. The first dimension is the process path that defines the flow of work from the origin to the terminus. The second is perpendicular to the process path and is the system for the *n*th process point. Survival, reliability, and maintainability functions are essential for planning and controlling operations and maintenance of each process path, as well as for planning and controlling operations and maintenance of each system employed by the process paths. The system of systems RBD for multiple process lines is illustrated in Figure 9.15.

Maintainability requirements for each process path are focused on the interface equipment with common skills mix, spare parts, and specialty tools for the paths. Sustainment requirements for each system (maintainability practices 1, 2, and 3) require an equipment-specific skills mix, spare parts, and specialty tools.

FIGURE 9.13
Pick-and-place robot.

System maintainability is driven by two parameters: $MTBDE_i$ and $MTTR_i$ for each system across the paths. The $MTBDE_i$ is the indefinite integral of the system survival function, $S_{sys}(t)_i$. In the maintainability logic figure (Figure 9.16), the parameters of maintainability for the pick-and-place

FIGURE 9.14
Multiple process line of pick-and-place robots.

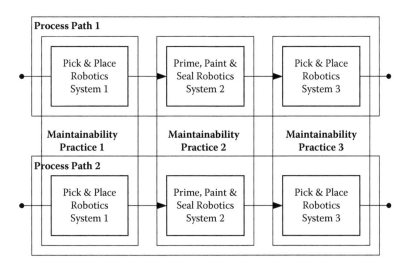

FIGURE 9.15
Multiple process line RBD.

robotics system 1 are the MTBDE and MTTR for the pick-and-place robotics across paths 1 and 2.

The survival function logic for a path shows that the path survival function, $S_{path}(t)$, is the product of the survival functions for the pick-and-place robotics system, $S_{sys}1(t)$; the prime, paint, and seal robotics system, $S_{sys}2(t)$; and the pick-and-place robotics system, $S_{sys}3(t)$. Process-path and total (all paths) capacity is characterized by the survival and reliability functions as it is for the simple single-mission system of systems.

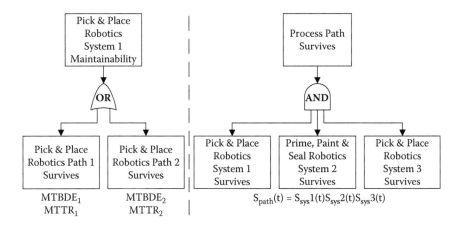

FIGURE 9.16
Complex single-mission system of systems maintainability and survival function logic.

TABLE 9.4

System of Systems and Conventional System Compared

| System of Systems Design Configuration | Conditions of Use | | | | |
	Mission	Duration	Ambient	Operational	Sustainment
Multiple mission	Different	Different	Different	Different	Different
Simple single mission	Different	Different	Common	Different	Different
Complex single mission	Different	Different	Common	Different	Different
Conventional system	Common	Common	Common	Common	Common

System of Systems Compared

The distinction between the three systems of systems design configurations and the conventional system is summarized in Table 9.4. Mission is the mission duration (*t*) for uninterrupted scheduled functionality. Duration is the time between scheduled maintenance actions. Sustainment is the sum of the logistical support resources.

The table illustrates that reliability system integration and sustainment best practices cannot be applied to a system of systems as if it were a conventional system. A system of systems does not have survival and reliability functions because system mission and durations differ. System of systems maintainability is driven by system ambient and operational conditions of use that differ. The system of systems sustainment for each system differs.

Notes

1. Design engineers are rarely confused by this distinction because they design parts and assemblies.
2. Information posted on the Web site for Federation of American Scientists.
3. Overburden removal is typically performed only during daylight hours for safety reasons. Mission durations can be scheduled up to two shifts/day or as extended day shifts during the summer.

10

Reliability-Centered Maintenance

System design reliability is an abstraction. System reliability is achieved by sustainment.

Bill Wessels

Introduction

System maintainability and sustainment happen one part at a time. Conventional maintainability fixes failed parts to restore system functionality; reliability-centered maintenance (RCM) controls part wear out to preserve system functionality.

> Conventional system sustainment reacts to demand for maintenance actions; RCM is proactive system sustainment that controls maintenance actions.

"Sustainment includes supply, maintenance, transportation, sustaining engineering, data management, configuration management, manpower, personnel, training, habitability, survivability, environment, safety, occupational health, protection of critical program information, anti-tamper provisions, and information technology," according to the *Defense Acquisition Guidebook*.[1] Industrial maintenance programs are defined by the same functions.

The Defense Acquisition University teaches that a successful and affordable system sustainment strategy includes understanding, measurement, and control of

- system availability
- system reliability
- system costs
- system downtime

The literature and sustainment practices suggest additional factors, including

- item unique identification (IUID)/serialized item management (SIM)
- failure reporting, analysis, and corrective action system (FRACAS)
- continuous process improvement (known as *kaizen*)
- maintenance planning
- obsolescence planning
- integrated supply chain management
- predictive modeling

Anecdotal evidence suggests that conventional maintenance programs are based on reactions to unscheduled maintenance events that prevent implementation of the systems sustainment strategy proposed by the Defense Acquisition University. Conventional maintenance practices vary from industry to industry as well as within industries, but there are two common activities:

repair/replacement of failed parts following an unscheduled system downing event during operations

periodic servicing, to include the following:

lubrication and fluid replenishment

inspection and replacement of consumable parts

inspection and adjustment of leaking/loose parts

Costs of maintenance practices include labor, materials, and overhead that are readily measured through conventional accounting methods. Well-documented system life-cycle costs include spare parts acquisition and storage—often the largest cost of system ownership. Excess spare parts consume operating capital and lack of spare parts incurs cost penalties. Less well-documented costs are those associated with safety hazards and lost opportunity from idle assets. Safety hazards incur tangible costs of damaged assets, lost-time employee accidents, regulatory agency penalties, and lawsuit settlements from fatal accidents. Intangible costs include reduced production by cautious employees and hazard inspections. Lost-opportunity costs are the result of production interruptions from repair or maintenance actions during scheduled operations and delays in maintenance actions waiting for maintenance resources and spare parts. Lost-opportunity costs can be prohibitively large and have an adverse impact on an organization's competitive position in its industry.

Reliability-centered maintenance is a departure from conventional maintenance that has been in practice for over 40 years. Its origin can be traced to the maintenance steering group formed by Boeing, the FAA, and United Airlines in the 1960s to create a maintenance program that would sustain the

747 aircraft. The need for RCM was the 747's technological leap over the preceding gradual evolution of aircraft design and development. The objectives of RCM were, and remain, safe and economically feasible system sustainment. Reliability-centered maintenance principles were rapidly adopted by NASA and the power-generation industry, especially nuclear power.

Aviation, NASA, and power-generation influences on the expansion of RCM to other public and private sector organizations are murky. Air Force and Navy aviation implemented RCM, but it is difficult to find documentation that describes their respective sources of RCM principles. The U.S. Army adoption of RCM lagged by over a decade after the other services and has been largely self-developed and focused on condition-based maintenance (CBM)—a single path of RCM. Private sector implementation of RCM has been a tough sell. Maintenance professionals view implementation of RCM as very expensive and difficult to perform. The flow down from the aviation origin of RCM to other industry sectors is illustrated in Figure 10.1.

Change does not occur easily in any organization without a dominant chief executive commitment that is well defined and backed up with resources. An excellent case study that is closely associated with implementation of RCM is Jack Welch at General Electric and the implementation of six-sigma quality. Management changes over the decades have been awash in fads that achieved

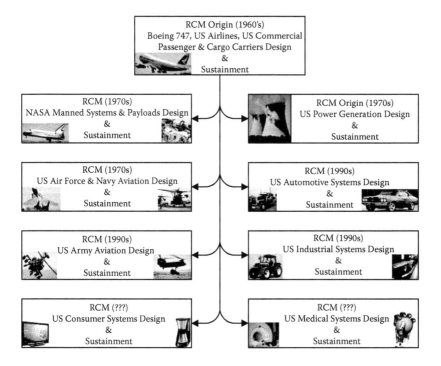

FIGURE 10.1
Historical RCM flow down.

little in economic return. Six-sigma transformed process quality. The conventional approach was to inspect process output for defects, which were then reworked or scrapped. Many organizations recognized an accepted quality level that included a proportion of defective output. Six-sigma demonstrated that world-class quality is achieved when each person in an organization owns his or her process step and is empowered by resources and authority to solve process problems that cause defects. The fundamental principle of six-sigma is that a process that is in control and capable is not able to produce a defective output.

The fundamental principle of RCM is to preserve system functionality.[2] This is in stark contrast to conventional maintenance, which is designed to repair a failed part after it fails.

Reliability-centered maintenance minimizes safety and lost-opportunity cost events to the point of total elimination and optimizes spare parts investment. The implementation of RCM demands a strong commitment from an organization's top executive level, but it need not demand a large investment.

Implementation of Reliability-Centered Maintenance

Reliability-centered maintenance is implemented through condition-based maintenance or time-directed maintenance (TDM). Condition-based maintenance uses understanding of part failure modes to measure degradation to use prognostic methods to schedule part replacement prior to a failure. Time-directed maintenance uses understanding of part failure modes to measure risk to use prognostic methods to schedule part replacement prior to a failure. A straightforward process leads from the identification of an RCM candidate part to the appropriate CBM or TDM path, as shown in Figure 10.2.

The RCM process may also yield a finding that no maintenance solution exists for the candidate part. This would include:

- Run to failure: the consequences of part failure incur costs lower than the cost to implement RCM.
- Engineering design: the consequences of part failure cannot be prevented by maintenance actions:
  - Design modification to system: the consequences of failure result from ambient and operational system conditions of use not anticipated by the system design (e.g., derated[3] design of original

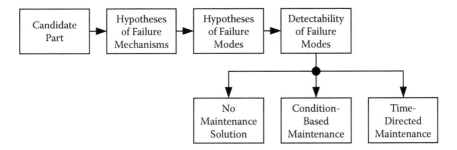

FIGURE 10.2
Logic for implementation of RCM.

equipment to increase materials properties that resist unique stress loads; insulation is designed to prevent thermal stresses or freezing unique to the organization's use; a limit switch is designed to prevent full range of operation of part functionality to prevent a power plant from overheating).

- Design modification to system interfaces: the consequences of failure result from ambient and operational interface conditions of use not anticipated by the system design (e.g., a barrier designed to prevent a haul truck from backing over the edge of a mine high wall while dumping overburden material; a high-pressure spray system designed to scrub liquid and particulate corrosive materials from a bulk materials handling machine; an overpack container designed to prevent exposure of precision machinery to vibration and humidity during transport from one job site to another).

- Administrative restrictions: the consequences of failure cannot be prevented by maintenance actions or engineering design:
  - limitations of operational conditions of use (e.g., impose speed and load limits on haul trucks to reduce failure mechanisms acting on brake and hydraulic parts)
  - limitation of ambient conditions of use (e.g., impose system isolation and shutdown during thunderstorms to prevent transient electrical shock)

The selection of RCM candidate parts begins with the part's failure effect on the system. An RCM candidate part is specified by the organization based on the system's conditions of use that are specific to the organization and the functional behavior of the systems that have an impact on the mission of the organization. It is not a task that can be performed in design, although design documentation is essential information that contributes to the RCM analysis.

Consider a model of a haul truck that is acquired by surface mining and construction operations. The operational and ambient conditions of use vary widely:

- Operational conditions of use differ within a location (i.e., removal of overburden requires different use of system assemblies than hauling ore—for example, distance of driving cycle and speed, frequency of backing up and braking, frequency of dumping).
- Ambient conditions of use differ by location (e.g., temperature, humidity, rain and ice, sand and salt exposure).
- Mission duration differs from 24/7 to 8 h/weekday shift operations.

Selection of an RCM candidate part is initiated by a review of maintenance records to identify the leading part failure events by

- effect
- cost impact
- frequency

This is a straightforward assessment—not a complex engineering analysis. The failure effect for a haul truck air brake can be catastrophic; the failure effect for a haul truck hydraulic bed lift hydraulic cylinder can be operational, and the failure effect for a haul truck radio can be a degraded mode. The failure effect for the previously mentioned hydraulic cylinder can occur infrequently yet incur high maintenance and lost-opportunity costs. On the other hand, a failed hydraulic hose may occur more frequently but have low maintenance and opportunity costs (Figure 10.3).

The findings of the selection of candidate parts are documented in the RCM critical items list (CIL). The organization develops an RCM critical items list to document part failures that have an impact on its economic health in its unique ambient and operational conditions of use—not to replicate the design CIL developed by the system manufacturer. The RCM critical items list ranks catastrophic part failure effects first; the ranking is identified in descending order:

personnel safety effects: fatal/permanent disability

personnel safety effects: lost-time injury

system safety effects: loss of system—capital asset

system safety effects: repairable damage beyond part maintenance— collateral damage

regulatory effects: compliance penalties—fines, shutdown, felony and tort legal actions

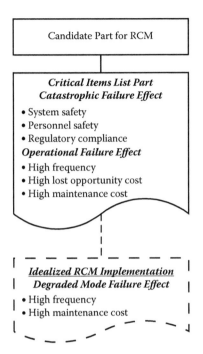

FIGURE 10.3
Logic diagram for selection of candidate part for RCM.

Part failure that has catastrophic effects must be mitigated through RCM analysis first. The outcome is that there is often no maintenance solution, other than run to failure.

Operational part failure effects are ranked in descending order of maintenance and lost-opportunity costs based on the organization's determination of priority:

- high system lost-opportunity cost from part failure: worst-case operational part failure scenario due to costs incurred
- low-frequency, high maintenance cost from part failure: often a low-visibility failure scenario, but incurs serious cost impact on the organization
- high-frequency, low maintenance cost part failure: high-visibility failure scenario, but least serious cost impact on organization

One may liken operational part failure scenarios with a rattlesnake bite, a bee sting, and gnat bites. The snake bite results in hospitalization and the bee sting is sore for days, but gnats are everywhere all of the time. The organization's determination of priorities is based on its unique exposure to rattlesnakes, bees, and gnats. Gnats will be an organization's priority if gnats are the only problem or if one breed of gnat is more prevalent than bees.

Catastrophic and operational failure effects cause unscheduled maintenance actions that interrupt scheduled mission operations; degraded mode failure effects do not. A degraded failure mode manifests reduced level or loss of functionality of a redundant part that will require maintenance action at the conclusion of the mission. Degraded modes do not cause damage to the system or a hazard to personnel. An organization applies RCM to degraded modes only if there are no remaining risks of a catastrophic or operational part failure.

Hypotheses of failure modes and mechanisms are performed using a modified fault tree analysis (FTA). The FTA is modified by not addressing the AND/OR logic of the failure modes and mechanisms. The objective is to identify failure modes and mechanisms that result in part failure regardless of whether they act individually or in combination with another mode and mechanism (Figure 10.4).

Emphasis is placed on failure modes that actually occur for the organization. This aspect of the analysis focuses on *relevant causes of failure* from the specific ambient and operational conditions of use experienced by parts for the organization's use of the system. Failure modes in a design failure analysis refer to changes in part geometry, material properties, conductivity, continuity, and other forms of degradation, as shown in Figure 10.5.

Loss of functionality must be sufficient for an RCM failure analysis. Knowing the distinction between hard and soft failures is appropriate for failures that may be mitigated through restart procedures for noncatastrophic failure effects. A soft failure that causes system shutdown is an RCM candidate when it causes collateral damage or unscheduled downtime. A drive pulley overload that causes a conveyor belt to stop will protect the drive motor and pulley from damage but will cause unscheduled downtime to dig out the belt load to restart the conveyor belt. A loss of a communications link

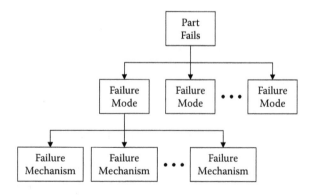

FIGURE 10.4
Modified FTA for RCM.

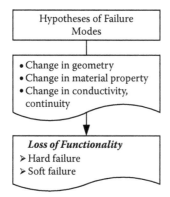

FIGURE 10.5
Logic for hypotheses of failure modes.

between haul truck and excavator operators that is restored by restarting the communication link is not an RCM candidate. Failure modes are documented in the RCM critical items list.

Failure mechanisms are hypothesized for each failure mode. Understanding the causes of failure of the line replaceable unit (LRU) is an absolute requirement that cannot be skipped. Failure mode, effects, and criticality analysis (FMECA) serve this purpose in design. Failure mechanisms must be understood in sustainment in the absence of an FMECA[4] or to supplement the work done in an FMECA. Evaluation of failure mechanisms remains necessary even when a design FMECA is available so as to include ambient and operating conditions not considered in the design analysis. Operating and ambient conditions of use that impose failure mechanisms realized by system users often differ from those considered in design (Figure 10.6).

A survey of historical failure records, brainstorming sessions with operations and maintenance personnel, and engineering judgment will identify realized failure mechanisms. The maintenance history is indicative of the failure mechanism's actions that are relevant to the failed part because they are the product of the actual conditions of use and the realized common causes and special causes of failure.

Common-cause failure describes intrinsic part material properties and defines how a part behaves in actual use. However, a part is subject to unique conditions of use by the using organization. Common-cause failure of the same part will differ based on an organization's use of the system. A portable generator used in Florida will be subject to common-cause failure different from that experienced by the same generator used in Wyoming. When it is used in a machine shop, the same generator will be subject to common-cause failures different from those experienced when it is used on a construction site. Common-cause variability must be understood and is an element in RCM failure analysis.

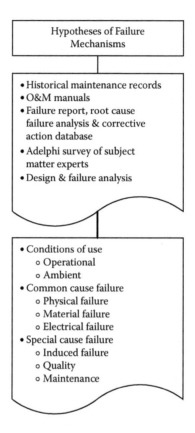

FIGURE 10.6
Logic for hypotheses of failure mechanisms.

Special-cause failure describes extrinsic part material properties and should not be allowed to occur. This type of failure is not explained by failure mechanisms; it is caused by doing the wrong thing through ignorance or negligence. Part design analysis assumes that special-cause failure does not occur. Special-cause failure can be summarized as the following:

- Induced failure results from poor workmanship, untrained personnel, and exceeding the specified ambient and operating conditions of use. Examples include fastener failure because threads were stripped by the maintainer, calibration performed by a trainee without supervision, and exceeding the load limits of a machine (Figure 10.7).

- Material failure results from use of the wrong part. Examples include fastener failure because it was the wrong grade, use of an ungrounded extension cord to supply AC power to a high-power tool, and using the wrong grade of lubricant in a machine (Figure 10.8).

FIGURE 10.7
Violation of operational conditions of use.

- Maintenance failure results from incorrect fault detection/fault isolation, use of incorrect maintenance procedures, and delaying maintenance actions. Examples include replacing a blown hydraulics hose when the overpressure was due to a failed valve, fastener failure because torque was not to specifications, and engine overheating due to lack of coolant inspection and replenishment (Figure 10.9).

| | Identification | | | Nominal Size | Mechanical Properties | | |
| | | Specification | Material | | Proof Load (psi) | Yield Strength Min (psi) | Tensile Strength Min (psi) |
	Grade Mark			Range (in.)			
Incorrect use of		ASTM A325 Grade 5	Low Carbon Martensitic Steel, Quenched and Tempered	1/2 thru 1	85,000	92,000	120,000
Instead of		SAE J429 Grade 8	Medium Carbon Alloy Steel, Quenched and Tempered	1/ 4 thru 1-1/2	120,000	130,000	150,000

FIGURE 10.8
ASTM-, SAE-, and ISO-grade markings and mechanical properties for steel fasteners.

Bolt Dia.	Thread per inch	Grade 5	Grade 8
1/2	20	90	120

USE GRADE 5 TORQU E
INSTEAD OF GRADE 8 TORQUE, OR
'HAND' TIGHTEN

FIGURE 10.9
U.S. bolt torque specifications (torque in pounds-foot).

Hypotheses of failure mechanisms are documented in the RCM critical items list. Detectability of RCM[5] candidate parts is evaluated to identify the RCM path. Two criteria are applied.

How evident is the failure mode? The question to be answered is whether the failure mode is evident to the operator as it occurs during a mission or is evident to maintenance personnel following completion of a mission. Operators of mobile and process machinery and equipment recognize departures from normal functionality from changes in system performance and the primary senses—sound, smell, visual, and tactile. Functional feedback may be provided through monitoring systems, condition indicators, and diagnostic systems.

Maintainers are able to recognize a failure mode between missions. They have the time and training to inspect a system for indicators that the operator cannot perceive. An important maintenance task performed between missions includes inspection and adjustment of interface hardware that are precursors to or indicators of pending failure modes.

Relationship between the failure mode and the maximum mission duration. The time between perception (P) of part failure and the failed (F) state is the P–F interval. There are two applications of the P–F interval[6]: (1) relationship between the P–F interval and maximum mission duration, and (2) relationship between the P–F interval and safe mission shutdown duration (Figure 10.10).

The P–F interval greater than maximum mission duration describes a scenario where perception of part failure happens while a mission is in progress for a failed state that will occur after the mission is completed. Tire tread degradation

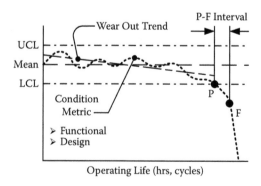

FIGURE 10.10
Moubrey's P–F interval.

on a haul truck is an example of perception of pending tire failure that will occur long after completion of a scheduled operation. This example illustrates a characteristic of failure perception that a maintenance action will be scheduled.

The P–F interval less than maximum mission duration describes a scenario where perception of part failure happens while a mission is in progress for a failed state that will occur before the mission is completed. Seeing and smelling steam from the haul truck engine compartment is an example of perception of a pending system downing event from a part failure in the engine coolant assembly that will occur before completion of a scheduled operation. This example illustrates the second application of the P–F interval: the relationship between the P–F interval and safe mission shutdown duration. In the preceding example, the perception of engine overheating provides a P–F interval of sufficient duration to permit safe system shutdown. Contrast the system downing event from engine overheating to a shock-induced downing of the pneumatic brake assembly. The P–F interval is instantaneous. The operator has no reaction time.

Detectability is documented in the RCM critical items list. The RCM implementation path for a candidate part is determined for common-cause failure mechanisms, modes, effects, and detectability, as shown in the logic flow chart in Figure 10.11. Detectability is the determinant for the appropriate RCM path. The determinant detectability factor for catastrophic failure effects is whether the failure mode is evident to the operator and maintainer.

Catastrophic safety failure modes that are not evident to the operator and maintainer have no maintenance solution regardless of the P–F interval. Catastrophic regulatory compliance failure modes that are not evident to operators and maintainers have a time-directed maintenance solution regardless of the P–F interval, as shown in Figure 10.12.

The absence of a maintenance solution yields the following operational criteria:

- The part design must be verified to be robust to a sufficient degree that the likelihood of failure under the operator's conditions of use will be less than an acceptable risk threshold.
- The part's exposure to failure mechanisms must be controlled to a level that assures that the likelihood of failure under the operator's conditions of use will be less than an acceptable risk threshold.
- External engineering solutions must be developed to protect the system from catastrophic failure effects.

Regulatory catastrophic effects may be mitigated through time-directed maintenance. The key evaluation criteria are economic feasibility and economic risk assessment. Time-directed and condition-based maintenance solutions are applicable for catastrophic part failure effects when the failure mode is evident to the operator and the maintainer, depending on the

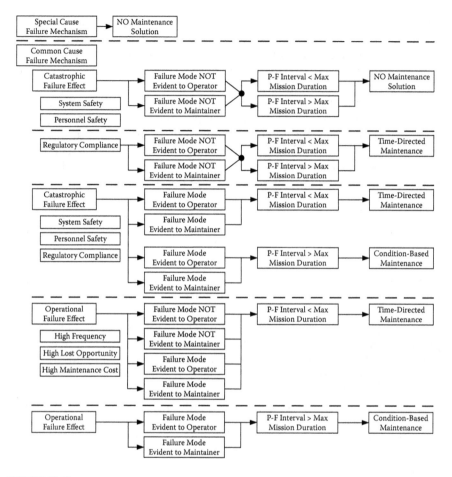

FIGURE 10.11
Reliability-centered maintenance logic: consequences analysis to selection of approach.

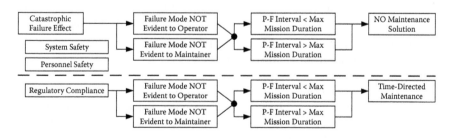

FIGURE 10.12
Catastrophic failure mode is *not* evident.

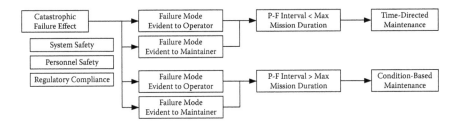

FIGURE 10.13
Catastrophic failure mode *is* evident.

relationship between the P–F interval and the maximum mission duration, as shown in Figure 10.13.

Time-directed maintenance may be appropriate for catastrophic failure modes when the P–F interval is less than the maximum mission duration. The operating risk is acceptable when the P–F interval allows sufficient time to perform a safe system shutdown. Similarly, maintenance inspections can be performed between missions specifically called out for the failure mode. Part replacement is performed prior to part degradation reaching the risk threshold.

Condition-based maintenance may be appropriate for catastrophic failure modes when the P–F interval is greater than the maximum mission duration. The operating risk is further reduced as more information is available to the operator and maintainer over the system's useful life.

Application of TDM and CBM for catastrophic failure effects is qualified as "may be appropriate." The judgment of whether to employ RCM rather than no maintenance solution is exclusively the prerogative of the organization. Application of TDM and CBM for catastrophic failure effects demands RCM analysis and understanding of the failure mechanisms, modes, effects, and detectability. "No maintenance solution" is the only acceptable maintenance approach absent RCM analysis.

The determinant detectability factor for operational failure effects is the relationship between the P–F interval and maximum mission duration. Time-directed maintenance is the appropriate RCM path when the P–F interval is less than the maximum mission duration, as shown in Figure 10.14. Condition-based maintenance is the appropriate RCM path

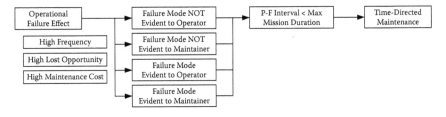

FIGURE 10.14
Operational failure effect: P–F interval *less than* maximum mission duration.

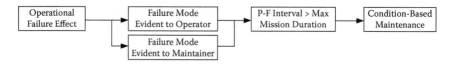

FIGURE 10.15
Operational failure effect: P–F interval *greater than* maximum mission duration.

when the P–F interval is greater than the maximum mission duration (see Figure 10.15).

The goal of RCM for catastrophic failure effects is to reduce risk to the operator, system, and organization. The goal of RCM for operational failure effects is to reduce the high costs of unscheduled maintenance and the disruption of the organization's operations.

Notes

1. Deputy Undersecretary of Defense for Logistics and Material Readiness Memorandum, March 10, 2007, Defense Acquisition University.
2. Moubray, J. 1997. *Reliability-centered maintenance,* 2nd ed. Oxford, England: Butterworth Heinemann.
3. Derating is the replacement of a weak part with a stronger part or an incapable part with a capable part, without changing the design "footprint" of the part (e.g., a higher grade fastener with the same geometry, a more efficient heat sink, etc.).
4. Users of systems do not typically have access to the system design FMECA.
5. Moubray, J. 1997. *Reliability-centered maintenance,* 2nd ed. Oxford, England: Butterworth Heinemann.
6. Ibid.

11

Reliability-Centered Failure Analysis

> Knowledge has to be improved, challenged, and increased constantly, or it vanishes.
>
> **Peter Drucker**

Introduction

The objective of reliability-centered failure analysis is to validate the hypotheses of failure mechanisms and characterize the reliability parameters that enable implementation of condition-based maintenance (CBM) and time-directed maintenance (TDM). Candidate parts for reliability-centered maintenance are selected from judgment and hypotheses, as described in the previous chapter. They are subjective, regardless of how much experience supports judgment or how much historical maintenance data support hypotheses. Reliability-centered failure analysis validates and quantifies the subjective judgments and hypotheses and adds characterization of the significance of the judgments and hypotheses.

The procedure for performing reliability-centered failure analysis is common for CBM and TDM, as shown in Figure 11.1. How the findings of reliability-centered analysis are applied helps to differentiate between CBM and TDM.

Reliability-centered failure analysis is the operations and maintenance (O&M) organization's responsibility. Sustainment personnel would be well served with access to the system design and failure analyses, but this is an unrealistic expectation. The system design and development organization treats this information as proprietary. System O&M organizations are typically separated by several levels of distribution organizations between them and the design and development organization, not to mention geopolitical boundaries.

An often expressed school of thought suggests that part CBM and TDM should, or must, be designed in. It is an idealization that may succeed when a system is exposed to the same operational and ambient conditions of use throughout its commercialization. Such is the case for medical diagnostic equipment, process equipment, and systems that share a requirement for controlled ambient conditions of use. However, their operational conditions of use differ between user organizations, including use rates (five daily shifts/week versus 24/7 continuous mission durations), proximity to other systems

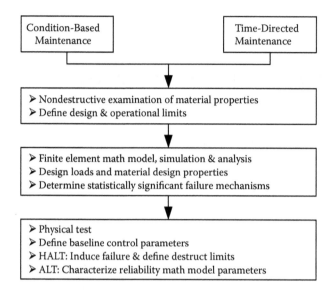

FIGURE 11.1
Reliability-centered failure analysis for CBM and TDM.

(isolated locations versus close exposure to other systems), and scheduled inspection and maintenance actions.

Reliability-centered failure analysis is an incremental approach that begins with nondestructive investigation and proceeds to physical testing until sufficient information is known. Sufficient information may be learned from nondestructive investigation coupled with field historical data and personnel experience to implement CBM, or CBM may require some degree of physical testing. Condition-based maintenance typically requires less investment and time than TDM. Time-directed maintenance must have physical tests to characterize failure parameters quantitatively.

Nondestructive Examination, Design, and Destruct Limits

Nondestructive examinations (NDEs) are inspection and evaluation methods that determine the design and functional utility of materials and parts. Nondestructive examination confirms the efficacy for installed design configuration, as well as materials and parts selection for components of rotating machinery, power transmission machinery, process machinery, diagnostic equipment, medical systems, pressure vessels and storage tanks, and reaction chambers. Such an examination focuses on machine parts, wiring harnesses, connectors, hoses and tubing, seals, welds, fasteners, linings, housing, shafts, and pulleys. Nondestructive examination is applied to identify and evaluate the following:

- surface and internal discontinuities and separations
- structural anomalies
- dimensions
- physical, mechanical, and chemical properties

Nondestructive examination serves several purposes: It can be used to establish and verify hypotheses of failure mechanisms and modes, to evaluate root causes of realized failures, and to design physical tests. Comparative evaluation between design specifications and the part with interfaces will identify potential failure mechanisms that are not evident in design documentation (design analysis, design art, and bills of materials). The condition of failed parts can identify wear-out mechanisms; for example, modal vibration tests identify natural harmonic frequencies that define deflection nodes used to identify accelerometer and thermocouple location in physical tests.

Visual inspection, including use of measuring devices, is a basic NDE method that can be performed quickly and inexpensively. Visual inspection ranges in sophistication from unaided viewing to use of a magnifying glass, a microscope, and an electron microscope (Figure 11.2). Visual inspection seeks to find leaks, surface cracks, fractures, changes in geometry, and corrosion. Hydraulic, brake, transmission, and engine oil leaks are prognostic of loose or failed connectors, fractured component housings, and seal wearout, which can be diagnosed by visual inspection. Tires along with measurement of tread depth and air pressure can be diagnosed by visual inspection.

Visual inspection is a powerful NDE method when performed by maintenance personnel who understand the operating and ambient conditions of use unique to the organization. Trend analysis of visual inspection findings for failure modes contributes to an understanding of part perception (P)–failure (F) intervals, and condition indicators are an essential element of reliability-centered maintenance (RCM).

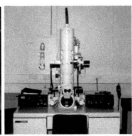

FIGURE 11.2
Magnifying glass, microscope, and scanning electron microscope.

Nondestructive examination includes various methods that are performed on the system between missions or after removal and replacement that serve to detect material property failure modes manifested by changes in geometry (strain, cracks, fracture, buckling, and misalignment) and properties (corrosion, hardness, and embrittlement). Commonly used NDE methods include:

- The modal test for harmonic frequency (Figure 11.3) provides a frequency sweep of a material to identify the natural frequency of the material and its harmonics. Analysis of modal results identifies maximum deflection for the design configuration and material geometry and locates nodes for finite element math modeling and simulation.

- Liquid penetrant testing (Figure 11.4) provides visual and microscopic investigation of cracks emanating from the surface into a material, delamination of composite materials, and gaps between joined parts. This method of testing is effective on a variety of materials (ferrous and nonferrous, homogeneous and composite) and is both quick and inexpensive.

- Radiographic testing (Figure 11.5) provides investigation of internal subsurface cracks, gaps, or geometric anomalies by use of penetrating radiation, including x-rays and electromagnetic eddy current.

- Ultrasonic testing (UT) (Figure 11.6) provides investigation of internal cracks, voids, or fissures by use of high-frequency sound energy.

FIGURE 11.3
Modal test inertial shaker, inertial hammer, and accelerometer.

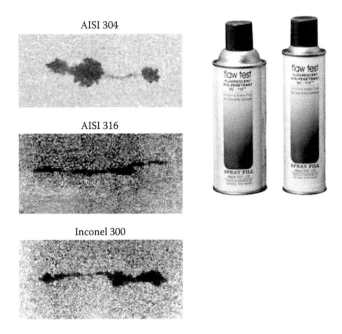

AISI 304

AISI 316

Inconel 300

FIGURE 11.4
Liquid penetrant test.

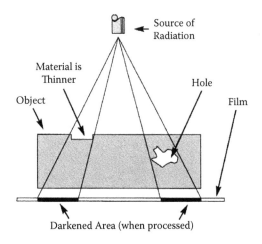

FIGURE 11.5
Radiographic test.

FIGURE 11.6
Ultrasonic test.

- Acoustic emission testing (Figure 11.7) provides investigation of cracks and voids, including delamination of composite materials, by introducing acoustic stress waves that reflect off anomalies.
- Leak testing or leak detection (Figure 11.8) provides investigation of liquid and gas leaking from precision machined actuators, pressure vessels, and linkages by introducing fluid pressure differentials across seals, joining seams, connections, and structure surfaces.

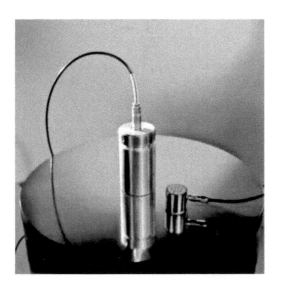

FIGURE 11.7
Acoustic emission test.

FIGURE 11.8
Leak test.

Many NDE test methods can be performed by the user organization or can be contracted with a test lab. Appropriate NDE test methods should be incorporated into the maintenance program to evaluate every critical part failure to continue to gain understanding of failure mechanisms. The findings of NDE are documented in the RCM critical items list.

Investigation of material design and destruct limits is performed concurrently with NDE to establish the baseline criteria to compare NDE findings. Design limits are the material properties of the part in excess of the expected maximum stress load, and the destruct limits are the material properties that cause a new, unaged part to fail. Theoretically, the failure condition described by NDE will begin at the design limits and approach the destruct limits. The NDE condition trend analysis will determine whether the part degradation is linear, nonlinear, cyclical, or abrupt. Abrupt degradation to failure disqualifies the hypothesis that a CBM solution exists; otherwise, the reliability-centered failure analysis can continue.

Condition-Based Maintenance NDE

Findings of CBM candidate part NDE are used to evaluate the conditions of part failure. Conditions of failure become the condition indicators that are measured to establish a CBM solution. The conditions of failure that can be used in a CBM solution include cracks, unrelaxed strain (elongation or compression deformation), and corrosion that can be visually detected and measured visually or microscopically or detected or measured by x-ray or strain gauge. The progression of cracks, strain, and corrosion can be trended as a function of operating time, load levels, or contaminate exposure to characterize the failed state and P–F interval.

A CBM solution can be defined and implemented when NDE provides sufficient information to define the condition indicator, the condition indicator measurement method, and the maintenance practices and inspection plan to perform the condition indicator measurement method. In such cases, no further reliability failure analysis is required; otherwise, physical tests are required.

Time-Directed Maintenance NDE

Findings of TDM candidate part NDE are used to validate the lack of existence of condition indicators of part failure. Conditions of failure used in a TDM solution are more complex in the methods required to detect the failure mode, do not behave in a trend, or both. Visual, microscopic, and x-ray detection is not technically or economically feasible. Strain gauge detection is not feasible or capable. Use of NDE for TDM does not provide sufficient information to preclude physical tests.

Finite Element Math Model, Simulation and Analysis, Design Loads and Material Design Properties, Statistically Significant Failure Mechanisms

Behavior of failure mechanisms and modes defined by strain due to physical loading (vibration and shock) and thermal loading (steady state and shock) can be understood using finite element modeling and simulation. The input to finite element math models includes part materials' design properties from NDE, boundary conditions for the part loading at rest, and introduction of design loads. Finite element analysis simulation and analysis describes strain for homogeneous materials and strain between two or more joined nonhomogeneous materials, and modal analysis for harmonic frequency nodes. Monte Carlo simulation of failure math models described by finite element math models and NDE findings determines the statistical significance of the failure mechanism and the correlation of the condition indicator measurement to the actual occurrence of the failed state.

Consider a hypothesis of failure mechanism—thermal strain—acting on a fastener that joins two plates of homogeneous material. The hypothesis of the failure mode is tensile stress that cracks the fastener. A finite element math model and simulation input the boundary conditions of the joined plates and the fastener. Design analysis shows the maximum allowable strain to be 15 µm. Findings of the finite element math model simulation are a mean strain of 11 µm with a standard deviation of 1 µm. The maximum strain is 14.25 µm. Assuming thermal strain to be normally distributed, MathCAD calculates the probability that thermal strain will exceed the allowable strain of 15 μm to be 0.0032%. The criticality of the part failure effect demands that the risk of fastener failure be less than 0.01%. The findings of the NDE and the finite element math model and simulation state that thermal strain is not a statistically significant failure mechanism and does not require a CBM solution or further reliability-centered failure analysis.

No Maintenance Solution

For situations is which there is no maintenance solution, findings of finite element math modeling and simulation identify harmonic frequency nodes or thermal strain locations that inform engineers how to implement design solutions for the failure mechanisms.

CBM Solution

For a CBM solution, findings of finite element math modeling and simulation validate hypothesis of failure of condition, identify locations for condition indicator measurements, estimate magnitudes of strain to specify condition indicator measurement products, and suggest the frequency of condition monitoring. A valid failure condition is statistically significant failure modes. Limits of statistical significance are determined by the risk of unscheduled part failure that an organization is willing to accept; no part is without risk of unscheduled failure.

Location for condition indicators is a combination of where the highest magnitude of the condition indicator occurs and access to that location. This typically is a "negotiation" between the ideal and the practical and carries trade-offs that must be understood to adjust the condition magnitude that initiates a maintenance action. Estimates of strain magnitude are an input specification to the selection of a commercially available product. Frequency of condition monitoring is based on operational exposure to the failure mechanism. The frequency can be periodic (e.g., end of each mission) or by exception (e.g., when an incident of the failure mechanism occurs).

TDM Solution

For a TDM solution, findings of finite element math modeling and simulation validate hypothesis of failure of condition and define the factors and levels of physical tests. Field historical failure data that describe complex behavior of failure mechanisms and modes will be simplified to identify failure mode limits and critical locations. Complexity includes failure modes that occur abruptly, are difficult to replicate in tear-down analysis of failed parts, and are the result of interactions between two or more failure mechanisms.

Physical Test

Physical tests are small-scale to assembly-level to full-scale experiments. Small-scale experiments are designed to induce failure in order to gain understanding of the failure mechanisms and modes and to confirm the findings of the NDI and simulation. Small scale is defined as a range of test articles from a material coupon to an entire part. A material coupon is a section of material that matches the geometry of one or more design axes and is large enough to be secured in a test fixture such that the interface to the test fixture does not introduce error in the application of the failure mechanism. The essential feature of a material coupon is that it acts in the test fixture just as it would under load in the part.

Design of a test fixture is required to mount the test article in the test chamber. The test fixture must not interfere with the exposure of the test article to the test stress. A minimal test fixture is a material used to fasten the test article to the test chamber. A more complex test fixture is a box fastened to the test chamber that provides exposure of the test article to corrosives. Design of a physical test includes inputs to the test article (e.g., power, fluids, and signals) from a source outside the test chamber and outputs of sensors from the test article.

Highly Accelerated Life Test

The highly accelerated life test (HALT) is performed to induce failure. The Hanse HALT chamber is shown in Figure 11.9. The objective is to validate the hypotheses of failure mechanisms. The tests are designed to subject the line replaceable unit (LRU) to the failure mechanisms that cause failure modes identified from the reliability-centered failure analysis, NDE, and finite element math model and simulation. The tests can take many forms: cycles of stresses applied at increasing levels (e.g., high- and low-temperature dwells, high- and low-temperature shock), vibration dwell and shock, exposure to corrosives, exposure to wear-out materials (e.g., salt spray and sand), and exposure to combined stresses to investigate the interactions of two or more stresses acting together.

Test methods to conduct failure analysis should be performed to a standard that provides credibility and validity to the test findings. Tests that are conducted ad hoc have no meaning to a system user, nor do such tests

FIGURE 11.9
Hanse HALT chamber system and chamber.

succeed in providing understanding of failure mechanisms. The following organizations promulgate standards for engineering tests:

- American National Standards Institute (ANSI)
- American Society for Testing and Materials (ASTM)
- European Committee for Electrotechnical Standardization (CENELEC)
- Electronic Industries Alliance (EIA)
- European Telecommunications Standard (ETS)
- International Electrotechnical Commission (IEC)
- International Organization for Standardization (ISO)
- Japanese Industry Standards (JIS)
- Military Standards and Handbooks (MIL)
- Society of Automotive Engineers (SAE)

MIL-STD-810

An excellent guideline for HALT experimental design is MIL-STD-810. Part 1 of this standard describes test management and planning. Test planning is an important aspect of failure analysis; all data cost money and no data are free, so direct labor and materials and test facility resources must be used efficiently. An effective test plan will assure that the right people are ready to perform the test, the necessary test materials are acquired and prepared for the test, and test fixtures are designed and ready for the test. A good test plan minimizes test setup and preparation time after the test facility and resources are made available; the first test article and test fixture will be ready to run at the beginning of access to the test facility resources. Test planning also provides guidelines to assure that the test will be performed on time and on budget.

Part 2 of MIL-STD-810 provides detailed test guideline methods and procedures for the failure mechanisms described in the following sections.

Method 501: High Temperature

This method exposes test articles to high temperatures that will be experienced in storage or operation of the system. High-temperature limits are determined from engineering judgment for the highest possible temperature extreme that the material or part will experience. Care should be taken to avoid considering only intended operational temperatures. A system designed to function in a sheltered, controlled environment may experience unplanned high temperatures during transport and storage.

A minimum of two thermocouples is used to monitor temperature: One thermocouple measures the air temperature in the chamber; at least one other thermocouple measures the temperature reached by the material (see Figure 11.10). A single test cycle exposes the test article to a baseline temperature from which the temperature is increased in step increments of small ramps and dwells to the maximum high temperature, or a single ramp and dwell, and then reduced in step increments or a single ramp to the baseline

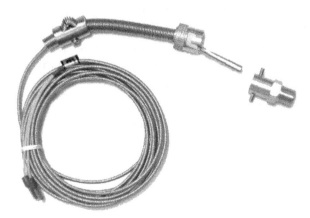

FIGURE 11.10
Thermocouple.

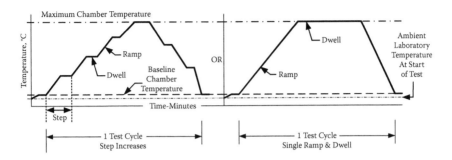

FIGURE 11.11
High temperature test profile.

temperature, as shown in Figure 11.11. High-temperature test cycles are plotted in the test plan and take one of the two forms shown in the figure. The increase in temperature, ramp slope, must be controlled so as not to induce thermal shock.

High-temperature exposure is combined with strength-of-materials tests between cycles to determine the degradation of the material strength properties (tensile, shear, compression, and torsion). High-temperature failures are often observed as an elongation strain or expansion of the geometry and a softening of the material structure. High-temperature exposure for dynamic parts and joining of materials uses strain gauges during the thermal test to measure thermal strain over repeated cycles (Figure 11.12). An empirical Young's modulus is found along with thermal failure limits.

Method 502: Low Temperature

This method exposes test articles to low temperatures that will be experienced in storage or operation of the system. A minimum of two thermocouples is used to monitor temperature: One thermocouple measures the air temperature in the chamber and at least One other thermocouple measures the temperature reached by the material.

As with high-temperature tests, a low-temperature single test cycle exposes the test article to a baseline temperature from which the temperature is decreased in step increments of small ramps and dwells to the minimum low temperature, or a single ramp and dwell, and then increased in step increments or a single ramp to the baseline temperature. The low-temperature profile is a mirror image of the high temperature. Test article failure analysis is the same as for the high-temperature test. Low-temperature failures are often observed as a compression strain or contraction of the geometry and a hardening or embrittlement of the material structure.

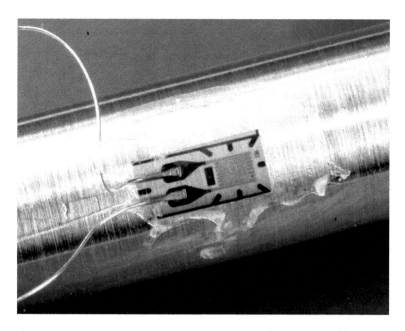

FIGURE 11.12
Strain gauge mounted on test article.

Method 503: Temperature Shock

This method exposes test articles to study successively larger changes of temperature that will be experienced in start up operation and due to changes of altitude. Two thermocouples and at least one strain gauge per strain axis are installed to capture data. One thermocouple measures the temperature of the chamber air; the second measures the material temperature, and the strain gauge measures thermal strain.

Engineering judgment is applied to determine the limits of the temperature shock exposure that the material and part will experience; indeed, engineering judgment determines whether temperature changes have the characteristics of shock. Most sources define "shock" (thermal or physical) to be a sudden or rapid change. Other sources define shock as a change sufficient to cause cracking or fracture. Suffice it to say that what might be thermal shock to a glass product might be a mild ramp to a steel product. HALT allows the engineer to resolve the confusion of how to define shock by testing a material or part to maximum expected temperature changes and verifying the existence or lack of thermal shock as a failure mechanism.

A system mission is defined in phases that include start-up, functionality, and shutdown, at a minimum. Part material temperature is the same as ambient temperature at commencement of the system start-up phase. Part material temperature changes during the system functionality phase due to part operational temperatures, exposure to temperature changes from proximate

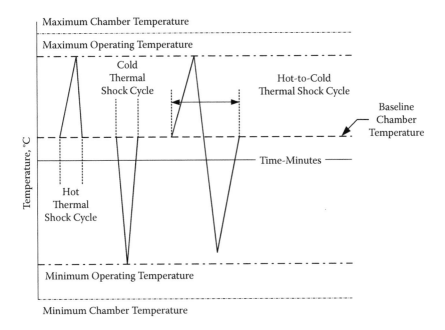

FIGURE 11.13
Temperature shock.

parts, and changes in ambient temperature. Part material temperature will change to ambient temperature at system shutdown. Systems can experience far more phases that are defined by specific changes in part material temperature. Temperature shock profiles can be plotted for single shock simulating mission start-up and shutdown and cyclical shock at varying magnitudes simulating multiphase changes in temperature (Figure 11.13).

Failure analysis for temperature shock should focus on cracking and fracture, unrelaxed strain, deformation, and embrittlement. Joining tensile fracture, cracking, and delamination of composite structures are common failure modes caused by temperature shock.

Method 507: Humidity

This method exposes test articles to high humidity. Humidity by itself is a slow-acting failure mechanism, but it can be accelerated in the presence of other failure mechanisms, including any combination of high temperature, temperature shock, salt, mineral sands, biological materials, lubricants, and organization-specific chemicals. Failure analysis for humidity as a main effect or in combination with other failure mechanisms includes corrosion moisture absorption, and changes in material properties resulting from chemical and biological reactivity.

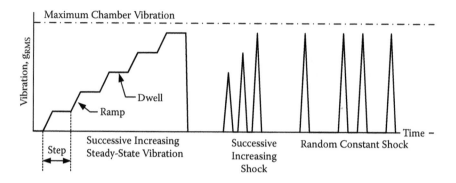

FIGURE 11.14
Vibration and physical shock.

Method 514: Vibration

This method exposes test articles to random vibration, forcing functions over a range of frequencies and amplitudes measured in g_{RMS}. Vibration levels are determined by engineering judgment, including steady-state vibration and physical shock—as one would expect from a truck driving over a consistently bumpy, undeveloped road and hitting random pot holes at full speed.

Test apparatus includes accelerometers and strain gauges in appropriate axes of strain. Vibration and random physical shock profiles are plotted in Figure 11.14. Failure analysis for vibration and physical shock includes material surface wear and displacement of joined faces, along with cracks and fracture.

Method 520: Combined Environments (Temperature, Vibration, and Humidity)

This method exposes test articles to high- and low-temperature step changes in combination with varying vibration forcing functions and varying humidity levels. The range of temperature changes and the ramp and dwell for step changes are determined by engineering judgment. The temperature range will normally include the extremes of seasonal weather variation as the starting point for each test cycle and the ambient or operational temperature extreme resulting from system functionality. Steady-state vibration cycles are nested within temperature cycles at operational levels determined by engineering judgment. Humidity is varied from high to low for alternating vibration cycles.

Combined environments are particularly informative for parts that comprise varied materials, fittings, joinings, and interfaces. Consider a valve seal: Temperature changes, temperature shock, vibration, physical shock, and humidity applied alone may not induce failure. However, the interactions of high temperature and humidity, or low temperature and vibration,

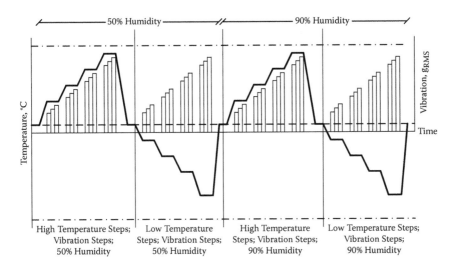

FIGURE 11.15
Combined environments test profile.

may very well cause a failure mode. Plots of various combined environment profiles are shown in Figure 11.15.

Accelerated Life Testing

The objective of accelerated life testing (ALT) is to age a material or part under a combination of stresses or a single stress at load levels at or above the maximum ambient and operational exposure to estimate the parameters of a reliability model or Weibull or normal distribution. As with HALT, ALT experiments induce failure in all or some test articles. Accelerated life testing equipment ranges from commercially available chambers to custom chambers (Figure 11.16).

FIGURE 11.16
Industrial oven, salt spray chamber, and corrosion test chamber.

Time Compression Accelerated Life Testing

The most straightforward ALT is time compression. Accelerated life testing can be designed for time compression where test cycles of stress repeatedly expose the test article to several hours of operating stress exposure in minutes of test time. Accelerated life testing applies continuous or step-increased stress at and above the maximum operating stress. Validity of ALT is based on the assumption that the same failure mechanisms will occur in less time. Accelerated stresses linearly transform the useful life of the material or part. The transformation is calculated by an acceleration factor (AF), a unit-less number. Existence of an acceleration factor is based on the assumption that material or part wear-out processes are dependent on the stress. An acceleration factor exists for each stress acting on a material or part. The time to fail (TTF) for a stress (TTF_{stress}) is transformed to calculate the estimator for time to fail in operations (TTF_{opn}), as shown in the following equation:

$$TTF_{opn} = AF \times TTF_{stress} \tag{11.1}$$

The general expression for the probability density function (pdf) of failure for a stress, $f_{stress}(t)$, transformed to characterize the pdf of failure in operations, $f_{opn}(t)$, is shown in the following equation:

$$f_{opn}(t) = \left(\frac{1}{AF} \right) f_{stress}(t) \tag{11.2}$$

The accelerated transformed exponential expression for the pdf is

$$f_{opn}(t) = \left(\frac{\lambda}{AF} \right) e^{-\left(\frac{\lambda}{AF} \right)t} \tag{11.3}$$

The accelerated transformed Weibull expression for the pdf is

$$f_{opn}(t) = \left(\frac{\beta}{AF \times \eta} \right) \left(\frac{t}{AF \times \eta} \right)^{\beta-1} e^{-\left(\frac{t}{AF \times \eta} \right)^{\beta}} \tag{11.4}$$

The general expression for survival function for a stress, $S_{stress}(t)$, transformed to characterize the survival function in operations, $S_{opn}(t)$, is shown in the following equation:

$$S_{opn}(t) = S_{stress} \left(\frac{t}{AF} \right) \tag{11.5}$$

The accelerated transformed exponential expression for the survival function is

$$S_{opn}(t) = e^{-\left(\frac{\lambda}{AF}\right)t}$$

(11.6)

The accelerated transformed Weibull expression for the survival function is

$$S_{opn}(t) = e^{\left(\frac{t}{AF \times \eta}\right)^{\beta}}$$

(11.7)

The general expression for hazard function for a stress, $h_{stress}(t)$, transformed to characterize the hazard function in operations, $h_{opn}(t)$, is shown in the following equation:

$$h_{opn}(t) = \left(\frac{1}{AF}\right) h_{stress}\left(\frac{t}{AF}\right)$$

(11.8)

The accelerated transformed exponential expression for the hazard function is

$$h_{opn}(t) = \frac{\lambda}{AF}$$

(11.9)

The accelerated transformed Weibull expression for the hazard function is

$$h_{opn}(t) = \left(\frac{\beta}{AF \times \eta}\right)\left(\frac{t}{AF \times \eta}\right)^{\beta-1}$$

(11.10)

Consider a fastener subject to two failure mechanisms—physical shock and thermal shock—and select to design and perform an ALT for physical shock. Physical shock is determined to act on the fastener randomly at a maximum frequency of four times per operating hour at 13 g_{RMS}. Fasteners are installed in a test fixture that loads each to the operating loads and torque. The test fixture is placed in a HALT chamber that is programmed to cycle one physical shock at 13 g_{RMS} over 3-min intervals (20 cycles/test hour; 480 cycles/day). The acceleration factor for physical shock ($AF_{physical}$ = 20/4 = 5) transforms one test hour to five operating hours. The HALT chamber can hold three test fixtures of 20 fasteners each for a total of 60 test articles. The test fixtures are inspected every hour for failed fasteners, which are replaced with new test articles. The failure data are tabulated in MS Excel, as shown in Figure 11.17.

Compressed time accelerated data for time to failure is fit to a Weibull distribution in MathCAD using a median ranks regression, as shown in Figure 11.18. The TTF data and index tables are imported from MS Excel in

| Time | Failured Fasteners | | | | | TTF FF | |
	Day 1	Day 2	Day 3	Day 4	Day 5	Test 1	Test 2
Hour 1	0	0	2	0	0	33	29
Hour 2	0	0	3	0	0	33	32
Hour 3	0	0	3	0	0	34	32
Hour 4	0	0	2	0	0	34	32
Hour 5	0	0	1	0	1	34	33
Hour 6	0	0	4	0	0	35	33
Hour 7	0	0	3	0	0	35	33
Hour 8	0	0	2	0	3	35	33
Hour 9	0	0	0	0	5	36	33
Hour 10	0	0	0	0	2	36	34
Hour 11	0	0	0	0	3	37	34
Hour 12	0	0	0	0	4	38	35
Hour 13	0	0	0	0	2	38	35
Hour 14	0	0	0	0	*	38	35
Hour 15	0	0	0	0	*	38	36
Hour 16	0	0	0	0	*	39	36
Test 1	0	0	20				39	36
Test 2			0	0	20		39 39	36
						40	37
						40	37

FIGURE 11.17
Accelerated life test TTF data: physical shock.

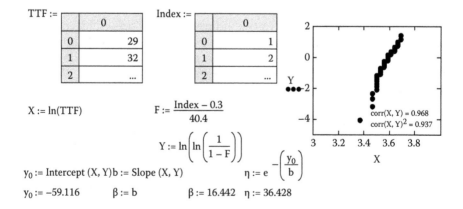

TTF :=

	0
0	29
1	32
2	...

Index :=

	0
0	1
1	2
2	...

$X := \ln(\text{TTF})$

$F := \dfrac{\text{Index} - 0.3}{40.4}$

$Y := \ln\left(\ln\left(\dfrac{1}{1-F}\right)\right)$

$y_0 := \text{Intercept}(X, Y)$ $b := \text{Slope}(X, Y)$ $\eta := e^{-\left(\frac{y_0}{b}\right)}$

$y_0 := -59.116$ $\beta := b$ $\beta := 16.442$ $\eta := 36.428$

corr(X, Y) = 0.968
corr(X, Y)^2 = 0.937

FIGURE 11.18
Median ranks regression TTF data: MathCAD.

$$fw(t) := \left(\frac{\beta}{AF \cdot \eta}\right) \cdot \left(\frac{t}{AF \cdot \eta}\right)^{\beta-1} \cdot e^{-\left(\frac{t}{AF \cdot \eta}\right)^{\beta}}$$

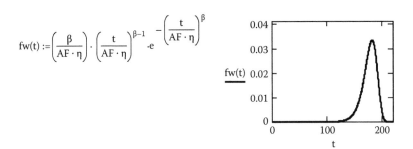

FIGURE 11.19
Fastener physical shock pdf and plot.

rank order. The independent variable of the median ranks regression (X) is calculated as the natural logarithm of TTF data. The cumulative distribution of the TTF data (F) is calculated using Bartlett's median ranks. The dependent variable of the median ranks regression (Y) is calculated as the natural log of the natural log of the inverse of $1 - F$.

The scatter plot of the TTF data is plotted with the coefficients of correlation and determination. The scatter plot illustrates the goodness of fit and the correlation of the TTF data. The lack of a wide spread and the positive increase indicate a good fit. The coefficients of the regression quantify the high positive correlation and the high predictive capacity of the regression model. The parameters of the Weibull reliability model are calculated from the slope (b) and y-intercept (y^0) of the regression line.

The expression for the Weibull pdf of failure, including the acceleration factor, $fw(t)$, and the plot of the PDF, is shown in Figure 11.19. The expression for the Weibull survival function, including the acceleration factor, $Sw(t)$, and the plot of the survival function, is shown in Figure 11.20. The expression for the Weibull hazard function, including the acceleration factor, $hw(t)$, and the plot of the hazard function, is shown in Figure 11.21.

$$Sw(t) := e^{-\left(\frac{t}{AF \cdot \eta}\right)^{\beta}}$$

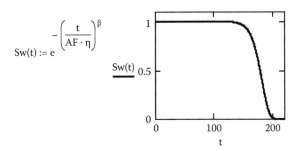

FIGURE 11.20
Fastener physical shock survival function and plot.

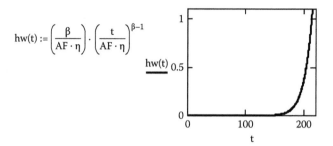

$$\text{hw}(t) := \left(\frac{\beta}{\text{AF} \cdot \eta}\right) \cdot \left(\frac{t}{\text{AF} \cdot \eta}\right)^{\beta-1}$$

FIGURE 11.21
Fastener physical shock hazard function and plot.

Life-versus-Stress Analysis Accelerated Life Test

Time compression is not always feasible. When it is not, the most common accelerated life test method is the life-versus-stress analysis, which exposes a test article to successively higher stress levels to back-trend the operating stress level capability of the part. Continuing with the example for a fastener, thermal shock is known to act on the fastener at the beginning of each mission at $\Delta T = 210°F$. Engineering judgment selects three temperature levels of accelerated stresses: level 1: 225°F; level 2: 240°F; and level 3: 255°F.

The design of the experiment includes operational vibration loading on the test articles of 5 g_{RMS}. A test fixture with 20 fasteners loaded and torqued to operational specifications is installed in the HALT chamber. The HALT chamber is cycled from –25 to 200°F in 5-min time intervals for the level 1 temperature shock exposure. Each fastener on the test fixture is inspected at the end of each cycle for failure modes, including cracking, loosening, shear, and tensile strain. The number of starts to failure is tabulated in MS Excel, shown in the table in Figure 11.22.

The experiment is repeated with a new test fixture mounted with new fasteners for the level 2 thermal shock, ranging from –40 to 200°F in 5-min intervals. Level 2 starts-to-failure data are tabulated along with the level 1 data. The experiment is repeated for the level 3 thermal shock, ranging from –40 to 215°F. Level 3 starts-to-failure data are tabulated along with level 1 and level 2 data.

The fifth percentile life points are calculated in MS Excel for each column of data. The fifth percentile life for level 1 data is 79 starts to failure. We can interpret that to mean that 95% of fasteners subjected to a level 1 thermal shock will survive beyond 79 starts. The natural logarithm of the fifth percentile life points is calculated and tabulated in Figure 11.23. It is plotted against that of the thermal shock magnitude in degrees Fahrenheit.

A regression line is fitted to the data using the MS Excel "Linear Trendline" routine. The regression equation is evaluated at a coefficient of determination equal to 0.9729 to be highly predictive. Substituting the

Starts-to-Failure		
DT Level 1	DT Level 2	DT Level 3
225°F	240°F	255°F
87	60	32
82	72	58
80	69	53
89	65	73
80	64	41
79	81	66
85	71	44
83	68	75
85	59	56
85	65	52
87	69	47
82	65	46
87	63	39
	54	70
	86	32
	77	29
		67
		48
		60
Count	13	16	19

FIGURE 11.22
Starts-to-failure data for thermal shock.

expected thermal shock of 210°F for y, we solve the regression equation for x to be 4.89. The antilogarithm of 4.89—$\ln^{-1}(4.89) = e^{4.89}$—is equal to 132 starts to failure. This is the fifth percentile life point for our maximum expected thermal shock. We can infer that 95% of all bolts will survive beyond 132 starts.

The value of small-scale physical tests extends well beyond understanding which failure mechanisms are valid and which are invalid and character-ization of the parameter estimators for the reliability functions for materi-als and parts. Small-scale physical tests that induce failure provide us with an intimate understanding of the conditions of the failure mechanisms and modes manifested by the material and part as they degrade to failure. As we understand the conditions of failure mechanisms, we can define metrics that measure the degradation; these metrics enable us to define condition indica-tors that will be a key element in condition-based maintenance.

Small-scale physical tests that reveal abrupt or complex failure mecha-nisms provide us with the initial data and understanding of the reliabil-ity parameters that must be down in order to characterize the probabilistic

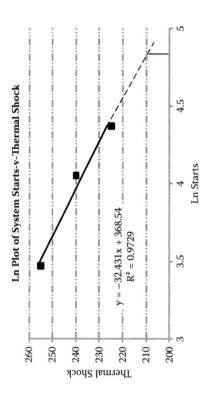

Ln Plot of System Starts-v-Thermal Shock

$y = -32.431x + 368.54$
$R^2 = 0.9729$

Ln Starts

Thermal Shock

5% Life Points		
DT Level 1	DT Level 2	DT Level 3
225°F	240°F	255°F
79	57	32
4.37	4.05	3.47

$\Delta T = 210°F$
$Ln(Starts) = 4.89$
$Starts = 132.76 \mid 132$

FIGURE 11.23
Fifth percentile life points and regression equation.

risk of failure manifested by our selection of materials and parts. This initial understanding guides us to define and implement more detailed experiments that will precisely characterize reliability functions of the materials and parts, including the hazard function. We will be able to compare the values of the hazard function with a risk threshold that enables us to implement time-directed maintenance.

12

Condition-Based Maintenance

> An organization's ability to learn, and to translate that learning into action rapidly, is the ultimate competitive edge.
>
> **Jack Welch**

Introduction

Condition-based maintenance (CBM) enables an organization to change the way it does business from the situation where system downing events control operations and maintenance to that where CBM controls system downing events.

Condition-based maintenance makes the transition from probabilistic risk assessment of populations of a material and part to direct evaluation of an individual material and part. The ideal maintenance scenario is one in which materials and parts have a failure mechanism that causes a failure mode that has an operational consequence evident to the operator and maintainer with a perception (P)–failure (F) interval greater than the maximum mission duration.[1] Such a material and part should rarely cause a system downing event during scheduled operations.

Consider a fleet of cars used by a taxi or courier service organization. Tires are identified by reliability-centered failure analysis to be a candidate part for reliability-centered maintenance. A failure hypothesis is a flat tire caused by wear out of the tire tread. The consequence of failure is operational; a flat tire ceases system operation and requires immediate corrective maintenance actions. Maintenance actions can be performed by

- the operator at the site of the flat tire
- organization maintenance personnel who travel to the failure site and perform the maintenance action at the site
- organization maintenance personnel at a maintenance facility after the downed system has been towed to the nearest maintenance facility

Each maintenance option incurs tangible maintenance costs associated with direct labor, direct materials, and direct overhead:

- Operator-performed maintenance action incurs direct labor cost for the operator, charged as maintenance direct labor, and direct labor and material costs for repair or replacement of the flat tire.

- Organization maintenance personnel incur direct labor costs from the time they depart the previous job site until completion of the tire replacement. The operator incurs direct labor costs that are charged to maintenance or to an "idle" account during the maintenance downtime. The maintenance vehicle used by maintenance personnel incurs direct overhead costs, and there are direct labor and material costs for repair or replacement of the flat tire.

- Organization maintenance personnel incur direct labor costs for the tire replacement at the maintenance facility. The operator incurs direct labor costs that are charged to maintenance or to an "idle" account during the maintenance downtime. The tow-truck driver incurs direct labor and the tow truck incurs direct overhead costs for the round-trip towing time, and there are direct labor and material costs for repair or replacement of the flat tire.

Each maintenance option incurs tangible lost-opportunity costs to the organization. A system generates revenue only when it is operating. Lost-opportunity costs equal the scheduled revenue that was not earned during the unscheduled maintenance downtime. Lost-opportunity costs should be viewed by sustainment engineers as the forfeiture of positive cash flow that is replaced by negative cash flow. Consider a system that earns \$1,000/operating hour. Assume that a flat tire replacement takes 1 h and costs the organization \$200. The organization forfeits \$1,000 and incurs \$200 in unscheduled costs for a lost-opportunity cost equal to \$1,200. Lost-opportunity costs change the gross margin of the organization by \$1,200—not just the maintenance costs incurred!

Unscheduled maintenance actions that interrupt scheduled operations incur intangible costs that are difficult to quantify. An organization profits not only from the service or product that it provides to a customer, but also from its reputation. A taxi or courier service customer who experiences a disruption of service will express displeasure in two ways: He or she will find another service provider and will also pass the bad experience by word of mouth to colleagues. Customers will be loyal to service providers who meet their needs and may recommend the service provider to their colleagues. But customers are far more disposed to express bad experiences than good ones. An organization must seek to avoid every bad customer experience; CBM enables that goal.

Returning to the CBM example for a taxi or courier service system, the condition indicator of the failure mechanism is identified by reliability-centered failure analysis to be tire tread depth. New tires have a tread depth of 7/16 in. Reliability-centered failure analysis defines a P–F interval where

- reliability-centered analysis suggests that flat tire failures (F) occur below tread depth of 1/16 in.
- engineering judgment and reliability-centered analysis recommend that a tire be perceived to depart from structural integrity at tread depth of 3/16 in.
- reliability-centered analysis suggests that operating time between P and F is 60 h
- maximum mission duration is 16 h, followed by a minimum of 4 h scheduled maintenance before the system is returned to scheduled operations
- replacement tires are received from the supplier no later than 18 h following an order[2]

A CBM solution is proposed that specifies that maintenance personnel will use a tread depth gauge at the completion of each system mission. The post-mission tread depth is entered in the system maintenance log book. A new tire is ordered when the tread depth reaches P. A time line for the maximum expected maintenance actions is shown in Figure 12.1.

The worst case is defined for tread depth occurring instantly following the commencement of a mission that lasts for 16 h. Postmission maintenance actions record and administratively process the tread depth that triggers a new tire order at the end of the maintenance period. The replacement tire arrives during the next postmission maintenance period. The CBM analysis provides the maintenance organization with the flexibility to replace the tire immediately or schedule the replacement at the next postmaintenance

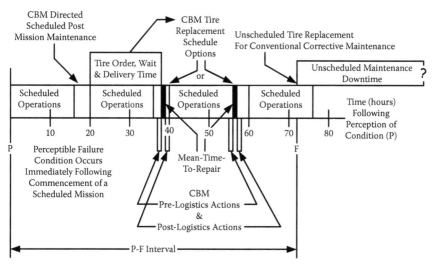

FIGURE 12.1
Condition-based maintenance P–F interval: tire.

opportunity. Either option assures tire replacement before the tire reaches the failed-state condition. Alternatively, not implementing CBM will result in a mid-mission tire failure with all of the tangible and intangible costs.

CBM Logic

The method to apply reliability-centered failure analysis to a CBM solution as illustrated in the previous tire example is presented in the flow chart in Figure 12.2. The logic integrates understanding of failure mechanisms and modes with knowledge of the P–F interval to determine technically and economically feasible condition indicators. This critical action must absolutely,

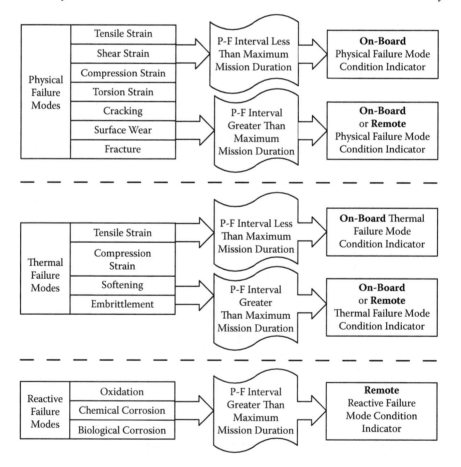

FIGURE 12.2
Condition indicator logic.

without exception, include participation of maintenance personnel and system operators who possess intimate understanding of the sound, feel, and smell of a system as it operates correctly and when it experiences downing events.

Selection of a condition indicator is part science (engineering) and part art (operator and maintainer intuition); both are required. An engineering solution absent operator and maintainer intuition effectively detects and isolates a part fault condition indicator only by random chance; more often, the results will be disappointing to the point of useless. Operator and maintainer intuition provide qualitative detection of a problem, but cannot evaluate the P–F interval or detect and isolate the fault of the failed part.

Technological and economic feasibility of the CBM solution is determined by selection of an effective condition indicator. Tire tread depth, measured by a depth gauge, is a technologically and economically effective condition indicator. Circumference of a tire, measured by a tape measure, is less technologically effective and costs more in direct labor and maintenance downtime to perform; it is an ineffective condition indicator. The condition indicator must be technically feasible to access and measure. Coolant temperature measured by a thermocouple is technically feasible to access and measure; internal engine block temperature is inaccessible and not economically feasible.

Equipment used to monitor and measure condition indicators must be technically and economically feasible. It should be intuitively obvious that monitoring equipment must be mature products with acceptable performance range that are accurate and unlikely to issue false alarms. They must have acceptable acquisition costs, including purchase price, installment costs, and operating costs. Acceptable costs are defined as an investment that has a rate of return in cost avoidance that meets or exceeds the organization's financial discount rate. This is non-negotiable; "gee-wizardry" technology must earn more than it costs by an acceptable margin. For this reason, organization financial personnel must be part of the CBM analysis team.[3] Therefore, a condition-indicator monitoring product that is not economically feasible to one organization may be feasible to another.

Condition indicators that define a perception (P) of imminent failure (F) that do not provide sufficient time to complete the mission must use onboard fault-detection/fault-isolation equipment. Onboard monitoring equipment measures the magnitudes of the condition indicator and reports those measures to the operator in real-time or short-time intervals. The engine coolant temperature gauge and dashboard read-out in a car perform this function for a condition indicator that defines a P–F interval that is less than mission duration. Operator feedback is essential for an onboard condition indicator monitor that allows sufficient time to consider operational actions that will minimize adverse safety and system consequences. The driver of the car will have sufficient time to exit the road in a controlled manner or to take actions to reduce the load causing the failure mechanism and mode by turning off auxiliary subsystems or slowing down.

Condition-indicator fault-detect/fault-isolate equipment for a P–F interval greater than mission duration can be either onboard or remote. Onboard monitoring equipment need not provide feedback to the operator; instead, it measures and saves the measures of the condition indicator for review by maintenance personnel. Onboard thermocouple monitors are used to measure the temperature profile of oil and fluids to determine whether viscosity may be materially changed; onboard strain gauge monitors are used to inspect the unrelaxed strain on a part.

Condition indicators measured remotely include monitoring equipment brought to the system following a mission or an inspection method. Using a tire tread depth gauge following a mission is an example of remote monitoring equipment that measures a condition indicator. Drawing an oil sample from the engine or transmission is an example of a remote inspection method that measures a condition indicator. The logic for an onboard and remote approach is provided in the flow chart in Figure 12.3. A shaft is known to experience physical loads in tension and high temperatures that can cause strain that causes the shaft to fail, resulting in a system downing event. The P–F interval is greater than the maximum mission duration. An organization can either employ an onboard strain gauge fault-detect/fault-isolate monitor or perform a remote maintenance inspection that measures the shaft with a micrometer. There is no prohibition to using two or more fault-detect/fault-isolate approaches; indeed, the inspection can be used to validate the strain gauge.

This example illustrates fault detection and isolation for a condition indicator that represents a single failure mode: strain. Strain on the shaft is the main effect—an experimental term for one response variable. The main effect has two failure mechanisms (physical load and thermal load) that are the factors. The factors have levels—the magnitudes of each load. The onboard and/or remote methods to measure the response variable are technically feasible and capable of accurately measuring the magnitudes and expected range of strain; they are mature technologies. Both are also economically feasible; the costs to acquire, install, operate, and maintain the strain gauge and micrometer are less than those for the problem they serve to prevent.

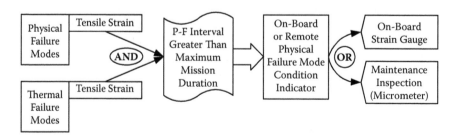

FIGURE 12.3
Main effect condition indicator logic.

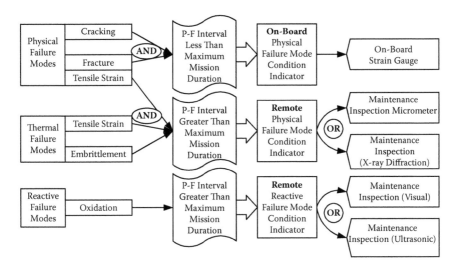

FIGURE 12.4
Interactive failure modes logic.

System downing events are also the result of interactions between two or more failure mechanisms that have two or more failure modes that result in a part failure that causes a system downing event. The logic for interactive failure mechanisms is presented in the flow chart in Figure 12.4. A tension bar is mounted between two dynamic subassemblies to maintain structural integrity. The tension bar is unloaded at rest and experiences physical loads during operation (vibration, tension, and shock), thermal loads at rest and during operation (seasonal extremes in high and low ambient temperature, high operating temperatures, thermal shock at start-up and shutdown), and reactive loads at rest and during operations (exposure to water from rain, snow, and condensation).

Physical failure modes include tensile strain, cracking, and fracture; thermal failure modes include thermal strain from high temperatures and embrittlement, and reactive failure mode is oxidation. Tensile strain, cracking, and embrittlement are interactive failure modes; tensile strain and oxidation are interactive, and embrittlement and oxidation are interactive. The failure of the part is no longer a main effect; rather, it is an interactive effect. The CBM solution must measure condition indicators for all main effect factors and levels and the interaction factors and levels.

Furthermore, cracking and fracture have P–F intervals less than mission duration and require onboard fault-detect/fault-isolate monitoring equipment. Feedback from the equipment alerts the operator to perform a system shutdown immediately. Tensile strain, embrittlement, and oxidation have P–F intervals that are greater than mission duration and can use remote maintenance inspection. Data from the condition indicator monitoring equipment are entered in a complete data full factorial analysis of variance

(ANOVA) algorithm that updates RCM failure analysis periodically to refine maintenance decisions for inspection intervals. Full and fractional experimental analysis with ANOVA is a specialty that adds another member to the CBM team.

Maintainability Demonstration Test, Validate Part Fault Detection, and P–F Interval

Implementation of CBM solutions is qualified by a maintainability demonstration before being incorporated into the organization's maintenance practices. Maintainability demonstrations are best performed by installing the fault-detect/fault-isolate monitoring equipment on a single system for a suitable time to verify that the equipment performs its function as designed. The P–F interval is verified as well as the behavior of the failure mechanisms and modes from the RCM failure analysis. Maintainability demonstration results are reviewed and evaluated by the CBM team and improvements on the application of the CBM solution are proposed and implemented.

Develop and Implement Maintenance Procedures and Practices

Successes of the maintainability demonstration are presented to the organization to develop CBM maintenance procedures and practices. Implementation of these procedures and practices is achieved through written guidelines and work instructions for maintainers and operators on the installation, operation, and sustainment of the fault-detect/fault-isolate equipment. Operators are trained to use the onboard information provided by the equipment and on the actions they may take. Maintainers are trained to perform inspection methods and procedures and on how to use the information provided by the equipment. Maintenance planners are trained to develop maintenance inspection intervals. Data acquisition and reporting methods are developed to distribute the information provided by the equipment to maintenance engineers, the ANOVA program, and management.

Periodically, the CBM solution is reviewed by the CBM team to evaluate successes and shortcomings in order to improve upon the CBM solution. Specifically, continual review of commercially available fault detection and isolation equipment is performed to update or upgrade to better technically and economically feasible alternatives; inspection intervals are adjusted to respond to the organization's specific conditions of use or to identify seasonal differences in ambient conditions of use, and financial analysis is trended to verify that the CBM solution remains acceptable.

The reader is correct to think that the transition RCM failure analysis to a CBM solution is vague. Two reasons account for this:

RCM failure analysis follows well-defined best practices to character-ize sample statistics of stress loads and material strength properties; CBM solutions are applied to a single individual part on an indi-vidual system.[4]

RCM failure analysis is performed by the body of knowledge for a single engineering discipline; CBM solutions are multidiscipline team projects crossing not only engineering disciplines (mechanical, electrical, civil, and chemical) but also management (maintenance supervision, middle management, financial management, purchas-ing, information technology) and skilled trades (system operators, maintenance technicians, vendor support).

CBM solutions force an organization to confront complex operating and ambient conditions of use that are unique to the system. The rewards in tangible and intangible financial returns will be large compared to the investment.

Notes

1. A material or part that never fails is not the ideal maintenance scenario; it is the null maintenance scenario.
2. The economic order quantity (EOQ) is assumed to be one tire for this example.
3. ABET engineering schools teach an engineering economists' course that cov-ers rate of return, net present value (NPV), and margin on operations. The PE exam has a mandatory economics question. This limited introduction to orga-nization finance does not qualify engineers to perform the project financial analysis, but rather enables engineers to understand the information needed by finance employees to perform project financial analysis and to understand the findings.
4. The periodic review of CBM data will show that a part will behave differently from one system to another.

13

Time-Directed Maintenance

> One would rather live with a problem one cannot solve than to employ a
> solution one does not understand.

Anonymous

Introduction

Time-directed maintenance (TDM) enables an organization to exchange lost
opportunity and intangible costs of an unscheduled system downing event
with maintenance costs to replace an unfailed part.

Time-directed maintenance uses the hazard function of part failure to
determine when a part is replaced based on the organization's definition of
allowable risk. Part replacement occurs when the part has been exposed to
a load that approaches or exceeds the likelihood of failure based on interference theory. There are no condition indicators, as are used in condition-based
maintenance (CBM). No part is risk free and parts will fail at different instantaneous failure rates under different operating and operational conditions
of use. The part mean time between failure (MTBF) is of little economical
value for TDM replacement intervals. The part hazard function is characterized from the reliability-centered maintenance (RCM) failure analysis and
describes the behavior of part failure resulting from loads and operational
and ambient conditions of use that are specific to the system.

The RCM hazard function is based on failure analysis of the loads that
cause failure and the cycles of loads. Calling the TDM hazard function an
instantaneous failure rate is not precise because it assumes the rate to be in
failures per units of time. The TDM hazard function is influenced more by
part exposure to failure mechanisms than to operating time. Consider a part
used on a system with an interference area defined as shown graphically
in Figure 13.1. Two systems are operated for 11 operational cycles and the
maximum load is measured and recorded below the interference area plot.
The part on system 1 experiences maximum loads on two operational cycles
that fall in the interference area; the part on system 2 experiences maximum
loads that fall within one standard deviation of the mean load. The TDM
solution would suggest that the part on system 1 requires at least a maintenance

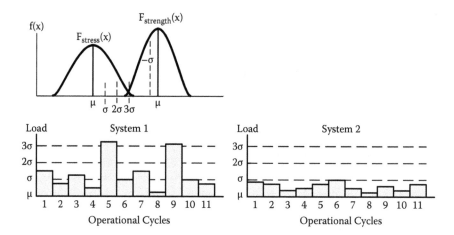

FIGURE 13.1
Example interference theory and load profile per cycle.

inspection following cycles 5 and 9 and that the part on system 2 does not require a maintenance inspection yet.

The TDM solution utilizes onboard failure mechanism measuring equipment to capture magnitudes and duration of exposure of part failure mechanisms. Remote maintenance inspection is an alternative, but it barely qualifies as TDM. Remote maintenance following an operator-reported excessive load serves the same purpose of maintenance inspection following an excessive load measurement but is general to the system or selected parts. The TDM solution is failure mechanism and part specific. Data captured by onboard failure mechanism measuring equipment are compared to the part stress-strength interference area to determine whether maintenance action is required. The logic for implementing a TDM solution is provided in the flow chart in Figure 13.2.

Reliability-centered failure analysis correlates failure mechanisms to failure modes. Maintenance inspection uses nondestructive examination (NDE) to evaluate failure modes when an excessive failure mechanism is reported from a mission. Maintenance actions are performed on the part on the system or the part is replaced and NDE is performed. Parts that pass the NDE are returned to service. Parts can be limited to a fixed number of exposures to excessive loads and are disposed following the limit.

Consider a reciprocating vane pump. Reliability-centered failure analysis finds that vibration and physical shock are failure mechanisms that cause the following failure modes: cracking, fracture, and surface wear, as shown in Figure 13.3. A condition indicator for failure modes is not technically or economically feasible. But an accelerometer can be mounted on the pump

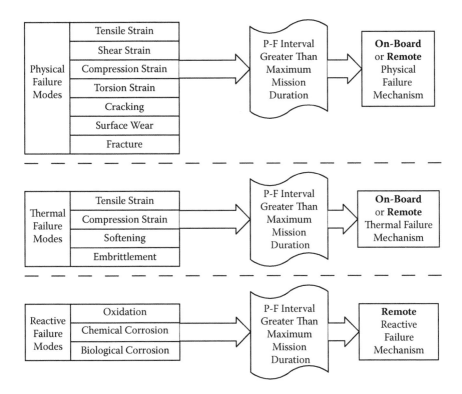

FIGURE 13.2
Time-directed maintenance logic.

to measure vibration levels. The physical loads are a combination of steady-state vibration and random shock.

Reliability-centered failure analysis finds that the material strength properties of the vane material and insertion design provide a 95th percentile stress (L_{95}) of 15 g_{RMS}. Any load that exceeds 11 g_{RMS} is defined as an acceptable load level risk for continued pump operation. The pump is replaced

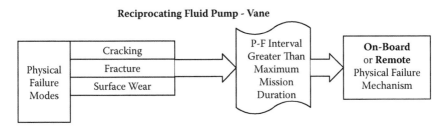

FIGURE 13.3
Time-directed maintenance logic: vane.

between missions when an accelerometer report exceeds 11 g_{RMS}. The pump data are analyzed to determine how the pump vane reached the load limit:

- A trend line shows that the pump reached the load limit over time as a wear-out degradation and requires NDE inspection for suitability for rebuilding or disposal.
- A data line with one or more shocks shows that the pump exceeded the load limit under discrete transient loads and requires NDE inspection for suitability of return to service or rebuilding.

The pump NDE maintenance inspection determines the extent of cracking, fracture, and surface wear. Pumps that pass the maintenance inspection are assembled and returned to inventory as a spare part. Pumps that fail the maintenance inspection are rebuilt and placed in spare parts inventory. Pumps that pass the maintenance inspection but have reached the limit of load exposure are rebuilt and placed in spare parts inventory.

Root cause failure analysis must be performed following maintenance actions in the TDM solution. The goal for TDM is much more than understanding part failure; it also includes understanding operation and ambient conditions of use that proactive maintenance is not just anticipation of failure occurrence to schedule maintenance actions before an unscheduled downing event. Rather, it is also evaluation of stress load that can be controlled to prevent part exposure to failure mechanisms.

Characterize Hazard Function

The preceding TDM approach assumes that failure mechanisms occur randomly and cannot be characterized by a function of time. Indeed, attempts to fit time-to-failure data will yield low coefficients of correlation and determination. Time-to-failure data are predictive for TDM when wear-out mechanisms are highly correlated to failure modes over operating time—not just operating time. Use of time-to-failure data to characterize a TDM hazard function assumes that the point estimate of the mean stress load and the measure of dispersion remain constant over time. Part strength is assumed to degrade over time, but the way it degrades is not described. There are three scenarios for strength degradation, as illustrated in Figure 13.4:

- Degradation scenario 1: mean strength degrades while the measure of dispersion for strength remains constant. This describes failure modes that cease to relax to initial conditions for strength following loading.

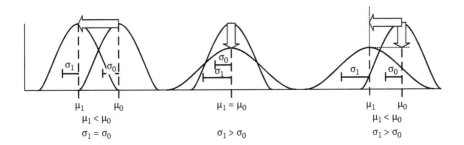

FIGURE 13.4
Part degradation scenarios.

- Degradation scenario 2: mean strength remains constant while the measure of dispersion for strength increases. This describes failure modes that relax to initial conditions for strength following loading or resist deformation, but experience changes in material strength properties and fail abruptly.
- Degradation scenario 3: mean strength degrades while the measure of dispersion for strength increases. This describes failure modes that experience complex failure mode interaction.

The current body of knowledge allows characterization of a TDM hazard function for the first degradation scenario. Research is in progress to characterize TDM hazard functions for the second and third degradation scenarios.

Define Hazard Threshold

An organization defines its risk threshold for part failure. A common practice is a risk threshold matrix, as shown in Figure 13.5. The risk threshold matrix defines the allowable risk that a part design must meet for initial acceptance and also defines the limits for part degradation in use until it must be removed from service. The example in Figure 13.5 defines the unacceptable risk for part failure rate for catastrophic failure consequences (highlighted in a bold border in the upper left section). The lower highlighted border defines risk that exceeds design requirements and part failure scenarios that do not require RCM solution. The RCM investment in low-risk parts would cost an organization far more than the expected financial return—a waste of engineering resources. The middle sectors define the organization's prerogative for risk assumption. Parts will be replaced when a risk threshold is reached based solely on the organization's willingness to incur the associated risk of failure.

Failure Consequences	Failure Category					
Catastrophic	**1A**	**1B**	**1C**	1D	1E	1F
Operational	**2A**	**2B**	**2C**	2D	2E	2F
Degraded Mode	**3A**	3B	3C	3D	3E	3F
Run-to-Failure	4A	4B	4C	4D	4E	4F

$$10^{-2} \quad 10^{-3} \quad 10^{-4} \quad 10^{-5} \quad 10^{-6} \quad 10^{-7}$$

h(x): Instantaneous Failure Rate

FIGURE 13.5
Risk threshold matrix.

Characterization of a hazard function was presented in previous chapters and follows the progression from empirical characterization of the part frequency distribution to fitting the Weibull reliability functions to characterization and plotting the hazard function, as illustrated in Figure 13.6.

The part risk threshold is plotted on the hazard function plot to define the maximum time in use the part will be allowed until it reaches the risk

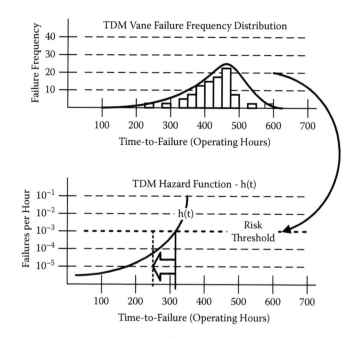

FIGURE 13.6
Part failure probability density function to hazard function.

limit. In this example, the organization will approve use of the part until it reaches a service life where wear-out is expected to result in degradation to one failure per 1,000 h. The example part will be replaced and either disposed or rebuilt at approximately 315 h. An organization may further reduce risk by reducing the service life limit based on engineering and maintenance judgment. Such a reduction must be accompanied by a financial analysis to assure that a cost benefit continues to justify the reduced service life. Cost benefit must be applied as a constraint; otherwise, one could reduce service life to a ridiculous "safe life" of 100 or 50 h, where risk is the only criterion.

Maintainability Demonstration Test, Validate Hazard Function

Implementation of TDM solutions is qualified by a maintainability demonstration before being incorporated into the organization's maintenance practices. Maintainability demonstrations are best performed on a single system for a suitable time to verify that the part risk threshold behaves as designed. Maintainability demonstration results are reviewed and evaluated by the TDM team and improvements on the application of the TDM solution are proposed and implemented.

Develop and Implement Maintenance Procedures and Practices

Successes of the maintainability demonstration are presented to the organization to develop TDM maintenance procedures and practices. Implementation of TDM maintenance procedures and practices is achieved through written guidelines and work instructions for maintainers to evaluate behavior of failure mechanisms measuring equipment and improvement in risk understanding and reduction. Data acquisition and reporting methods are developed to distribute the TDM information to maintenance engineers, the analysis of variance (ANOVA) program, and management. Periodically, the TDM solution is reviewed by the TDM team to evaluate successes and shortcomings to improve upon the TDM solution.

Bibliography

Abernathy, R. B. 2006. *The new Weibull handbook,* 5th ed. North Palm Beach, FL: Robert B. Abernathy.

AMCP 706-196. 1976. *Engineering design handbook: Development guide for reliability.* HQ U.S. Army Materiel Command.

Azarkhail, M., and M. Modarres. 2007. Markov chain simulation for estimating accelerated life model parameters. *Proceedings of the Reliability and Maintainability Symposium,* Orlando, FL.

Bayoumi, A., N. Goodman, R. Shah, T. Roebuck, A. Jarvie, L. Eisner, L. Grant, and J. Keller. 2008. Conditioned-based maintenance at USC—Part I: Integration of maintenance management systems and health monitoring systems through historical data investigation. *Proceedings of the American Helicopter Society Specialists Meeting on Condition-Based Maintenance,* Huntsville, AL.

———. 2008. Conditioned-based maintenance at USC—Part III: Aircraft components mapping and testing for CBM. *Proceedings of the American Helicopter Society Specialists Meeting on Condition-Based Maintenance,* Huntsville, AL.

Bazargan, M., and R. N. McGrath. 2003. Discrete event simulation to improve aircraft availability and maintainability. *Proceedings of the Reliability and Maintainability Symposium,* Tampa, FL.

Bazovsky, I. 1961. *Reliability theory and practice.* Upper Saddle River, NJ: Prentice Hall.

Birolini, A. 1994. *Reliability engineering: Theory and practice,* 4th ed. Zurich: Swiss Federal Institute of Technology.

Black, P. H., and O. E. Adams. 1968. *Machine design,* 3rd ed. New York: McGraw–Hill.

Blanchard, B. S. 2004. *Logistics engineering and management,* 6th ed. Upper Saddle River, NJ: Prentice Hall.

Brall, A., W. Hagen, and H. Tran. 2007. Reliability block diagram modeling—Comparisons of three software packages. *Proceedings of the Reliability and Maintainability Symposium,* Orlando, FL.

Briand, D., and J. E. Campbell. 2007. Real-time consequence engine. *Proceedings of the Reliability and Maintainability Symposium,* Orlando, FL.

Carter, C. M., and A. W. Malerich. 2007. The exponential repair assumption: Practical impacts. *Proceedings of the Reliability and Maintainability Symposium,* Orlando, FL.

Collins, J. A. 1993. *Failure of materials in design analysis prediction prevention,* 2nd ed. New York: John Wiley & Sons.

Condra, L. W. 1993. *Reliability improvement with design of experiments.* New York: Marcel Dekker.

Cook, J. 2009. System of systems reliability for multistate systems. *Proceedings of the Reliability and Maintainability Symposium,* Ft. Worth, TX.

Distefano, S., and A. Puliafito. 2007. Dynamic reliability block diagrams vs. dynamic fault trees. *Proceedings of the Reliability and Maintainability Symposium,* Orlando, FL.

Dodson, B. 1994. *Weibull analysis.* Milwaukee, WI: ASQ Quality Press.

Dovich, R. A. 1990. *Reliability statistics.* Milwaukee, WI: ASQ Quality Press.

Draper, N. R., and H. Smith. 1981. *Applied regression analysis,* 2nd ed. New York: John Wiley & Sons.

Farquharson, J. A., and J. L. McDuffee. 2003. Using quantitative analysis to make risk-based decisions. *Proceedings of the Reliability and Maintainability Symposium,* Tampa, FL.

Goel, H., J. Grievink, P. Herder, and M. Weijnen. 2003. Optimal reliability design of process systems at the conceptual stage of design. *Proceedings of the Reliability and Maintainability Symposium,* Tampa, FL.

Hartog, J. P. 1984. *Mechanical vibrations,* 4th ed. New York: Dover Publications.

Hauge, B. S., and B. A. Mercier. 2003. Reliability centered maintenance maturity level road map. *Proceedings of the Reliability and Maintainability Symposium,* Tampa, FL.

Hicks, C. R. 1993. *Fundamental concepts in the design of experiments,* 4th ed. New York: Saunders College Publishing.

Ireson, W. G., C. F. Coombs, and R. Y. Moss. 1996. *Handbook of reliability engineering and management,* 2nd ed. New York: McGraw–Hill.

Kapur, K. C., and L. R. Lamberson. 1977. *Reliability in engineering design.* New York: John Wiley & Sons.

Krasich, M. 2003. Accelerated testing for demonstration of product lifetime reliability. *Proceedings of the Reliability and Maintainability Symposium,* Tampa, FL.

———. 2007. Realistic reliability requirements for the stresses in use. *Proceedings of the Reliability and Maintainability Symposium,* Orlando, FL.

———. 2009. How to estimate and use MTTF/MTBF. Would the real MTBF please stand up? *Proceedings of the Reliability and Maintainability Symposium,* Ft. Worth, TX.

Krishnamoorthi, K. S. 1992. *Reliability methods for engineers.* Milwaukee, WI: ASQC Quality Press.

Lambeck, R. P. 1983. *Hydraulic pumps and motors: Selection and application for hydraulic power control systems.* New York: Marcel Dekker.

Lanza, G., P. Werner, and S. Niggeschmidt. 2009. Adapted reliability prediction by integrating mechanical load impacts. *Proceedings of the Reliability and Maintainability Symposium,* Ft. Worth, TX.

———. 2009. Behavior of dynamic preventive maintenance optimization for machine tools. *Proceedings of the Reliability and Maintainability Symposium,* Ft. Worth, TX.

Lapin, L. L. 1998. *Probability and statistics for modern engineering,* 2nd ed. Prospect Heights, IL: Waveland Press.

Leemis, L. 1995. *Reliability probabilistic models and statistical methods.* Upper Saddle River, NJ: Prentice Hall.

Lefebvre, Y. 2003. Using equivalent failure rates to assess the unavailability of an ageing system. *Proceedings of the Reliability and Maintainability Symposium,* Tampa, FL.

Liddown, M., and G. Parlier. 2008. Connecting CBM to the supply chain: Condition-based maintenance data for improved inventory management and increased readiness. *Proceedings of the American Helicopter Society Specialists Meeting on Condition-Based Maintenance,* Huntsville, AL.

Liu, Y., H.-Z. Huang, and M. J. Zuo. 2009. Optimal selective maintenance for multistate systems under imperfect maintenance. *Proceedings of the Reliability and Maintainability Symposium,* Ft. Worth, TX.

Luo, M., and T. Jiang. 2009. Step stress accelerated life testing data analysis for repairable system using proportional intensity model. *Proceedings of the Reliability and Maintainability Symposium,* Ft. Worth, TX.

Mannhart, A., A. Bilgic, and B. Bertsche. 2007. Modeling expert judgment for reliability prediction—Comparison of methods. *Proceedings of the Reliability and Maintainability Symposium,* Orlando, FL.

MIL-HDBK-472. Maintainability prediction.

MIL-HDBK-781. Reliability test methods, plans and environments for engineering development, qualification and production.

MIL-STD-470. Maintainability program for systems and equipment.

MIL-STD-471. Maintainability verification/demonstration/evaluation.

MIL-STD-690. Failure rate sampling plans and procedures.

MIL-STD-756. Reliability modeling and prediction.

MIL-STD-781. Reliability testing for engineering development, qualification and production.

MIL-STD-785. Reliability program for systems and equipment development and production.

MIL-STD-810. Environmental test methods and engineering guidelines.

MIL-STD-1629. Procedures for performing a failure mode, effects and criticality analysis.

MIL-STD-2155. Failure reporting, analysis and corrective action system (FRACAS).

Misra, R. B., and B. M. Vyas. 2003. Cost effective accelerated testing. *Proceedings of the Reliability and Maintainability Symposium,* Tampa, FL.

Montgomery, D. C., G. C. Runger, and N. F. Hubele. 2006. *Engineering statistics,* 3rd ed. New York: John Wiley & Sons.

Moubray, J. 1997. *Reliability-centered maintenance,* 2nd ed. Oxford, England: Butterworth Heinemann.

Murphy, K. E., C. M. Carter, and R. H. Gass. 2003. Who's eating your lunch? A practical guide to determining the weak points of any system. *Proceedings of the Reliability and Maintainability Symposium,* Tampa, FL.

Murphy, K. E., C. M. Carter, and A. W. Malerich. 2007. Reliability analysis of phased-mission systems: A correct approach. *Proceedings of the Reliability and Maintainability Symposium,* Orlando, FL.

Nachlas, J. A. 2005. *Reliability engineering: Probabilistic models and maintenance methods.* Boca Raton, FL: Taylor & Francis.

O'Connor, P. D. T. 2002. *Practical reliability engineering,* 4th ed. New York: John Wiley & Sons.

Pipe, K. 2008. Engineering the gateway for implementing prognostics in CBM. *Proceedings of the American Helicopter Society Specialists Meeting on Condition-Based Maintenance,* Huntsville, AL.

Pukite, J., and P. Pukite 1998. *Modeling for reliability analysis.* New York: IEEE Press.

Ramakumar, R. 1993. *Engineering reliability fundamentals and applications.* Upper Saddle River, NJ: Prentice Hall.

Sage, A. P. 1992. *Systems engineering.* New York: John Wiley & Sons.

Sautter, F. C. 2008. A systems approach to condition-based maintenance. *Proceedings of the American Helicopter Society Specialists Meeting on Condition-Based Maintenance,* Huntsville, AL.

Shanley, F. R. 1967. *Mechanics of materials.* New York: McGraw–Hill.

Shigley, J. E. 1977. *Mechanical engineering design,* 3rd ed. New York: McGraw–Hill.

Singh, J., S. Vittal, and T. Zou 2009. Modeling strategies for reparable systems having multi-aging parameters. *Proceedings of the Reliability and Maintainability Symposium,* Ft. Worth, TX.

Smith, A. M. 1993. *Reliability-centered maintenance.* New York: McGraw–Hill.

Snook, I., J. M. Marshall, and R. M. Newman. 2003. Physics of failure as an integrated part of design for reliability. *Proceedings of the Reliability and Maintainability Symposium*, Tampa, FL.

Tebbi, O., F. Guerin, and B. Dumon. 2003. Statistical analysis of accelerated experiments in mechanics using a mechanical accelerated life model. *Proceedings of the Reliability and Maintainability Symposium*, Tampa, FL.

van den Bogaard, J. A., J. Shreeram, and A. C. Brombacher. 2003. A method for reliability optimization through degradation analysis and robust design. *Proceedings of the Reliability and Maintainability Symposium*, Tampa, FL.

Vaughan, R. E., and D. O. Tipps. 2008. A condition-based maintenance approach to fleet management 1. *Proceedings of the American Helicopter Society Specialists Meeting on Condition-Based Maintenance*, Huntsville, AL.

Wang, W., and J. Loman. 2003. A new approach for evaluating the reliability of highly reliable systems. *Proceedings of the Reliability and Maintainability Symposium*, Tampa, FL.

Warrington, L., and J. A. Jones. 2003. A business model for reliability. *Proceedings of the Reliability and Maintainability Symposium*, Tampa, FL.

Wessels, W. R. 2003. Cost-optimized scheduled maintenance interval for reliability-centered maintenance. *Proceedings of the Reliability and Maintainability Symposium*, Tampa, FL.

———. 2007. Use of the Weibull versus exponential to model part reliability. *Proceedings of the Reliability and Maintainability Symposium*, Orlando, FL.

———. 2008. Reliability functions of flight-critical structural materials from stress-strength analysis. *Proceedings of the American Helicopter Society 64th Annual Forum*, Montréal, Canada.

Wessels, W. R., W. Roark, and S. Hardy. 2004. Application of modeling and simulation to predict impact on system availability due to logistical decisions for sparing and resource allocations. *Proceedings of the Huntsville Simulation Conference*, Huntsville, AL.

Wessels, W. R., and F. C. Sautter. 2009. Reliability analysis required to determine CBM condition indicators. *Proceedings of the Reliability and Maintainability Symposium*, Ft. Worth, TX.

Wolstenholme, L. C. 1999. *Reliability modeling: A statistical approach*. London: Chapman & Hall/CRC.

Xing, L., P. Boddu, and Y. Sun. 2009. System reliability analysis considering fatal and nonfatal shocks in a fault tolerant system. *Proceedings of the Reliability and Maintainability Symposium*, Ft. Worth, TX.

Zhang, Y., R. Rogers, and T. Skrzyszewski. 2003. Reliability prediction of hydraulic gasket sealing. *Proceedings of the Reliability and Maintainability Symposium*, Tampa, FL.

Index